Reflectionless Filters

For a complete listing of titles in the
Artech House Microwave Library,
turn to the back of this book.

Reflectionless Filters

Matthew A. Morgan

ARTECH HOUSE
BOSTON | LONDON
artechhouse.com

Library of Congress Cataloging-in-Publication Data
A catalog record for this book is available from the U.S. Library of Congress.

British Library Cataloguing in Publication Data
A catalogue record for this book is available from the British Library.

Cover design by John Gomes

ISBN 13: 978-1-63081-327-7

685 Canton Street
Norwood, MA 02062

10 9 8 7 6 5 4 3 2 1

Contents

Preface

I think it is not a stretch to say that anyone who works in electronics — no matter what their specialty — has had to think about filters at one time or another, quite possibly on a daily basis. In fact, I would argue that filtering is so common in the practice of our field that most degreed electrical engineers, even those (like myself) that are not trained filter experts, are required to design a filter occasionally. At the very least, it has surely fallen upon engineers to specify a filter for construction or purchase so that it may clean up the spurious signals and excess noise generated by their mixers, multipliers, amplifiers, modulators, power supplies, or other components, or else to reject the unwanted signals that may be incident upon their system from external sources (usually, the product of other engineers' electronic creations).

If that has not been your experience, and yet you are reading this, then I can only conclude that you have a long career yet ahead of you. May it be a good one!

For the rest of us, I would wager that more than half have found our efforts occasionally frustrated by something that our undergraduate courses did not fully prepare us for — the undesirable effects that a filter may have in its stop-band upon neighboring components. At the risk of sounding overly critical of our educational institutions, I think the way filters are initially presented is too simplistic — filters, it is said, pass signals at some frequencies, while attenuating others. "Attenuation" here implies that the undesired signal, or at least most of it, is somehow removed from the system. However, since most conventional filters are made up of nominally lossless components, it bears asking what happens to the energy in these stop-bands, as it has to go somewhere. Some might argue that it doesn't matter; it is outside the band of interest after all. Out of band, out of mind. If the out-of-band signals didn't matter, why filter them in the first place? Experienced designers know that out-of-band signals can have disastrous effects upon system performance.

Mixers, for example, change their performance in response to the sometimes unpredictable and often rapidly varying reactive terminations presented to them in their image bands. Ripples, or in extreme cases a "suckout" — a sharp, narrow-band drop in gain or efficiency[1] — may appear in the conversion loss. Intermodulation products that the filter was intended to attenuate may in fact be bounced back into the mixer to multiply. These issues are all the more troublesome because immediately before and after a mixer are two of the most common places that a filter is needed. Seasoned engineers have learned to "pad" the mixer Radio Frequency (RF), Intermediate Frequency (IF), and Local Oscillator (LO) ports with moderately valued attenuators to mitigate these effects.[2]

Amplifiers may also have their dynamic range diminished by standing waves built up between their outputs and an adjacent filter when the gain has not sufficiently diminished outside the operational band. Unintended self-biasing may result, and in the worst cases, instability and oscillations. Multipliers too may develop excessive conversion ripples due, at least in part, to the trapping of their harmonics in standing waves by a filter trying to block them.

Ironically, even filters themselves are not immune to the problems of stop-band interaction. It is widely recognized that a cascade of filters is not equal to the sum of its parts. Reactive elements in adjacent filters will almost certainly find frequencies at which they resonate with one another, leading to spurious pass-band spikes where there should be none. Perhaps — because my own background lies in a different arena than filter design — I am able to ask the uncomfortable question: If filters cannot even be cascaded with each other, why do we expect them to be cascadable with other components?

It is with this question that the story of this book begins for me. I am a microwave engineer. Among other things, that means that I think principally in terms of waves. Not voltages between nodes or currents along wires but waves propagating along transmission lines (a good microwave engineer, of course, is adept at switching his point of view between these different models as befits the situation). In the jargon of microwave engineering, conventional filters in their stop-bands are poorly impedance-matched. That, as every microwave engineer knows, can only lead to one thing...

Reflection.

No, conventional filters do not attenuate; they reflect. That is a much more descriptive term for what filters do, and it is damning. With the benefit of hindsight, I feel it surprising now that — in a field where impedance-matching is second nature — we should so readily accept the intentional mismatching of one of our primary building blocks.

Why can't a filter be matched in its stop-band as well as its pass-band? Something in my gut told me that this should be possible. A review of the available literature confirmed my suspicion that I wasn't the first to consider this question. Over the years, a number of gifted researchers had taken on the

---

[1] Should any of the terms used here be unfamiliar to the reader, a useful general reference is "Microwaves 101" available online as www.microwaves101.com.

[2] There are other reasons for doing this, such as to improve return loss that is often relatively poor for mixers, or to ensure adequate isolation between signal paths that share a common LO source.

challenge of designing so-called absorptive filters. A few of these approaches are well-grounded in classical theory — directional filters, for example, which are four-port networks analogous to directional couplers in which the coupling response is strongly frequency-dependent. However, almost all directional-coupling structures that I was able to find did not lend themselves well to high-order implementations. This limited their usefulness in practice. Some tried employing balanced filters where two identical and potentially high-order filters are combined at both ports with a quadrature hybrid. This too failed to satisfy me, as an ideal quadrature hybrid with unlimited frequency response does not exist, and those with sufficient bandwidth to be useful in this context are lossy and difficult to realize in practice.

Yet another well-understood approach is simply to design a diplexer and terminate one of the outputs; however, this would only be matched in the stop-band from one side, unless two diplexers were placed back to back. Moreover, those with anything other than a Butterworth response are usually poorly matched in their transition-bands.

To my mind, however, diplexers suffer from an even bigger problem: they are complicated. At least, they seem that way to me, if the reader will permit the confession. I don't mean conceptually; the idea of a diplexer is straightforward enough, but the details are hard to get right. They're difficult to design and finicky to properly tune, so much so that the average electronics engineer who is not a filter specialist (again, like myself) might find that their time is better spent solving problems that are more in their wheelhouse and resign themselves to simply adding an attenuator to their filter. Certainly, there are filter professionals out there who are more than up to the challenge — and to them, I take my hat off — but I think it is fair to say that the labor required to generate a custom diplexer makes them prohibitively expensive for all but the most demanding cases.

Not finding an adequate solution in theory, I, like many researchers before me, began resorting to a more empirical approach. "What if I add some loss here in these resonators, or at the end of these stubs?" While this yielded some limited success, for the most part it resulted in compromised performance. Historically, I think these explorations (while useful) had the negative side-effect of further convincing engineers (perhaps subconsciously) that absorptive filters were, by definition, lossy filters in the sense that absorption in the stop-band must come at the cost of extra loss in the pass-band, especially for filters with sharp rejection. It does not, as I hope to show in this book.

The word *reflectionless*, as I will use it in these pages, refers to the property that a filter may have, at least in principle, given ideal components, whereby its reflection coefficient is identically zero at all ports and at all frequencies from zero to infinity. This is in contrast to many other filters that are merely absorptive, having some finite, if nevertheless small, reflection coefficient at all but discrete frequency points.

I have stated a couple of times now that I am not a filter expert. In some sense, I think this gave me an advantage; I wasn't smart enough to know what couldn't be done. Had I known just a little more (but not too much) I might have concluded too quickly, as many have, that the solution I was seeking could not be found. Classical synthesis techniques for what are called *constant-resistance networks* — that is, networks that present a fixed, real impedance at their inputs, which I would call *reflectionless* if that were made equal to the characteristic impedance — focus on realizing prescribed transfer functions using a relatively small number of canonical topologies comprising the first and second order. Higher-order transfer functions were to be implemented by cascades. Giants in the field, such as Bell Labs' Hendrik Bode, proved that these topologies are sufficient to realize any pass-band shape that is theoretically possible for a finite, passive, causal network — but only within a constant loss factor. In that sense (and only in that sense), it was claimed that more sophisticated topologies were not needed. This gives classically trained circuit engineers the impression that anything that can be done in this arena has already been done. As I hope to show with this book, however, the unspecified loss factor in those claims often represents the glaring difference between a filter implementation that is both practical and useful, and another possible implementation that is not. We will see that the prior existing topologies of constant-resistance networks often fall very short of providing filter realizations of much value. More to the point, topologies that perform much better do exist and are waiting to be found.

Therefore the design of reflectionless filters is, to a large extent, a search for topologies. Of the countless ways in which simple elements such as resistors, inductors, and capacitors may be combined and connected together, we seek those relatively few that permit a solution whereby the impedance seen looking into one of its ports is constant and matches the prescribed characteristic impedance of the system. Once discovered, the process of assigning values to the elements within these topologies, and then coming up with practical realizations, is a matter of fairly routine engineering.

This book will primarily cover the tools needed to arrive at those topologies. At the present time, this is not a mechanical process. We have no way of writing down an equation for what we want, and then turning the mathematical crank to make the topology that achieves it appear on our sheet of paper.[3] Rather, what we have is a set of rules, a "bag of tricks," that if applied will guarantee that the end product is reflectionless. The manner and sequence in which those rules are applied is left up to the instinct and creativity of the designer. Once finished, the transmission response of the final result can then be analyzed, and if it is found to be useful, recorded for future reference. Otherwise, it is discarded, and the process repeats.

In the first chapter, some background is provided regarding microwave concepts, classical lumped-element filters, and the boundaries of what constitutes

[3] Coming up with just such an algorithm is a subject of active research. One thing we do know for certain is that the solution will often not be unique. For example, there is a class of seventh-order transfer function for which there are at least eight different reflectionless topologies that we know of, not including the near-limitless trivial variations.

a transfer function that is realizable using a finite number of passive elements. We also introduce basic transmission-line concepts to better elucidate what we mean when referring to reflections. In Chapter 2, we review some of the previous attempts at making absorptive filters and discuss their trade-offs. In Chapter 3, we introduce the foundational rules for deriving the symmetric, reflectionless filter that is the cornerstone of this work, and show how it leads us to the most basic, nontrivial topology, with which one can realize a third-order inverse Chebyshev (or Chebyshev Type II) frequency response. In Chapters 4–6, we build upon these techniques to derive topologies of increasing sophistication and flexibility, culminating in the realization of optimal Chebyshev Type I and Type II filters of arbitrary order with a totally reflectionless topology. In Chapter 7, we leave the realm of lumped elements to explore how the same basic principles can be applied to transmission line structures. Chapter 8 reveals how our repertoire of reflectionless filters may aid us in the development of multiplexers with 3 dB crossovers and identically zero reflection coefficient at all frequencies and ports. Chapter 9 discusses application scenarios in which reflectionless filters can improve the performance of our electronic systems — including nonlinear frequency-domain effects where out-of-band interactions corrupt in-band performance, and time-domain applications where low-reflections combine with time-delay characteristics to minimize distortion. Finally, in Chapter 10, we present some of the most promising techniques for creating the next generation of reflectionless topologies.

This book is intended for students, professionals, and hobbyists alike. It is thought that the material contained in this subject is new and refreshing enough to capture the interest of even the most seasoned experts, while still being accessible to the layman. As a result, advanced training in mathematics or engineering is not required, though it is assumed throughout that the reader has some casual familiarity with analog circuits, signal analysis in the frequency domain, and high frequency concepts.

My favorite spiritual writer once advised, "be grateful that this universe is not so shallow that our little plummets can touch bottom."[4] Above all, my hope with this book is to inspire the same sense of wonder that I feel when I realize that no field of endeavor is ever truly closed. No matter how much we learn and discover and prove, there is always more. I have no illusions that these reflectionless filters will ever become more than a niche within the massive canon of literature that is modern filter theory, but it is a niche that I firmly believe is well worth exploring, to the very bottom.

---

[4] Harry Emerson Fosdick, *Riverside Sermons*.

Acknowledgments

Completion of a work such as this is impossible without the generous support of others. Although the author was privileged to have the assistance of many individuals during the writing of this book, he is particularly indebted to the following people.

First, the author is grateful to his many associates and new friends at Mini-Circuits, including especially Ted Heil and Harvey Kaylie, for their faith in the technology that is reported within. Undoubtedly, without that belief to fuel continuing developments, this book would not exist.

Humble appreciation similarly goes out to Minghao "Martina" Yan, also of Mini-Circuits, for her enthusiastic endorsement of the writing project, without which the author would not have begun, and for the role she played in characterizing the first prototypes of these reflectionless filters as well as the systems that use them. The measurements and feedback she provided figure prominently in the examples found within these pages.

Also a debt of gratitude is owed to Aaron Vaisman, whose mastery and appreciation of the subtle intricacies of filter design and the elegant mathematics behind it are infectious. His critical insights into the deeper meaning of the symmetries, physical principles, and equations we all use has been an inspiration.

Finally, the author offers his most sincere thanks to his friend and colleague, Dr. Shing-Kuo Pan, for his encouragement to begin writing when this was little more than a gleam in the author's eye, and for the many hours he spent reading the manuscript, checking its calculations, and meticulously reviewing the explanations for technical accuracy. His insightful comments and suggestions have greatly enhanced the clarity and readability of this work.

Chapter 1

Passive Microwave Networks

Before attempting to derive a reflectionless filter, it is useful to review some of the fundamental principles of microwave networks. This chapter begins with a brief explanation of terms regarding scattering parameters and their mathematical relationship to impedance. Beginning in section 1.2 is a summary of the conditions for realizability of a driving-point immittance or transfer function assuming networks comprising a finite number of lumped elements. The most commonly used optimal filter transfer functions, their mathematical basis, and their topologies are then reviewed. The chapter concludes with a review of basic transmission-line concepts, which further underscores the role and importance of reflections in high-frequency systems.

1.1 TWO-PORT NETWORKS

The microwave engineer's point of view of a two-port network is shown in Figure 1.1(a). In this diagram, the ports are represented by two terminals to which the input and output transmission lines are connected; the ground terminal is implicit. The arrows represent waves propagating into and away from those ports, where the labels represent the normalized wave amplitudes. For example, the voltage and current phasors from the incoming wave at port 1 may be written as

$$v_{a1} = a_1\sqrt{Z_0} \tag{1.1a}$$

$$i_{a1} = \frac{a_1}{\sqrt{Z_0}} \tag{1.1b}$$

and those from the outgoing wave may be written as

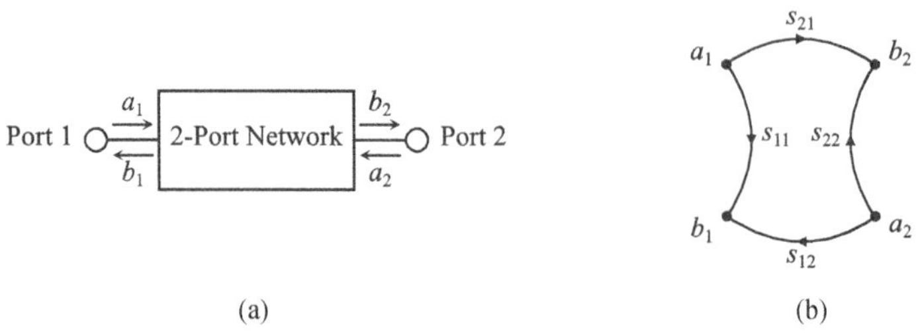

Figure 1.1 (a) A two-port network from a wave propagation point of view. (b) Signal flow graph for the two-port network.

$$v_{b1} = b_1\sqrt{Z_0} \tag{1.2a}$$

$$i_{b1} = -\frac{b_1}{\sqrt{Z_0}} \tag{1.2b}$$

where Z_0 is the characteristic impedance of the input/output transmission line [1]. It is important to realize that these are complex numbers, where the argument of the complex number defines the relative phase of the quantity specified. Note that the squares of the wave amplitudes in (1.2) above have units of power. Additionally, we have adopted the convention that positive current is in the direction of propagation, hence the sign reversal in (1.2b). The input impedance of the network at port 1 is then given by the ratio of the total voltage and current at this terminal,

$$Z_{in} = \frac{v_{a1} + v_{b1}}{i_{a1} + i_{b1}} = \frac{(a_1 + b_1)\sqrt{Z_0}}{(a_1 - b_1)/\sqrt{Z_0}} = Z_0\frac{a_1 + b_1}{a_1 - b_1} \tag{1.3}$$

In Figure 1.1(b), a signal flow diagram is shown for the two-port network. The scattering parameter, s_{ij}, is the ratio of the outgoing wave amplitude at port i divided by the ingoing wave amplitude at port j. For example, $s_{11} = b_1/a_1$. Note that if a_1 and b_1 have the same phase or opposite phase, then s_{11} is real; otherwise s_{11} is complex, where its argument defines the relative phase of the incoming and outgoing waves. Substituting this into (1.3),

$$Z_{in} = Z_0\frac{a_1 + s_{11}a_1}{a_1 - s_{11}a_1} = Z_0\frac{1 + s_{11}}{1 - s_{11}} \tag{1.4}$$

Therefore, solving for s_{11} in terms of the impedance, we have

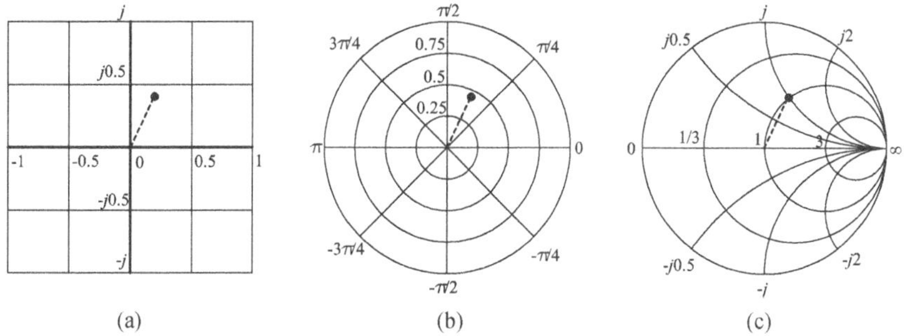

Figure 1.2 Three ways to plot the reflection coefficient in the complex plane. (a) Cartesian coordinates. (b) Polar coordinates. (c) Smith chart. The normalized impedance plotted in all cases is $z = 1 + j$, corresponding to a reflection coefficient of $s_{11} = 0.2 + j0.4$.

$$s_{11} = \frac{Z_{in} - Z_0}{Z_{in} + Z_0} = \frac{z-1}{z+1} \tag{1.5}$$

where the latter expression is given in terms of the normalized impedance, $z = Z_{in}/Z_0$. This relationship between the impedance and complex reflection coefficient is very often illustrated on what is known as a Smith chart. See Figure 1.2 for an example. The reflection coefficient is plotted as a point in the complex plane, but the axes, rather than representing lines of constant real and imaginary part (as in the Cartesian plot of Figure 1.2(a)) or of magnitude and angle (as in the polar plot of Figure 1.2(b)) comprise instead lines of constant resistance and reactance. The resulting graph, or impedance Smith chart, is shown in Figure 1.2(c). Alternatively, lines of constant conductance and susceptance may be drawn instead, in which case the graph is referred to as an admittance Smith chart [2,3] — and appears the same, but for a 180 degree rotation.

In general, the full set of scattering parameters, **S**, may be derived from the network impedance matrix, **Z**, or the admittance matrix, **Y**, as follows [4],

$$\begin{aligned}\mathbf{S} &= (\mathbf{Z} + Z_0\mathbf{I})^{-1}(\mathbf{Z} - Z_0\mathbf{I}) \\ &= (Y_0\mathbf{I} + \mathbf{Y})^{-1}(Y_0\mathbf{I} - \mathbf{Y})\end{aligned} \tag{1.6}$$

Conservation of energy requires that the magnitude of any row or column vector of the scattering parameter matrix (whatever its order) is less than or equal to 1. If the network is lossless, the matrix **S** is required to be unitary. That is,

$$\sum_{k=1}^{N} s_{ki} s_{kl}^{*} = \delta_{il} = \begin{cases} 1 & i = l \\ 0 & i \neq l \end{cases} \tag{1.7}$$

Furthermore, reciprocity requires that $s_{ik} = s_{ki}$ for networks of this type.

Finally, it is common in microwave engineering to express ratios of powers in logarithmic units known as decibels (dB), defined as ten times the base-ten logarithm. For example,

$$s_{ij}(\mathrm{dB}) = 10\log_{10}\left(\left|s_{ij}\right|^2\right) \tag{1.8}$$

In (1.8), if $i = j$, then s_{ij} (dB) is typically referred to as reflected power; otherwise it is referred to as transmitted power. For a more thorough discussion of network and scattering parameters, the reader is referred to [4].

1.2 PROPERTIES OF IMMITTANCE AND TRANSFER FUNCTIONS

1.2.1 The Laplace Frequency Domain

It is useful to express the foregoing immittance (impedance or admittance) quantities and scattering parameters as functions of the complex frequency variable, $s = \sigma + j\omega$, where ω is the angular frequency in radians per second, and σ controls the rate of exponential growth or decay of the sinusoidal amplitude as a function of time.[1] Mathematically, this derives from the Laplace transform of a given time-domain signal [5],

$$F(s) = L\{f\}(s) = \int_0^\infty e^{-st} f(t)dt \tag{1.9}$$

If the voltage applied to a capacitor, C, is given by $v(t)$, then by definition the current passing through it is proportional to the time derivative of the voltage, or

$$i(t) = C\frac{dv(t)}{dt} \tag{1.10}$$

Expressed in the complex frequency domain, this becomes

$$I(s) = \int_0^\infty e^{-st} i(t)dt = C\int_0^\infty e^{-st}\frac{dv(t)}{dt}dt = sCV(s) \tag{1.11}$$

where the last step follows from integration by parts (and we have assumed for

[1] Like many electrical engineers, I prefer to use j instead of i as the imaginary constant ($\sqrt{-1}$) to avoid confusion with the current, $i(t)$.

simplicity that the initial value of the voltage is zero). We thus find that the impedance of the capacitor in the complex frequency domain is $(sC)^{-1}$. In a similar fashion, one may show that the impedance of an ideal inductor is sL, and that of a resistor is simply R. The admittances are, of course, the reciprocals of these quantities.

The value of the Laplace transform is that it converts a differential equation (or, equivalently, an integral equation) relating time-variable parameters (such as voltage and current) into an algebraic equation, thus simplifying the analysis of circuits and systems. Through the power of the Laplace frequency-domain representation, the mathematical treatment of reflectionless filters throughout the remainder of this book shall consist solely of algebra.

1.2.2 Immittance Functions

As one may discover by applying the well-known rules for series and parallel combinations of these elements (as well as delta-wye transformations and other techniques), the driving-point impedance of a lumped-element network must always be a rational function comprising the ratio of two polynomials in s. Moreover, it is easy to see that the coefficients of those polynomials must always be positive.

To be a little more rigorous, it has been proven that the immittance functions of finite, passive, lumped-element networks must be positive-real (p.r.) [6]. Positive-real functions are those that map the right half of the complex s-plane to the right half, and the real axis to the real axis. For rational p.r. functions, the following properties must hold:

1. If $F(s)$ is p.r., then $1/F(s)$ is also p.r.
2. The sum of two p.r. functions is also p.r. (but the difference, in general, may not be).
3. There are no poles or zeros in the right half plane.
4. Only simple poles with real positive residues may exist on the $j\omega$ axis.
5. All poles and zeros are either real or exist in conjugate pairs.
6. The highest powers of the numerator and denominator may differ by at most unity.
7. The lowest powers of the numerator and denominator may differ by at most unity.

1.2.3 Transfer Functions

The immittance functions listed above are a subset of an even broader class of rational polynomials[2] corresponding to the transfer functions, or any ratio of

[2] For brevity, I will use the term *rational polynomial* to refer to the ratio of two polynomial functions. It should be clear from context that I do not mean polynomials with rational coefficients.

currents and/or voltages in an electrical network. This, by definition, includes the driving-point immittances above, but also adds the scattering parameters. Most typically, transfer functions relate the output response of a filter to its input.

The principal concerns here are causality and stability — the requirements that the excitation (or input) must always precede the response, and the response remains bounded for bounded excitations. Without going into the details (the reader is referred to [6] for a more complete derivation), to satisfy these conditions, the denominator of a transfer function must be what is known as a *Hurwitz polynomial*. A Hurwitz polynomial satisfies the following conditions:

1. The coefficients are all real and nonnegative.
2. Between the highest and lowest order terms, none of the coefficients may be zero (missing) unless the polynomial is even (having all even powers) or odd (having all odd powers).
3. Both the even and odd parts of the polynomial have imaginary roots only.
4. A continued fraction expansion of the odd part divided by the even part (or the reciprocal) yields only positive quotients.

Note that because a driving point impedance and its inverse (admittance) are both also transfer functions, the numerator and denominator of all immittance functions must then be a Hurwitz polynomial as well.

1.3 DUALITY

The concept of duality is a recurring theme in the subject of reflectionless filters, and it will reappear several times throughout this book. Its origins trace back to the symmetrical relationship between the electric and magnetic fields in Maxwell's equations, and duals result from interchanging the two [7]. Consequently, for almost every quantity or concept in electromagnetic theory and electronic circuits, there exists a dual, such as

1. Voltage is the dual of current.
2. Inductance is the dual of capacitance.
3. Resistance is the dual of conductance.
4. Impedance is the dual of admittance.
5. Parallel circuit structure is the dual of serial circuit structure.
6. Kirchoff's voltage law is the dual of Kirchoff's current law.
7. Thévenin's theorem is the dual of Norton's theorem.

and many more of these dualities exist. Take, as an example, the two networks in Figure 1.3. Recall that in terms of the Laplace frequency, $s = \sigma + j\omega$, the impedance of an inductor is sL, the admittance of a capacitor is sC, and the impedance of a resistor is simply R. The combined impedance of the first network,

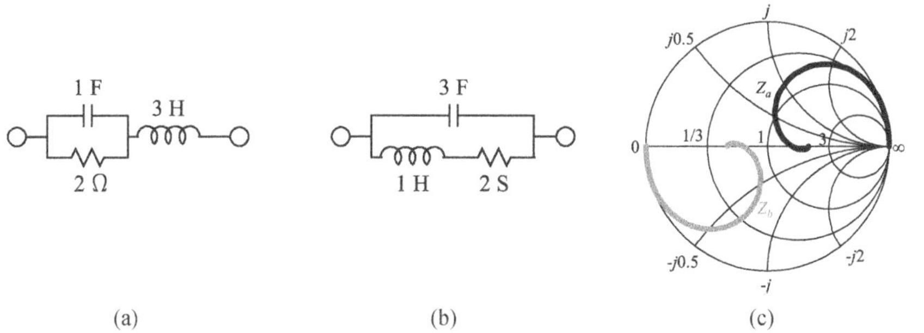

Figure 1.3 (a) An example of an impedance network, (b) its dual, and (c) their corresponding impedance/reflection coefficient plotted on a Smith chart.

then, is given by

$$Z_a = 3s + \left(1s + 2^{-1}\right)^{-1} = 3s + \frac{2}{2s+1} = \frac{6s^2 + 3s + 2}{2s+1} \tag{1.12}$$

and the impedance of the second network is given by

$$Z_b = \left(3s + \left(1s + 2^{-1}\right)^{-1}\right)^{-1} = \left(3s + \frac{2}{2s+1}\right)^{-1} = \frac{2s+1}{6s^2 + 3s + 2} \tag{1.13}$$

The second network was formed by replacing the 1F capacitor with a 1H inductor, the 3H inductor with a 3F capacitor, the 2Ω resistor with a 2S conductor (or 0.5Ω resistor), and all series combinations with parallel combinations, or vice versa. One can see that the impedance functions are inverses of one another. Moreover, when plotted on a Smith chart, Figure 1.3(c), it is evident that the reflections from dual circuits are 180 degrees out of phase.

The concept may be generalized for any desired characteristic impedance, Z_0, in which case the following substitutions apply,

$$L' = Z_0^2 C \tag{1.14a}$$

$$C' = Y_0^2 L \tag{1.14b}$$

$$R' = Z_0^2 G \tag{1.14c}$$

and, as before, the series and parallel connectivity is reversed.

1.4 FILTER TRANSFER FUNCTIONS

1.4.1 Pole-Zero Configuration

Because transfer functions of lumped-element networks take the form of rational polynomials, it is useful to classify different types of transfer functions by the arrangement of their poles and zeros — that is, the complex frequencies in the Laplace domain where the transfer function becomes infinite or zero, respectively.

Take, for example, the following transfer function,

$$H(s) = \frac{s^2+1}{2s^3+3s^2+2s+1} \tag{1.15}$$

The frequency response is given by the magnitude of this transfer function in terms of pure frequency, ω. Thus we substitute $s = j\omega$ (i.e., let $\sigma = 0$) into the above expression and take the magnitude,

$$|H(j\omega)| = \left|\frac{-\omega^2+1}{-j2\omega^3-3\omega^2+j2\omega+1}\right| = \frac{|1-\omega^2|}{\sqrt{(1-3\omega^2)^2+4\omega^2(\omega^2-1)^2}} \tag{1.16}$$

This is plotted in Figure 1.4(a). In the terminology of filters, the range of frequencies for which the magnitude of the transfer function is close to 1 is referred to as the *pass-band*, whereas the frequency range for which the transfer function is near zero is called the *stop-band*. Frequencies that lie in between these two extremes may be called the *transition band(s)*. The particular thresholds for these definitions will depend on the specific application.

To find the poles and zeros, we return to the original transfer function and factor both the numerator and denominator,

$$H(s) = \frac{s^2+1}{(2s^2+s+1)(s+1)} = \frac{(s+j)(s-j)}{2\left(s+\frac{1}{4}+j\frac{\sqrt{7}}{4}\right)\left(s+\frac{1}{4}-j\frac{\sqrt{7}}{4}\right)(s+1)} \tag{1.17}$$

This reveals that the transfer function has zeros at $s = \pm j$, and poles at $s = -1$ and $s = (-1 \pm j\sqrt{7})/4$. Additionally, there is a zero at infinity because the degree of the numerator is one less than that of the denominator. The degree of the denominator, and thus the number of poles, is referred to as the *order* of the filter. In this case, the filter is third order.

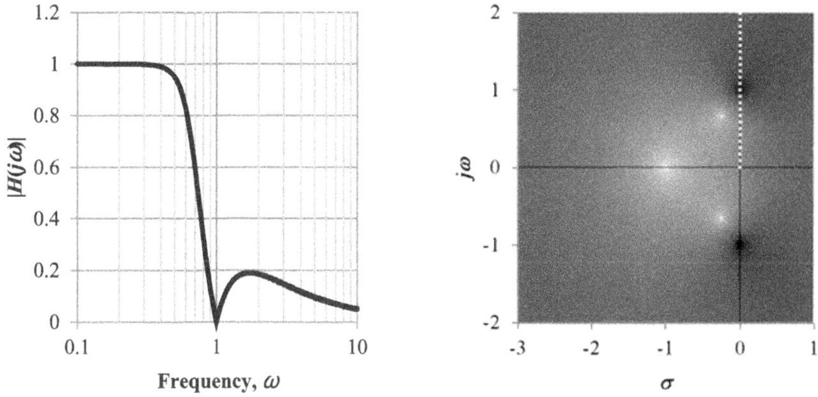

(a) (b)

Figure 1.4 (a) Frequency response and (b) complex frequency space plot of the transfer function in (1.15). The white dashed line identifies the region of the complex plane from which the true frequency response is obtained.

This list of poles and zeros is sufficient to completely describe the transfer function, and consequently the frequency response, as well as phase characteristics — though a constant factor such as the 2 in the denominator above may be needed to ensure passivity. A complex frequency space plot of the transfer function over the complex plane is shown Figure 1.4(b). White in the complex frequency space plot corresponds to amplitude gains greater than 100, while black corresponds to less than $1/100^{th}$. The poles are readily visible as the white spots in the left half-plane, while the zeros are apparent as black spots on the vertical axis. The white dashed line indicates the region of the complex plane from which the frequency response is obtained. Note that since the zero on the positive imaginary axis corresponds to a pure frequency, $\omega = 1$, it is also visible as a null in the frequency response of Figure 1.4(a). On the other hand, the poles do not lie on the pure frequency axis, so the frequency response never becomes infinite — as it shouldn't, since that would clearly violate stability. Only complex frequencies in the left half-plane corresponding to exponentially decaying sinusoids are allowed to have infinite gain.

In general, any realizable filter may be completely described by a list of its finite poles and zeros, p_k and z_k. The transfer function may then be written

$$H(s) = a_0 \frac{\prod_{k \le N}(s - z_k)}{\prod_{k \le N}(s - p_k)} \tag{1.18}$$

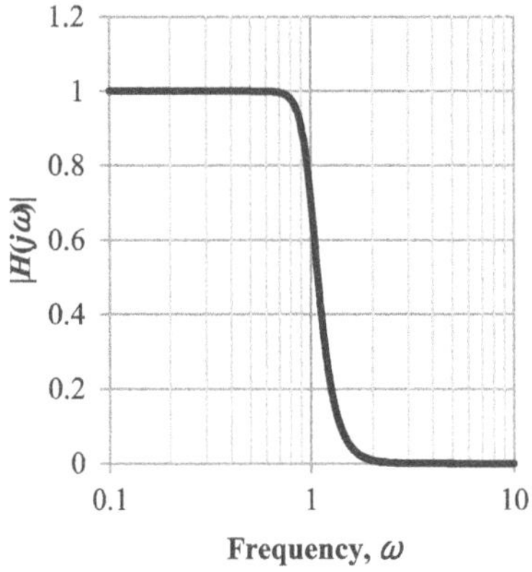

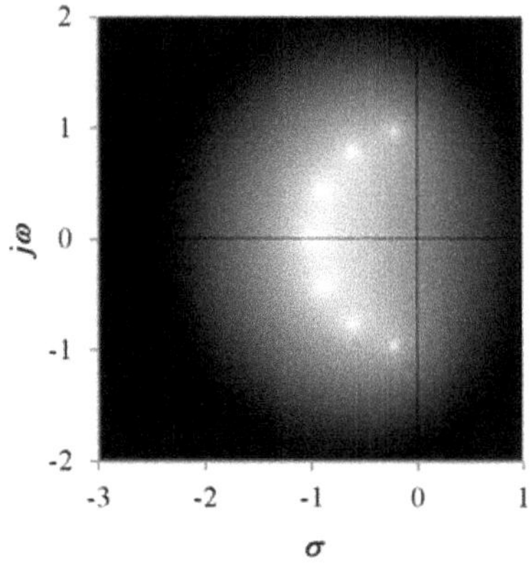

Figure 1.5 (a) Frequency response and (b) complex frequency space plot of a seventh-order low-pass Butterworth filter.

where N is the order of the filter. Zeros at infinity are not included in (1.18); they manifest as a smaller order in the numerator than in the denominator. (Poles at infinity would violate passivity for the transmission characteristic of a filter, and hence are not allowed.) The scale factor a_0 is sometimes not specified, but determined instead as the largest permissible value that still limits the magnitude of the transfer function to less than unity at all points on the imaginary (pure frequency) axis.

An ideal low-pass filter would have a rectangular frequency response in ω, with a value of unity up to the cutoff frequency and zero beyond. However, such a transfer function is not causal; its impulse response would be a sinc function having nonzero energy for time, t, less than zero (and indeed for all times, t, from minus infinity to plus infinity). The well-known optimized filter responses, such as Butterworth, Chebyshev, and elliptic, are designed to approximate this response given certain other constraints.

1.4.2 Butterworth Filter Response

For example, Butterworth filters are designed to be as flat as possible in the pass-band, meaning that the first $2N$-1 derivatives of the frequency response of an Nth-order filter are identically zero at $\omega = 0$. Butterworth filters are monotonic in the pass-band and stop-band. They are characterized by poles which are equally distributed around a unit half-circle in the left half of the s-plane,

$$p_k = e^{\frac{j(2k+n-1)\pi}{2n}} \tag{1.19}$$

for k = 1...N, as shown in Figure 1.5. The zeros are all located at infinity. The transfer function is then given by substituting these values into (1.18), such that

$$H(s)=\frac{a_0}{(s-p_1)(s-p_2)\cdots(s-p_N)}=\frac{1}{(s-p_1)(s-p_2)\cdots(s-p_N)} \tag{1.20}$$

Note that a_0 = 1 in this case. The frequency response associated with this transfer function is shown in Figure 1.5(a) for the seventh order, and a complex frequency space plot in Figure 1.5(b).

1.4.3 Chebyshev Type I Filter Response

A steeper transition may be achieved by allowing some ripple in either the pass-band or the stop-band. The Chebyshev Type I filter is optimized to provide the steepest transition for a given order and maximum allowable ripple in the pass-band, while the stop-band is constrained to be monotonic. The magnitude of the transfer function is given by

$$|H(j\omega)|=\frac{1}{\sqrt{1+\varepsilon^2 T_N^2(\omega)}} \tag{1.21}$$

where ε is a free parameter known as the ripple factor, and T_N is the Nth-order Chebyshev polynomial of the first kind [8]. The Chebyshev polynomials are an orthogonal set with the unique property of having the largest leading coefficient for any polynomial subject to the condition that their absolute value is limited to less than unity over an interval $-1 \leq \omega \leq 1$. This makes them ideal building blocks for filters that achieve the steepest possible transition slope with bounded ripples in the pass- or stop-bands. They are defined indirectly by the trigonometric relationship,

$$T_N(\omega)=\cos\left(n\cos^{-1}\omega\right)=\cosh\left(n\cosh^{-1}\omega\right) \tag{1.22}$$

which, despite appearances, is in fact reducible to a polynomial in the variable ω. They are equivalently defined by the recurrence relation,

$$T_0(\omega)=1 \tag{1.23a}$$

$$T_1(\omega)=\omega \tag{1.23b}$$

$$T_{N+1}(\omega)=2\omega T_N(\omega)-T_{N-1}(\omega) \tag{1.23c}$$

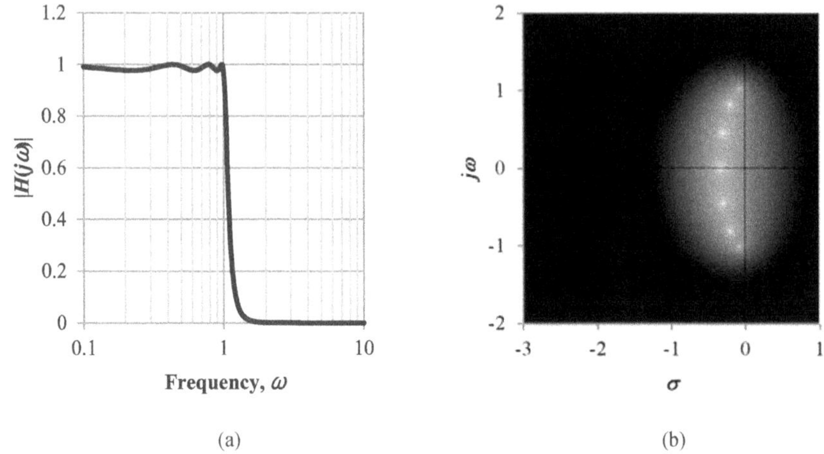

Figure 1.6 (a) Frequency response and (b) complex frequency space plot of a seventh-order Chebyshev Type I filter.

An example of the Chebyshev Type I filter response is shown in Figure 1.6. The pole-zero configuration needed to achieve such a response is similar to that of a Butterworth, except that the poles are arranged along an ellipse in the left half of the plane instead of in a circle. Specifically, the poles are given by

$$p_k = -\sinh\left(\tfrac{1}{N}\sinh^{-1}\left(\tfrac{1}{\varepsilon}\right)\right)\sin\left(\tfrac{\pi}{2}\tfrac{2k-1}{N}\right) + j\cosh\left(\tfrac{1}{N}\sinh^{-1}\left(\tfrac{1}{\varepsilon}\right)\right)\cos\left(\tfrac{\pi}{2}\tfrac{2k-1}{N}\right) \tag{1.24}$$

for $k = 1...N$. The transfer function scaling factor for passivity is given by

$$a_0 = \frac{1}{2^{N-1}\varepsilon} \tag{1.25}$$

The amplitude in the pass-band oscillates between a value of 1 and $1/(1+\varepsilon^2)^{1/2}$.

1.4.4 Chebyshev Type II Filter Response

Another example is the Chebyshev Type II filter. In this case, the pass-band is monotonic while the stop-band is allowed to have ripple. The magnitude of its transfer function is also based on the Chebyshev polynomials,

$$|H(j\omega)| = \frac{1}{\sqrt{1+\varepsilon^{-2}T_N^{-2}\left(\omega^{-1}\right)}} \tag{1.26}$$

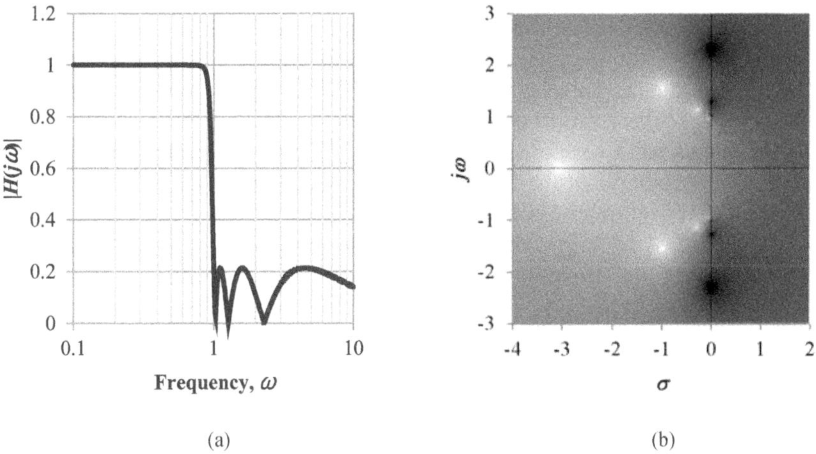

Figure 1.7 (a) Frequency response and (b) complex frequency space plot of a seventh-order Chebyshev Type II filter.

The poles of a Chebyshev Type II filter are simply the inverses of the poles of the Chebyshev Type I filter,

$$p_k = \frac{1}{-\sinh\left(\frac{1}{N}\sinh^{-1}\left(\frac{1}{\varepsilon}\right)\right)\sin\left(\frac{\pi}{2}\frac{2k-1}{N}\right) + j\cosh\left(\frac{1}{N}\sinh^{-1}\left(\frac{1}{\varepsilon}\right)\right)\cos\left(\frac{\pi}{2}\frac{2k-1}{N}\right)} \tag{1.27}$$

for $k = 1...N$. The zeros, no longer all located at infinity, are given by

$$z_k = j\sec\left(\frac{\pi}{2}\frac{2k-1}{N}\right) \text{ for } k = 1...N \tag{1.28}$$

Note that if N is odd, one of the zeros in the above list will be infinite, and therefore does not appear as a term $(s-z_k)$ in (1.18). Instead, the zero at infinity results from the numerator having a lower order than the denominator. The scaling factor in this case is

$$a_0 = \begin{cases} N\varepsilon & \text{if } N \text{ is odd} \\ \dfrac{\varepsilon}{\sqrt{1+\varepsilon^2}} & \text{if } N \text{ is even} \end{cases} \tag{1.29}$$

An example of this filter response in shown in Figure 1.7. The amplitude of the stop-band ripples is given by $\varepsilon/(1+\varepsilon^2)^{1/2}$.

Although Butterworth filters are described as *maximally flat*, defined by the lowest order derivatives being equal to zero, the Chebyshev Type II can actually be flatter in the pass-band. This is because it has less slope very near the cutoff at the cost of having ripples in the stop-band. It is worth pointing out that the very simple transfer function given at the beginning of this section in (1.15) and plotted in Figure 1.4 is also that of a Chebyshev Type II filter; in this case, it is a third-order response where the frequency and ripple factor have been scaled such that the coefficients are all simple integers. We will return to this transfer function at several points throughout the book as an easy test case for various principles and properties of reflectionless filters.

1.4.5 Elliptic Filter Response

A final example is the elliptic filter, also known as a Cauer filter (after the mathematician Wilhelm Cauer). It achieves greater selectivity than any of the previous filter types by allowing ripple in both the pass-band and stop-band. Its frequency response is given by

$$|H(j\omega)| = \frac{1}{\sqrt{1+\varepsilon^2 R_N^2(\xi,\omega)}} \tag{1.30}$$

where R_N is the Nth-order elliptic rational function with selectivity factor ξ, and ε is again called the ripple factor [9]. The elliptic rational functions in turn are generally defined by

$$R_N(\xi,\omega) = \mathrm{cd}\left(N\frac{K(1/L_N)}{K(1/\xi)}\mathrm{cd}^{-1}(\omega,1/\xi),1/L_N \right) \tag{1.31a}$$

$$L_N = R_N(\xi,\xi) \tag{1.31b}$$

where cd() is the Jacobi elliptic cosine function, and $K()$ is the complete elliptic integral of the first kind. L_N is known as the discrimination factor, and is the minimum value of $|R_N(\xi,\omega)|$ for $|\omega| \geq \xi$. The circular nature of the definition of R_N and L_N (1.31b) will usually require that they be evaluated by other means. One method is to construct R_N as a rational polynomial whose poles and zeros (not to be confused with the poles and zeros of the elliptic filter transfer function itself) can be determined without prior knowledge of L_N [10]. The zeros of the elliptic rational function, R_N, are given by

$$z_k^{(R_N)} = \mathrm{cd}\left(K(1/\xi)\tfrac{2k-1}{N},1/\xi\right) \tag{1.32}$$

and the poles are given by

$$p_k^{(R_N)} = \frac{\xi}{z_k^{(R_N)}} \tag{1.33}$$

for $k = 1...N$. The elliptic rational function may then be calculated as

$$R_N(\xi, \omega) = r_0 \frac{\prod_k \left(\omega - z_k^{(R_N)}\right)}{\prod_k \left(\omega - p_k^{(R_N)}\right)} \tag{1.34}$$

where r_0 is a scaling factor that ensures $R_N(\xi,1) = 1$. If N is odd, one of the poles will be infinite, and may be omitted from (1.34).

In this manner, (1.34) may be used along with (1.31b) to calculate $L_N(\xi)$. The poles and zeros of the elliptic filter transfer function are then given by

$$p_k = j\,\mathrm{cd}\left(\frac{K(1/\xi)}{N \cdot K(1/L_N)}\mathrm{cd}^{-1}\left(\frac{j}{\varepsilon}, \frac{1}{L_N}\right) + \frac{2(k-1) \cdot K(1/\xi)}{N}, \frac{1}{\xi}\right) \tag{1.35a}$$

$$z_k = j \cdot p_k^{(R_N)} = \frac{j\xi}{\mathrm{cd}\left(K(1/\xi)\frac{2k-1}{N}, 1/\xi\right)} \tag{1.35b}$$

for $k = 1...N$. Note that if N is odd, one of the zeros will be infinite. Evaluation of the special functions above and their inverses can be difficult without software coded to use them, but one free online tool that is capable of it is Wolfram Alpha [11]. Be aware that some software packages (including Wolfram Alpha) are coded such that the elliptic integrals and Jacobi elliptic cosine functions take the square of the arguments above — that is, $K(x)$ above becomes $K(x^2)$, $\mathrm{cd}(x,y)$ becomes $\mathrm{cd}(x,y^2)$, and $\mathrm{cd}^{-1}(x,y)$ becomes $\mathrm{cd}^{-1}(x,y^2)$.

For example, assuming a seventh-order filter with $\varepsilon = 0.2$ and $\xi = 1.05$, we find from (1.34) and (1.31b) that the discrimination factor $L_7(1.05) = R_7(1.05,1.05) = 218.6219$. The poles are given by (1.35a) as $p_k = -0.01719 \pm j1.0094$, $-0.09136 \pm j0.9681$, $-0.3281 \pm j0.7589$, and -0.6089. The zeros are given by (1.35b) as $\pm j1.05713$, $\pm j1.1438$, $\pm j1.6475$, and ∞.

The seventh-order elliptic filter frequency response associated with these poles and zeros is shown in Figure 1.8. The pass-band amplitude oscillates between a value of 1 and $1/(1+\varepsilon^2)^{1/2}$, the same as a Chebyshev Type I filter. The stop-band ripples have amplitude $1/(1+\varepsilon^2 L_N^2)^{1/2}$.

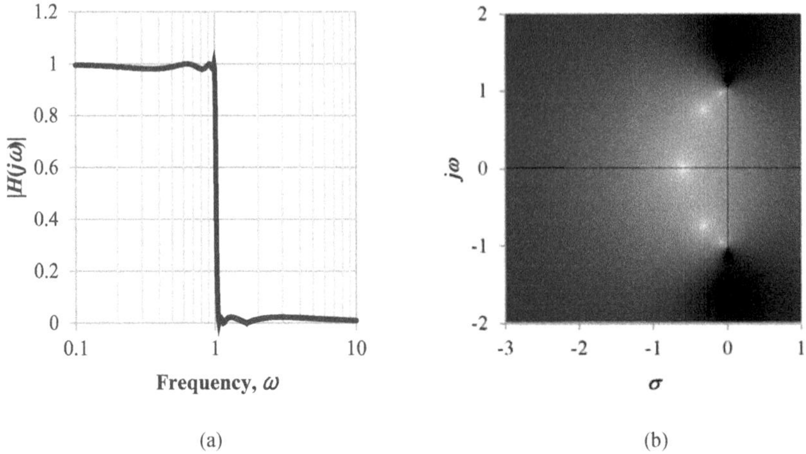

Figure 1.8 (a) Frequency response and (b) complex frequency space plot of a seventh-order elliptic filter.

Elliptic filters achieve the greatest frequency selectivity of any of the filter types presented, but they are difficult to make and are less common than, say, Chebyshev Type I filters. True elliptic filters will not be discussed any further in this book. However, sometimes filters that do have high selectivity and ripples in both the pass-band and stop-band are designed that do not strictly adhere to the equal-ripple constraint or fail to possess the full complement of (finite) stop-band zeros commensurate with the order of the filter. These are often referred to as quasi-elliptic or pseudo-elliptic filters. Examples of reflectionless pseudo-elliptic filters will be derived in Sections 4.2.3, 6.4.2, and 10.1.3.

1.4.6 Group Delay

In some filter applications, the frequency-dependent phase or time delay of a passing signal is an important concern. There are principally two different kinds of delay that are of interest to system designers, the phase delay,

$$\tau_\phi(\omega) = -\frac{\phi(\omega)}{\omega} \tag{1.36}$$

and the group delay,

$$\tau_g(\omega) = -\frac{d\phi(\omega)}{d\omega} \tag{1.37}$$

where

$$\phi(\omega) = \angle H(j\omega) \tag{1.38}$$

In a modulated waveform, the phase delay is the time required for the underlying carrier to propagate from one port of the filter to the other. Conversely, the group delay is the time required for the modulation envelope, or wave packet, to propagate from one port to the other.[3] The group delay is of primary concern, as it determines how the shape of a waveform comprising multiple frequencies is distorted beyond a simple scaling factor in amplitude or offset in time. For various applications in audio, communication, and digital data transmission, the goal is to provide a group delay that is as near to flat as possible across the pass-band of the filter.

The theoretical group delay for the first three filter types discussed in this section (Butterworth, Chebyshev Type I, and Chebyshev Type II) is shown in Figure 1.9. The Butterworth is the smoothest in general, while the Chebyshev Type II has the lowest group delay throughout the pass-band for filters of a given order. The Chebyshev Type II also has the rather curious property that its group delay actually decreases with increasing order, as seen in Figure 1.9(b).

1.5 CONVENTIONAL FILTER TOPOLOGIES

The most common topology used for lumped-element filters is the ladder-network, or Cauer topology, comprising alternating series elements and shunt elements, as shown in Figure 1.10 (not to be confused with *Cauer filters* which is another name for the elliptic filter frequency response introduced in Section 1.4.5). The element values required to realize a Chebyshev Type I filter, for example, using this topology have been derived [13],

$$\beta = \ln\left(\coth\left(\tfrac{1}{4}\ln\left(1+\varepsilon^2\right)\right)\right) \tag{1.39a}$$

$$\gamma = \sinh\left(\tfrac{\beta}{2N}\right) \tag{1.39b}$$

$$a_k = \sin\left(\tfrac{(2k-1)\pi}{2N}\right), \text{ for } k = 1 \ldots N \tag{1.39c}$$

$$b_k = \gamma^2 + \sin^2\left(\tfrac{k\pi}{N}\right), \text{ for } k = 1 \ldots N \tag{1.39d}$$

[3] At one time, it was thought that the group delay corresponded to the propagation time of information in a modulated signal. Although approximately true in most cases, this was a cause for some concern, as under certain conditions the propagation velocity associated with this delay could take on values that were negative or even superluminal (faster than light), which would be a violation of causality under special relativity. Such concerns were unjustified, as the seemingly anomalous velocities had already been explained in prior literature, and the actual speed at which information travels is the so-called *front velocity*, which is exactly *c*, the speed of light in a vacuum, regardless of the medium. For a thorough discussion of these topics, the reader is referred to [12].

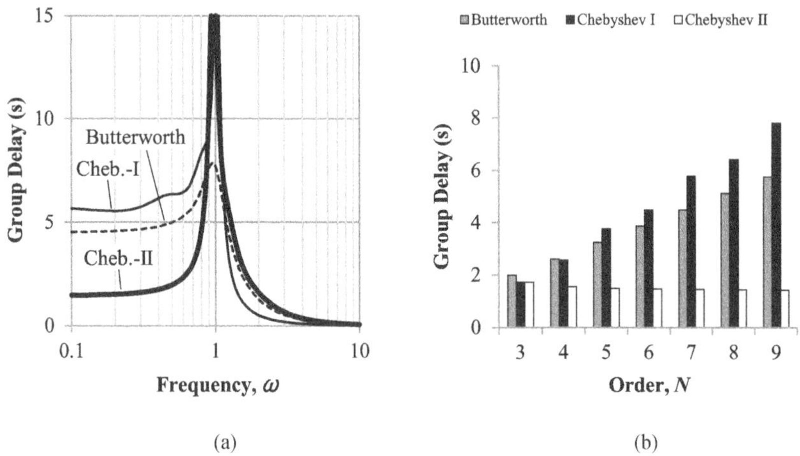

Figure 1.9 Theoretical group delay for the first three optimal filters discussed in this section. (a) As a function of frequency for order $N = 7$. (b) As a function of order, N, at $\omega = 0$. The curves are normalized such that the cutoff frequency is 1 rad/s.

$$g_1 = \frac{2a_1}{\gamma} \tag{1.39e}$$

$$g_k = \frac{4a_{k-1}a_k}{b_{k-1}g_{k-1}}, \text{ for } k = 2...N \tag{1.39f}$$

where g_k are the normalized values of the elements from left to right. For a filter with cutoff frequency, ω_c, designed to be matched in the pass-band to a characteristic impedance of Z_0, we must set each capacitor to $C_k = g_k Y_0/\omega_c$ and each inductor to $L_k = g_k Z_0/\omega_c$. Table 1.1 gives the normalized element values for a number of filters having different numbers of poles and ripple factors. More extensive tables are available in [13]. The parameters in Table 1.1 were chosen here because they correspond to filters of special significance when implemented with a reflectionless topology, as will be shown in Chapter 6. The third row for which $N = 7$ represents the filter that is plotted in Figure 1.6.

Butterworth filters may also be implemented using the Cauer topology, where different element values are assigned so as to move the poles further away from the $j\omega$ axis. The Chebyshev Type II filter, however, cannot be implemented using the Cauer topology shown in Figure 1.10. This is because this simple arrangement of inductors and capacitors cannot produce any transmission zeros except at infinity, regardless of what element values are chosen. To create zeros at finite frequencies, one must add capacitors in parallel with the series inductors or inductors in series with the shunt capacitors. This added complexity is one reason

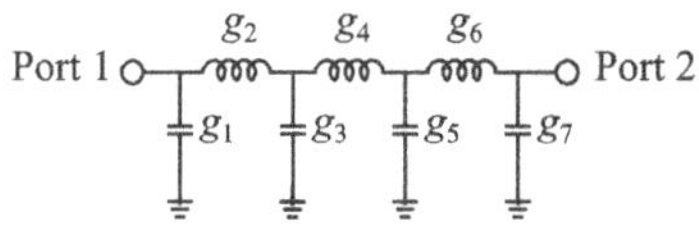

Figure 1.10 Low-pass Cauer topology. The elements are labeled with their normalized values.

Table 1.1
Prototype Element Values for Chebyshev Type I Filters

N	ε	g_1	g_2	g_3	g_4	g_5	g_6	g_7	g_8	g_9
3	0.1925	1.155	1.155	1.155						
5	0.2164	1.337	1.337	2.164	1.337	1.337				
7	0.2187	1.377	1.377	2.280	1.498	2.280	1.377	1.377		
9	0.2192	1.392	1.392	2.315	1.531	2.378	1.531	2.315	1.392	1.392

why Type II filters are not more common, despite having superior pass-band characteristics (vis à vis ripple and group delay) and the same overall slope in the transition-band (as evidenced by Figures 1.6 and 1.7). Interestingly, we will find that the opposite is true for reflectionless filters; in that case, the Type II filter is the easiest to realize, and Type I is more difficult.

1.5.1 Frequency Transformations

Filters are generally categorized as being one of four frequency response types: low-pass, high-pass, band-pass, and band-stop. These are illustrated in Figure 1.11, using the Chebyshev Type II response as a basis.

Thus far, all of the transfer functions discussed have been the low-pass type. To convert any of the given low-pass filters (or for that matter, any low-pass filter) to high-pass, band-pass, or band-stop, one may employ well-known frequency transformations [4]. For example, the transformation that converts a low-pass to a high-pass design is defined by the following substitution,

$$\frac{\omega}{\omega_c} \rightarrow -\frac{\omega_c}{\omega} \tag{1.40}$$

Therefore, the inductors within the low-pass filter topology that have reactance $\omega L_k = g_k Z_0 \omega / \omega_c$ are replaced with an element having reactance $-g_k Z_0 \omega_c / \omega$. The latter expression is that for a capacitor having value $Y_0/(g_k \omega_c)$. In a similar fashion, one finds that the capacitors in a low-pass prototype become inductors in the high-pass transformation.

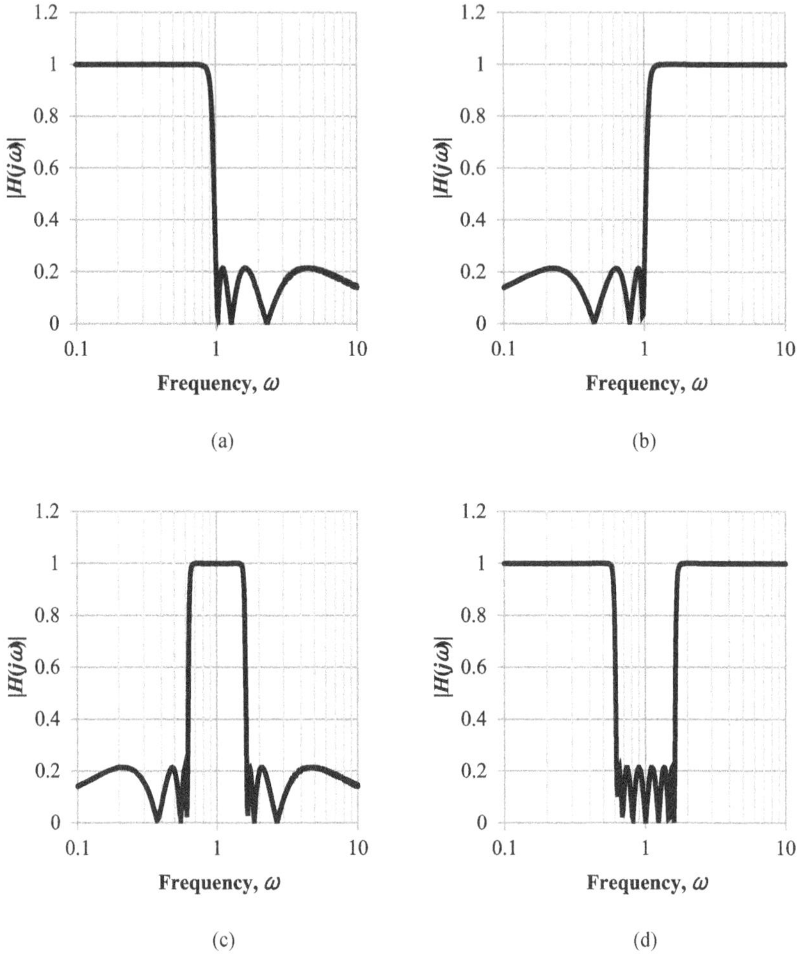

Figure 1.11 Illustration of the four general filter frequency response types. (a) Low-pass, (b) high-pass, (c) band-pass, and (d) band-stop. In all relevant cases, $\omega_c = \omega_0 = 1$, and $\Delta = 1$.

On the other hand, to convert from a low-pass prototype to band-pass, one may employ the alternate transformation,

$$\frac{\omega}{\omega_c} \rightarrow \frac{1}{\Delta}\left(\frac{\omega}{\omega_0} - \frac{\omega_0}{\omega}\right) \tag{1.41a}$$

Table 1.2

Frequency Transformation and Scaling

Low-Pass	*High-Pass*	*Band-Pass*	*Band-Stop*
$\dfrac{g_k Z_0}{\omega_c}$	$\dfrac{Y_0}{g_k \omega_c}$	$\dfrac{g_k Z_0}{\omega_0 \Delta}, \dfrac{Y_0 \Delta}{g_k \omega_0}$	$\dfrac{g_k Z_0 \Delta}{\omega_0} // \dfrac{Y_0}{g_k \omega_0 \Delta}$
	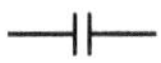		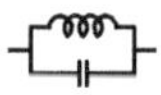
$\dfrac{g_k Y_0}{\omega_c}$	$\dfrac{Z_0}{g_k \omega_c}$	$\dfrac{Z_0 \Delta}{g_k \omega_0} // \dfrac{g_k Y_0}{\omega_0 \Delta}$	$\dfrac{Z_0}{g_k \omega_0 \Delta}, \dfrac{g_k Y_0 \Delta}{\omega_0}$
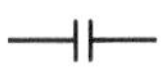			

$$\Delta = \frac{\omega_2 - \omega_1}{\omega_0} \tag{1.41b}$$

$$\omega_0 = \sqrt{\omega_1 \omega_2} \tag{1.41c}$$

where ω_1 and ω_2 are the pass-band edges, or the start and stop frequencies of the pass-band. In this case, an inductor in the low-pass prototype with reactance $g_k Z_0 \omega / \omega_c$ becomes an element with reactance $g_k Z_0 \omega / \omega_0 \Delta - g_k Z_0 \omega_0 / \omega \Delta$, which one may recognize as an inductor in series with a capacitor.

All the necessary element scalings and transformations required to implement low-pass, high-pass, band-pass, and band-stop filters are tabulated in Table 1.2.

1.6 EVEN- AND ODD-MODE ANALYSIS

Most of the reflectionless filters described in this book are symmetric. Those that aren't are constructed of smaller building blocks that are themselves symmetric. As such, the analysis of symmetric structures is pivotal in the development of reflectionless filters, as will become clear in Chapter 3. The mechanism for doing so is called even- and odd-mode analysis [4], related to an earlier concept known as Bartlett's bisection theorem [14]. Starting with a symmetric network, Figure 1.12(a), we imagine dividing the circuit along its symmetry plane, and pulling the two halves apart to expose the wires that occupy the central nodes of the circuit, as shown in Figure 1.12(b). We may then define two different kinds of excitation for

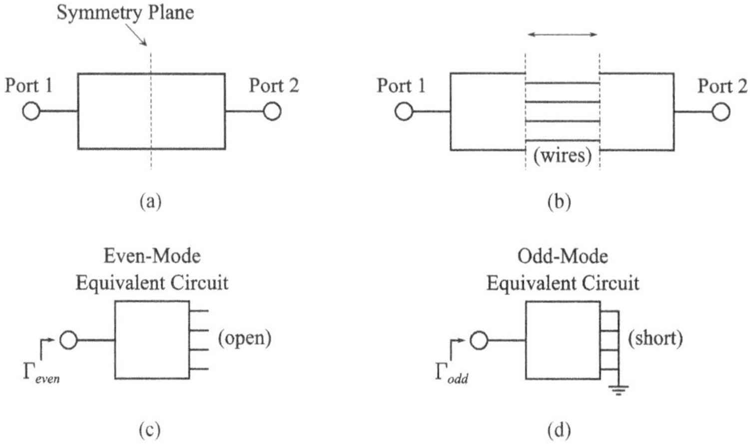

Figure 1.12 (a) A symmetric two-port network. (b) Exposed wires after separating along symmetry plane. (c) Even-mode equivalent circuit. (d) Odd-mode equivalent circuits.

this two-port, an even-mode and an odd-mode.

The even-mode excitation is characterized by the insertion of two identical signals into both ports simultaneously, with equal phase. It is intuitive that no current may cross the wires on the symmetry plane given such an excitation, for the direction of that crossing would be indeterminate. Therefore, it is permissible to draw only half of the network, replacing the symmetry plane with an open-circuit boundary, Figure 1.12(c). This half-circuit is called the *even-mode equivalent circuit*. It is equivalent in that it has the exact same port behavior from its single port as the full circuit does from either of its ports when stimulated with the even-mode excitation. The reflection coefficient seen looking into the single port of this equivalent circuit shall be called Γ_{even}.

The odd-mode excitation is characterized by the insertion of two identical signals into both ports simultaneously, but 180 degrees out of phase. Thus the signal at port 2 is the negative of the signal at port 1. In this case, one may logically deduce that no voltage with respect to ground may develop on any node touching the symmetry plane because the polarity of that voltage could not be determined. Therefore, one may draw only half of the network again, but this time the symmetry plane will become a virtual short to ground, Figure 1.12(d). Let the reflection coefficient seen looking into the odd-mode equivalent circuit be Γ_{odd}.

Consider how the behavior of the two equivalent half-circuits relates to that of the full two-port circuit. Since the two-port network is linear, the output from multiple excitations is simply given by the superposition of the outputs from each excitation individually. As described above, the even-mode excitation corresponds to simultaneous, identical excitations at both ports. Therefore, the output at port 1 under these conditions is given by the response of the network to identical inputs at both port 1 and port 2, or equivalently, the reflection from port 1 added to the transmission from port 2.

$$\Gamma_{even} = s_{11} + s_{12} \tag{1.42}$$

and, by symmetry,

$$\Gamma_{even} = s_{22} + s_{21} \tag{1.43}$$

On the other hand, the odd-mode excitation corresponds to simultaneous, identical excitations at both ports with opposite polarity. Thus, the output at port 1 is the superposition of the reflection from port 1 with the transmission from port 2 *from the equivalent negative input*, or

$$\Gamma_{odd} = s_{11} - s_{12} \tag{1.44}$$

and, again by symmetry,

$$\Gamma_{odd} = s_{22} - s_{21} \tag{1.45}$$

The full two-port scattering parameters of the original circuit may therefore be found by solving the above set of equations as,

$$s_{11} = s_{22} = \tfrac{1}{2}\left(\Gamma_{even} + \Gamma_{odd}\right) \tag{1.46a}$$

$$s_{21} = s_{12} = \tfrac{1}{2}\left(\Gamma_{even} - \Gamma_{odd}\right) \tag{1.46b}$$

In this way, the even- and odd-mode equivalent half-circuits are sufficient to completely describe the behavior of a symmetric two-port network.

It is important to note that the even- and odd-mode half-circuits are equivalent circuits only. The open- and short-circuit boundary conditions will typically result in some immediate and obvious simplifications that alter the original topology (e.g., some elements one might find to be shorted out and can be omitted, while others are left dangling with one end open and removed just as trivially). It may also be useful to rearrange elements (via exchanges of series elements, delta-wye transformations, etc.) so that analysis of the even- and odd-mode circuit behaviors is easier. Ultimately, what matters is the equivalence of the port parameters, regardless of what alterations to the topology have occurred.

Let us consider the simple example shown in Figure 1.13(a). A fairly tedious circuit to solve by conventional means, as it would have required the application of a delta-wye transformation yielding nonphysical elements, it becomes very easy with even- and odd-mode analysis. The circuit is redrawn in Figure 1.13(b)

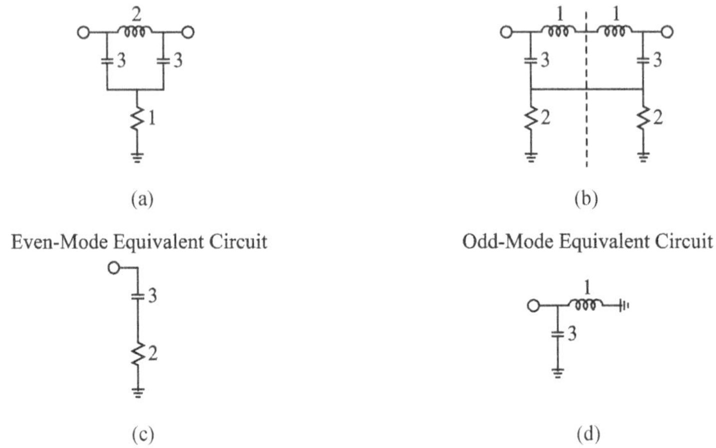

Figure 1.13 Example circuit for using even- and odd-mode analysis. (a) Original network. (b) Shown with explicit symmetry plane. (c) Even-mode equivalent circuit. (d) Odd-mode equivalent circuit. Elements are labeled with their normalized values.

with the symmetry plane explicit. To facilitate division of this network into half-circuits, we have split the elements that lie on the symmetry plane into pairs. The inductor has been omitted in the even-mode circuit (Figure 1.13(c)) as it would be an open element, whereas the resistor is omitted in the odd-mode circuit (Figure 1.13(d)) as it is shorted out.

The normalized input impedance of the even-mode equivalent circuit is given by

$$z_{even} = (3s)^{-1} + 2 = \frac{6s+1}{3s} \tag{1.47}$$

thus providing an even-mode reflection coefficient of

$$\Gamma_{even} = \frac{z_{even} - 1}{z_{even} - 1} = \frac{3s+1}{9s+1} \tag{1.48}$$

The odd-mode equivalent circuit has normalized input admittance

$$y_{odd} = s^{-1} + 3s = \frac{3s^2+1}{s} \tag{1.49}$$

thus providing an odd-mode reflection coefficient of

$$\Gamma_{odd} = \frac{1 - y_{odd}}{1 + y_{odd}} = \frac{-3s^2 + s - 1}{3s^2 + s + 1} \tag{1.50}$$

By (1.46), we have for the full two-port scattering parameters,

$$s_{11} = s_{22} = \frac{1}{2}\left(\frac{3s+1}{9s+1} + \frac{-3s^2 + s - 1}{3s^2 + s + 1}\right) = \frac{-9s^3 + 6s^2 - 2s}{(9s+1)(3s^2 + s + 1)} \tag{1.51a}$$

$$s_{12} = s_{21} = \frac{1}{2}\left(\frac{3s+1}{9s+1} - \frac{-3s^2 + s - 1}{3s^2 + s + 1}\right) = \frac{18s^3 + 6s + 1}{(9s+1)(3s^2 + s + 1)} \tag{1.51b}$$

1.7 TRANSMISSION LINES

The theoretical underpinning of lumped-element inductors and capacitors — and indeed of the whole of circuit theory embodied in Kirchoff's voltage and current laws — is an approximation of Maxwell's equations that results from assuming the physical extent of the circuit is very small compared to the wavelength of the signals passing through it. As the frequency increases and the wavelength shrinks, the circuit elements also must become smaller in order to continue behaving as the mathematical components they were intended to be. While the examples in this book will show lumped elements operating successfully as high as 40 GHz, eventually the scientific justification for lumped element design fails and a new paradigm is needed to extend the reflectionless filter concepts to higher frequencies.

Transmission lines offer a solution. In fact, transmission lines are a fundamental aspect of a microwave engineer's perspective of electronic circuits, as the definition of the scattering parameters at the beginning of this chapter worked on the assumption that transmission lines were connected to the ports. While a comprehensive treatment of transmission line theory is beyond the scope of this book, it will be useful to provide an overview of the most basic concepts. The reader is referred to [4, 15–17] for a more complete description.

1.7.1 The Telegrapher's Equations and Wave Solutions

Just as it was with lumped elements, the mathematical foundations of transmission lines derive from an approximation of Maxwell's equations, in this case that the transmission line structure is small in cross section, or the lateral direction, but potentially large in the longitudinal direction — "long and skinny," to put it in colloquial terms. The coaxial cable is an excellent example to bear in mind, having diameters that are much smaller than a wavelength, but extending

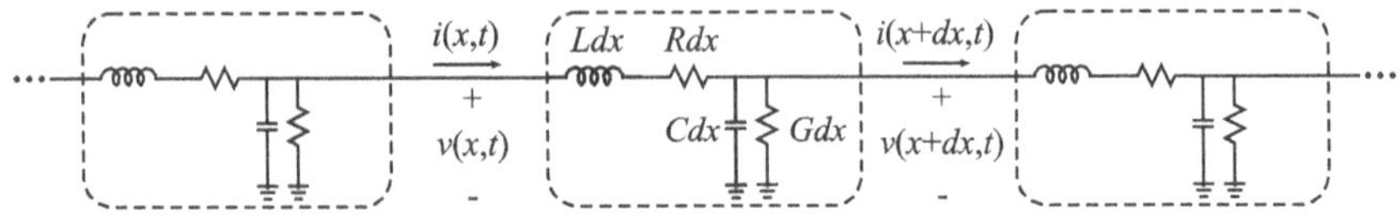

Figure 1.14 Model of a transmission line as a cascade of infinitesimal unit cells.

potentially many wavelengths long. To calculate the behavior of such a structure, we may first consider its behavior on an infinitesimal section of the line along its length, an electrically short piece that allows classical lumped-element approximations to apply. This viewpoint is shown in Figure 1.14, where L is the inductance per unit length, C is the capacitance per unit length, R is the series resistance per unit length, and G is the shunt conductance per unit length. R accounts for any Ohmic loss in the conductors of the transmission line (such as the inner and outer circular conductors of the coaxial cable), and G accounts for the dielectric loss or any leakage between the conductors.

Notionally, we model a transmission line as an ensemble of infinitely many, small, lumped elements. It is worth pointing out that this invalidates some of the conclusions of the previous sections with regard to realizable immittance and transfer functions, as they assumed a finite number of lumped elements. Still, the basic principles of causality and stability do apply.

Assuming a voltage, v, applied to the input terminals, and a current, i, injected into the unit cell, we may write a pair of equations for the same quantities at the output terminals that are the inputs of the next unit cell,

$$\frac{\partial v}{\partial x} = -iR - L\frac{\partial i}{\partial t} \tag{1.52a}$$

$$\frac{\partial i}{\partial x} = -vG - C\frac{\partial v}{\partial t} \tag{1.52b}$$

Known as the telegrapher's equations [13], these govern the time-dependent voltage and current, $v(x,t)$ and $i(x,t)$, at any point, x, along the line, and at any instant in time, t. These take on a somewhat simpler form in the Laplace domain,

$$\frac{dv}{dx} = -Z(s)i \tag{1.53a}$$

$$\frac{di}{dx} = -Y(s)v \tag{1.53b}$$

where

$$Z(s) = R + sL \tag{1.54a}$$

$$Y(s) = G + sC \tag{1.54b}$$

Differentiating (1.53a) with respect to x, and then substituting in from (1.53b), we have,

$$\frac{d^2 v}{dx^2} = \gamma^2 v \tag{1.55}$$

or similarly,

$$\frac{d^2 i}{dx^2} = \gamma^2 i \tag{1.56}$$

where

$$\gamma = \sqrt{Z(s)Y(s)} = \sqrt{(R + sL)(G + sC)} \tag{1.57}$$

Equations (1.55) and (1.56) are forms of the Helmholtz equation, a time-independent form of the wave equation. Its solutions take the form of exponentials, $e^{\pm\gamma x}$, corresponding to forward- and backward-traveling waves on the transmission line, where γ is known as the propagation constant.

1.7.2 The Standard Transmission-Line Model

We thus envision electromagnetic waves propagating from one end of the transmission line to the other and reflecting off of the terminating impedances at either end. The conceptual picture that materializes from this analysis is that shown in Figure 1.15.

The waves are characterized by voltage and current amplitudes on the conductors of the transmission line (the inner and outer conductors in the case of a coaxial cable). The wave amplitudes v^+ and i^+ correspond to the voltage and current phasors for a wave propagating in the forward/positive direction (from port 1 to port 2) and v^- and i^- correspond to the voltage and current phasor amplitudes for a wave propagating in the reverse/negative direction (from port 2 to port 1). In both cases, we define the current as being positive in the direction of propagation. The ratio of voltage to current for a given wave is determined by the characteristic impedance, given by

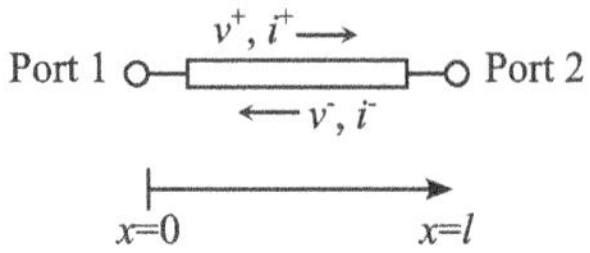

Figure 1.15 A transmission line with forward- and backward-travelling waves of voltage and current.

$$Z_0 = \sqrt{\frac{Z(s)}{Y(s)}} = \sqrt{\frac{R + sL}{G + sC}} \tag{1.58}$$

Like the propagation constant, γ, the characteristic impedance, Z_0, depends on cross-sectional dimensions of the transmission line (e.g., the diameters of the coaxial conductors) and any dielectric material that may be filling it. We thus have

$$v^+ = i^+ Z_0 \tag{1.59a}$$

$$v^- = i^- Z_0 \tag{1.59b}$$

The length of the transmission line is accounted for by the exponential solutions of the Helmholtz equation. Let us define the longitudinal coordinate system such that port 1 is located at $x = 0$, and port 2 at $x = l$, where l is the length of the line. At a given moment in time, the phase of the waveform at point x along the line is delayed, or lagging, by an amount γx for a positive travelling wave, but is advanced/leading by an equal amount for a negative travelling wave. The total voltage and current at any point along the line in phasor amplitude is thus given by

$$v(x) = v^+ e^{-\gamma x} + v^- e^{\gamma x} \tag{1.60a}$$

$$i(x) = i^+ e^{-\gamma x} - i^- e^{\gamma x} \tag{1.60b}$$

Here we have defined the current in the positive x direction. Since the negative travelling wave has its current directed in the negative direction (the direction of its propagation) we have used a minus sign in (1.60b).

1.7.3 Special Cases

Let us assume port 2 is terminated with a load impedance, Z_L. This constrains the ratio of the total voltage and current at the far end of the line as follows,

$$\frac{v(l)}{i(l)}=\frac{v^{+}e^{-\gamma l}+v^{-}e^{\gamma l}}{i^{+}e^{-\gamma l}-i^{-}e^{\gamma l}}=Z_L \tag{1.61}$$

Applying (1.59), we have

$$\frac{v^{+}e^{-\gamma l}+v^{-}e^{\gamma l}}{i^{+}e^{-\gamma l}-i^{-}e^{\gamma l}}=\frac{v^{+}e^{-\gamma l}+v^{-}e^{\gamma l}}{v^{+}e^{-\gamma l}-v^{-}e^{\gamma l}}Z_0=Z_L \tag{1.62a}$$

$$\frac{v^{-}}{v^{+}}=\frac{Z_L-Z_0}{Z_L+Z_0}e^{-2\gamma l} \tag{1.62b}$$

The impedance seen looking into port 1 is then

$$Z_{in}=\frac{v(0)}{i(0)}=\frac{v^{+}+v^{-}}{i^{+}-i^{-}}=Z_0\frac{v^{+}+v^{-}}{v^{+}-v^{-}}=Z_0\frac{(Z_L+Z_0)+(Z_L-Z_0)e^{-2\gamma l}}{(Z_L+Z_0)-(Z_L-Z_0)e^{-2\gamma l}} \tag{1.63a}$$

$$=Z_0\frac{Z_L\cosh(\gamma l)+Z_0\sinh(\gamma l)}{Z_0\cosh(\gamma l)+Z_L\sinh(\gamma l)} \tag{1.63b}$$

The above solution for the impedance looking into a transmission line is quite general. Some useful simplifications are worth noting for easy reference. First, if the transmission line is effectively lossless, then the propagation constant is purely imaginary, $\gamma = j\beta$. If the lossless propagation constant, β (not to be confused with the prototype parameter β from (1.39a)) is directly proportional to the frequency, ω, as is at least approximately true in most cases, then the transmission line is said to be free of dispersion. Specifically,

$$\beta=\frac{\omega}{v_p} \tag{1.64}$$

where v_p is known as the phase velocity.

Substituting the lossless propagation constant $j\beta$ into (1.63b), we have,

$$Z_{in}=Z_0\frac{Z_L+jZ_0\tan(\beta l)}{Z_0+jZ_L\tan(\beta l)} \tag{1.65}$$

Further, if the load impedance at port 2 is an open- or short-circuit (i.e., the transmission line is a *stub*),

$$Z_{in} = \begin{cases} -jZ_0 \cot(\beta l) & Z_L = \infty \\ jZ_0 \tan(\beta l) & Z_L = 0 \end{cases} \tag{1.66}$$

On the other hand, if the load impedance is finite, but the line is a quarter wavelength long at some frequency ($\beta l = \pi/2$),

$$Z_{in} = \frac{Z_0^2}{Z_L} \tag{1.67}$$

This is sometimes referred to as a quarter-wave transformer or impedance-inverter, since it inverts the load impedance, Z_L, relative to the characteristic impedance of the transmission line, Z_0.

1.7.4 Impedance and Scattering Parameters

Let us consider a transmission line having characteristic impedance Z_c embedded in a network with a system characteristic impedance of Z_0. The impedance parameters of the transmission line section may be derived by first assuming a current injected into port 1, while port 2 is open-circuited. Thus, the total current at the end of the transmission line at port 2 must be zero,

$$i(L) = i^+ e^{-\gamma l} - i^- e^{\gamma l} = 0 \tag{1.68a}$$

$$\therefore \frac{i^-}{i^+} = \frac{v^-}{v^+} = e^{-2\gamma l} \tag{1.68b}$$

The impedance parameters are then found by considering the total voltage developed at either port relative to the total current injected at port 1,

$$Z_{11} = \frac{v(0)}{i(0)} = \frac{v^+ + v^-}{i^+ - i^-} = \frac{v^+\left(1 + e^{-2\gamma l}\right)}{i^+\left(1 - e^{-2\gamma l}\right)} = Z_c \coth(\gamma l) \tag{1.69a}$$

$$Z_{21} = \frac{v(l)}{i(0)} = \frac{v^+ e^{-\gamma l} + v^- e^{\gamma l}}{i^+ - i^-} = \frac{v^+\left(2e^{-\gamma l}\right)}{i^+\left(1 - e^{-2\gamma l}\right)} = \frac{Z_c}{\sinh(\gamma l)} \tag{1.69b}$$

By symmetry, $Z_{22} = Z_{11}$ and $Z_{12} = Z_{21}$, therefore,

$$\mathbf{Z} = Z_c \begin{pmatrix} \coth(\gamma l) & \operatorname{csch}(\gamma l) \\ \operatorname{csch}(\gamma l) & \coth(\gamma l) \end{pmatrix} \tag{1.70}$$

The scattering parameters may then be found by application of (1.6),

$$\mathbf{S} = (\mathbf{Z} + Z_0\mathbf{I})^{-1}(\mathbf{Z} - Z_0\mathbf{I}) \tag{1.71a}$$

$$= \begin{pmatrix} z\coth(\gamma l)+1 & z\operatorname{csch}(\gamma l) \\ z\operatorname{csch}(\gamma l) & z\coth(\gamma l)+1 \end{pmatrix}^{-1} \begin{pmatrix} z\coth(\gamma l)-1 & z\operatorname{csch}(\gamma l) \\ z\operatorname{csch}(\gamma l) & z\coth(\gamma l)-1 \end{pmatrix} \tag{1.71b}$$

$$= \frac{1}{2\cosh(\gamma l)+(z+y)\sinh(\gamma l)} \begin{pmatrix} (z-y)\sinh(\gamma l) & 2 \\ 2 & (z-y)\sinh(\gamma l) \end{pmatrix} \tag{1.71c}$$

where $z = y^{-1} = Z_c/Z_0$ is the normalized characteristic impedance of the line.

1.7.5 Coupled Transmission Lines

Although coaxial cable is a good example to bear in mind for basic transmission line concepts, its modern usage is more often as a generic interconnect between subassemblies than as a circuit element in a filter. For the latter usage, printed transmission lines (such as microstrip or stripline) are far more common. Aside from the obvious geometric differences, the fundamental equations governing their behavior is essentially the same. Their *open* structure, however, does permit a different type of circuit element that is not easily realized with coax. Specifically, if two microstrips or striplines are brought very close together in parallel, waves propagating on one line can induce waves propagating in the other. The two lines are said to be *coupled.* The electromagnetic mechanism of this coupling will not be explored in detail here, but in principle, it results from mutual inductance and capacitance between them.

Analytically, coupled lines may be characterized via even- and odd-mode analysis, with the symmetry plane drawn down the middle of the gap between the two lines, as shown in Figure 1.16. Though the symmetry plane doesn't contact any nodes, it may be thought of as an electric or magnetic boundary condition very near to the transmission line in the even- or odd-mode equivalent circuit, respectively. This has the effect of distorting the fields around the transmission line, altering the characteristic impedance and potentially the propagation constant. Therefore, the even- and odd-mode equivalent circuits have the same scattering parameters as an isolated transmission line given in (1.71), but with line characteristics of Z_e and γ_e, or Z_o and γ_o, respectively. That is,

$$\mathbf{S}_e = \begin{pmatrix} s_{11,e} & s_{12,e} \\ s_{21,e} & s_{22,e} \end{pmatrix} = \frac{\begin{pmatrix} (z_e - y_e)\sinh(\gamma_e l) & 2 \\ 2 & (z_e - y_e)\sinh(\gamma_e l) \end{pmatrix}}{2\cosh(\gamma_e l) + (z_e + y_e)\sinh(\gamma_e l)} \tag{1.72a}$$

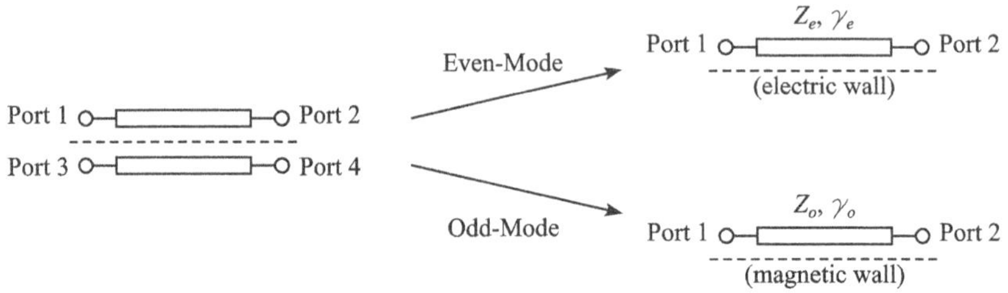

Figure 1.16 Even- and odd-mode decomposition of coupled transmission lines.

$$\mathbf{S}_o = \begin{pmatrix} s_{11,o} & s_{12,o} \\ s_{21,o} & s_{22,o} \end{pmatrix} = \frac{\begin{pmatrix} (z_o - y_o)\sinh(\gamma_o l) & 2 \\ 2 & (z_o - y_o)\sinh(\gamma_o l) \end{pmatrix}}{2\cosh(\gamma_o l) + (z_o + y_o)\sinh(\gamma_o l)} \tag{1.72b}$$

To find the full four-port scattering parameters of the coupled transmission line, we must apply superposition to the responses of the even- and odd-mode equivalent circuits given above. (This was described earlier in Section 1.6 in the context of two-port networks.) For example, the port 1 reflection coefficient, s_{11}, is seen as the combined response of the even- and odd-mode reflection coefficients,

$$s_{11} = \tfrac{1}{2}\left(s_{11,e} + s_{11,o}\right) \tag{1.73}$$

The coupling response to port 3, called s_{31}, is the difference of these two,

$$s_{31} = \tfrac{1}{2}\left(s_{11,e} - s_{11,o}\right) \tag{1.74}$$

These equations are essentially equivalent to (1.46a) and (1.46b), given earlier for two-port networks. To derive the coefficients of scattering from port 1 to ports 2 and 4 at the far end of the coupled lines, we must consider the even- and odd-mode equivalent circuit *transmission* coefficients, namely,

$$s_{21} = \tfrac{1}{2}\left(s_{21,e} + s_{21,o}\right) \tag{1.75a}$$

$$s_{41} = \tfrac{1}{2}\left(s_{21,e} - s_{21,o}\right) \tag{1.75b}$$

The remainder of the four-port scattering parameters may be derived by observing symmetry. Putting this all together, we have,

$$\mathbf{S} = \begin{pmatrix} s_{11} & s_{12} & s_{13} & s_{14} \\ s_{21} & s_{22} & s_{23} & s_{24} \\ s_{31} & s_{32} & s_{33} & s_{34} \\ s_{41} & s_{42} & s_{43} & s_{44} \end{pmatrix} \tag{1.76}$$

where

$$s_{11} = s_{22} = s_{33} = s_{44} = \frac{\frac{1}{2}(z_e - y_e)\sinh(\gamma_e l)}{2\cosh(\gamma_e l) + (z_e + y_e)\sinh(\gamma_e l)} + \frac{\frac{1}{2}(z_o - y_o)\sinh(\gamma_o l)}{2\cosh(\gamma_o l) + (z_o + y_o)\sinh(\gamma_o l)} \tag{1.77a}$$

$$s_{21} = s_{12} = s_{43} = s_{34} = \frac{1}{2\cosh(\gamma_e l) + (z_e + y_e)\sinh(\gamma_e l)} + \frac{1}{2\cosh(\gamma_o l) + (z_o + y_o)\sinh(\gamma_o l)} \tag{1.77b}$$

$$s_{31} = s_{13} = s_{42} = s_{24} = \frac{\frac{1}{2}(z_e - y_e)\sinh(\gamma_e l)}{2\cosh(\gamma_e L)l + (z_e + y_e)\sinh(\gamma_e l)} - \frac{\frac{1}{2}(z_o - y_o)\sinh(\gamma_o l)}{2\cosh(\gamma_o l) + (z_o + y_o)\sinh(\gamma_o l)} \tag{1.77c}$$

$$s_{41} = s_{14} = s_{32} = s_{23} = \frac{1}{2\cosh(\gamma_e l) + (z_e + y_e)\sinh(\gamma_e l)} - \frac{1}{2\cosh(\gamma_o l) + (z_o + y_o)\sinh(\gamma_o l)} \tag{1.77d}$$

References

[1] Gonzalez, G., *Microwave Transistor Amplifiers: Analysis and Design*, 2nd ed., Upper Saddle River, NJ: Prentice Hall, pp. 1–61, 1996.

[2] Smith, P., "Transmission line calculator," *Electronics*, vol. 12, no. 1, pp. 29–31, January 1939.

[3] Smith, P., "An improved transmission line calculator," *Electronics*, vol. 17, no. 1, p. 130, January 1944.

[4] Pozar, D., *Microwave Engineering*, 4th ed., New York: Wiley, November 2011.

[5] Arfken, G. and Weber, H., *Mathematical Methods for Physicists*, 4th ed., San Diego, CA: Academic Press, pp. 877–916, 1995.

[6] Kuo, F., *Network Analysis and Synthesis*, 2nd ed., New York: Wiley, pp. 290–308, 1962.

[7] Balanis, C., *Advanced Engineering Electromagnetics*, New York: Wiley, pp. 310–312, 1989.

[8] https://en.wikipedia.org/wiki/Chebyshev_polynomials.

[9] https://en.wikipedia.org/wiki/Elliptic_filter.

[10] https://en.wikipedia.org/wiki/Elliptic_rational_functions.

[11] https://www.wolframalpha.com/.

[12] Milonni, P., *Fast Light, Slow Light and Left-Handed Light*, New York: Taylor & Francis, pp. 26–58, 2005.

[13] Matthaei, G., Young, L., and Jones, E., *Microwave Filters, Impedance-Matching Networks, and Coupling Structures*, Norwood, MA: Artech House, 1980.

[14] Guillemin, E., *Synthesis of Passive Networks*, New York: Wiley, pp. 196–198, 1957.

[15] Ramo, S., Whinnery, J., and Van Duzer, T., *Fields and Waves in Communications Electronics*, 3rd ed., New York: Wiley, 1993.

[16] Miano, G. and Maffuci, A., *Transmission Lines and Lumped Circuits: Fundamentals and Applications*, London: Academic Press, 2001.

[17] Franceschetti, G., *Electromagnetics: Theory, Techniques, and Engineering Paradigms*, New York: Plenum Press, 1997.

Chapter 2

Classical Absorptive Filters

This chapter discusses some of the preexisting methods of constructing absorptive filters, including terminated diplexers, directional filters, constant-resistance networks, and a broad sample of empirical transmission-line and waveguide filtering structures. It explores the limitations of these approaches, both theoretical and practical, and reveals some of the common misconceptions about them that have propagated in academic literature.

2.1 DIPLEXERS

One fairly obvious way of constructing an absorptive filter is to create a diplexer and terminate the branch associated with the unwanted portion of the spectrum with a resistor. Though straightforward in principle, there are many details to consider. For one, a conventional diplexer will only have low reflections from one side. To have low reflections from both sides, a bidirectional diplexer is needed.

Supposing that low reflection is only needed from one side, the diplexer still presents many challenges. To construct a diplexer from conventional high- and low-pass filters, as shown in Figure 2.1(a), the two subfilters must present a high-impedance to the branch-point in their respective stop-bands in order to avoid loading down one another. Note that the core topology of the low-pass prototype in this case starts with a series element in order to provide that high impedance, unlike the topology in Figure 1.10, which would have the opposite effect. (The low-pass section here is, in fact, the dual of Figure 1.10.) Moreover, if the filter is to be truly reflectionless — which we shall define throughout this book as having theoretically zero reflection at all frequencies and not simply lossy at some frequencies — the transition regions of the two subfilters must balance in just the right way so that sum of the admittances at the branch-point is constant. Otherwise, at the very least, there will be a peak in reflection at the transition-band of the diplexer.

Unfortunately, the high- and low-pass versions of the Chebyshev and elliptic filter responses presented in the previous chapter do not behave in this way,

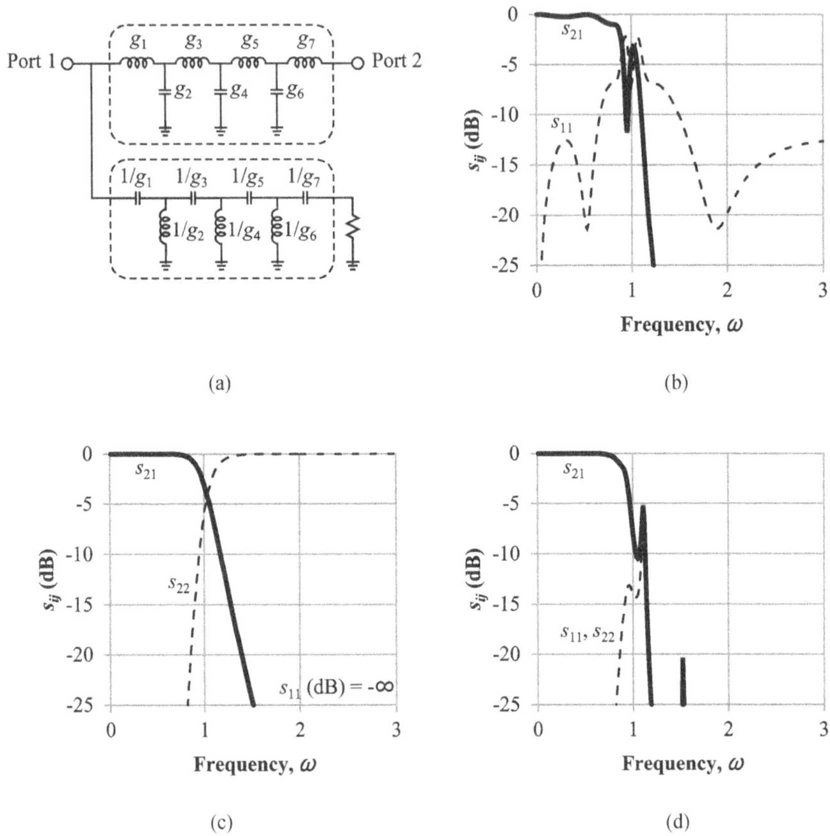

Figure 2.1 (a) Absorptive filter implemented as a diplexer. Note that the normalized values of the high-pass filter elements have been inverted in accordance with the high-pass transformation described in the previous chapter. (b) Frequency response when the prototype element values are taken from Table 1.1 for a seventh-order Chebyshev Type I filter. The admittances at the branch-point do not properly add up to a constant value. (c) Frequency response when the prototype element values correspond to a singly terminated Butterworth filter. (d) Frequency response when two Butterworth diplexers are used back to back. Resonances occur as a result of interactions between their mismatched ports.

regardless of their implementation. Figure 2.1(b), for example, shows what happens when the prototype element values are taken from Table 1.1 for a seventh-order Chebyshev Type I filter. Due to the interaction of the admittances of the two subfilters, not only is the reflection very high in the transition-band, but the overall pass-band shape has been severely corrupted.

In fact, only one of the optimal filter functions described thus far readily combines to produce low transition-band reflections — namely, the Butterworth filter. (Strictly speaking, what is required is a singly terminated Butterworth filter

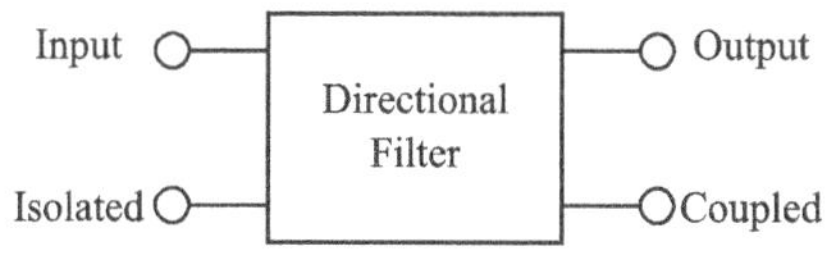

Figure 2.2 Directional filter concept.

prototype, the details of which are beyond the scope of this section [1]). This is illustrated in Figure 2.1(c), where it is evident that the return loss from the input side of the diplexer (s_{11}) is infinite, but the return loss from the output side (s_{22}) is the same as that for a nonabsorptive Butterworth filter. Simply cascading two diplexers back to back is not a solution, as the interaction between the two at their unmatched ports creates resonances that corrupt the frequency response near cutoff. Figure 2.1(d) is a simulation of the result when two singly terminated Butterworth diplexers are used in a back-to-back configuration. Sharp resonances are evident in all scattering parameters. This is an example of how conventional filters (or diplexers) do not cascade well with one another.

Despite these examples, it is possible to create a diplexer that approximates all the desired properties through empirical optimization; as a theoretical problem, however, it is only weakly solved.

In any case, combining subfilters to create diplexers is a delicate business, requiring careful consideration of the layout to ensure that reflections from the subfilters do not arrive back at the branch points with unanticipated phase errors.

2.2 DIRECTIONAL FILTERS

A lesser-known solution is to use a directional filter structure, illustrated by Figure 2.2. These intrinsically four-port networks operate similarly to directional couplers, except that the through- and coupled-response is strongly frequency-dependent. That is, signals at the input are divided so that power primarily from the pass-band is routed to the output-port, while power primarily from the stop-band is routed to the coupled-port. The fourth port is nominally isolated from signals entering the network from the input side. Typically, such structures are symmetric, such that the coupled- and isolated-ports exchange labels when the input and output are reversed. As such, a termination present on the coupled-port absorbs stop-band energy from the input side, while a termination on the isolated port absorbs stop-band energy incident from the output side.

Although it is feasible to realize a directional filter using lumped elements, the most common forms are usually first conceived in transmission lines or waveguides. Several commonly cited examples appear in Figure 2.3. The first two stripline filters employing capacitively coupled half-wave and full-wave resonators are only capable of yielding first-order responses. It was once thought

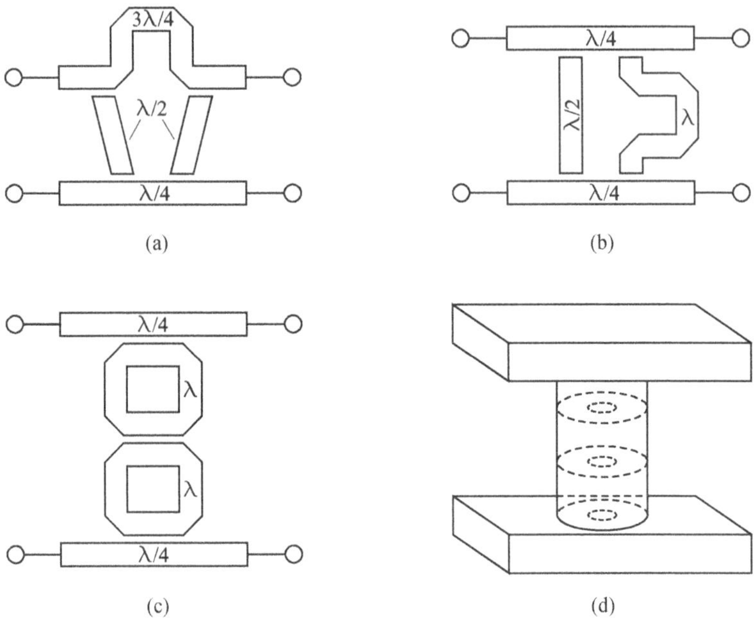

Figure 2.3 Examples of directional filter structures. (*After*: Matthaei, G., Young, L., and Jones, E. Artech House, 1980 [1].)

that the filter in Figure 2.3(a) could be extended to higher orders by cascading additional half-wave resonators in the shunt signal paths [2–3], but this was later proven to be erroneous [1].

The traveling wave ring directional filter in Figure 2.3(c) is capable of a higher-order response with additional rings, but it is found to be exceptionally sensitive to the directivity in the quarter-wavelength coupled-line sections, and by extension, to reflections from the mitered bends. It has not found widespread use in practice.

The final example in Figure 2.3(d), a waveguide filter employing direct-coupled cylindrical cavity resonators, is capable of higher-order responses with low-loss, and has found some use in applications for which the size of a waveguide implementation is not an issue. It is usually restricted to very narrow bandwidths, less than 1 percent unless special coupling apertures are used, in which case 2 percent is the maximum for a multiresonator response.

2.3 HYBRID-COUPLED FILTERS

One can also synthesize a directional filter by combining an identical pair of

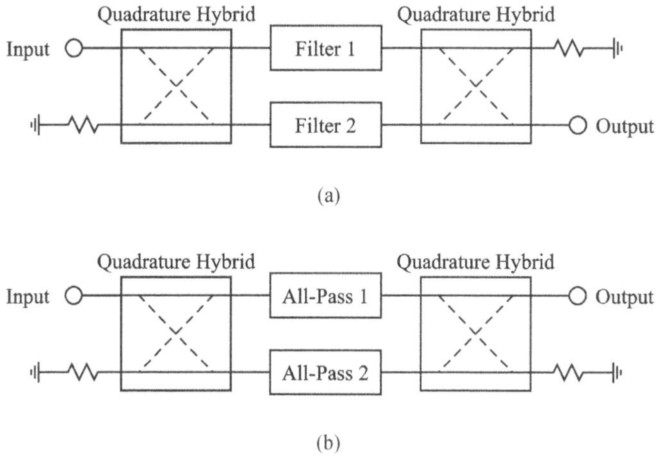

Figure 2.4 Two examples of hybrid-coupled absorptive filter designs. (a) Directional "balanced" filter. (b) All-pass parallel-cascade filter exploiting phase-cancellation.

filters with input and output quadrature hybrids, as shown in Figure 2.4(a). Any type of filter may be used. Its operation is similar to that of a balanced amplifier. The quadrature hybrid at the input splits the signal into two equal parts, each passing to one of the filters. The part that takes the diagonal path through the hybrid experiences a 90-degree phase shift relative to the horizontal path. For pass-band frequencies, the two signals combine in the output quadrature hybrid in-phase with one another at the output port. Reflections in the stop-band, however, cancel at the input as the signal that reflects from Filter 2 undergoes 180 degrees of phase change compared to the signal that reflects from Filter 1. These reflections instead add constructively at the isolated port of the hybrid where a termination resistor absorbs them.

Alternatively, one may use all-pass networks in place of the filters, as in Figure 2.4(b). The two all-pass networks differ in this arrangement, but provide for a differential phase delay in combination with the quadrature hybrids that cancels the signal out in the stop-band while adding constructively at pass-band frequencies [4].

The shortfall of these types of absorptive filters is a practical one — namely that an ideal quadrature hybrid with infinite bandwidth does not exist. In many applications, the bandwidth of the hybrids is little better than that of the filters themselves and is certainly not as wide as the stop-band that the user may require. As a consequence, this method is really only useful for providing band-pass or band-stop filtering with absorption locally near the transition-bands. Hybrids are also bulky and often more expensive than other filter components.

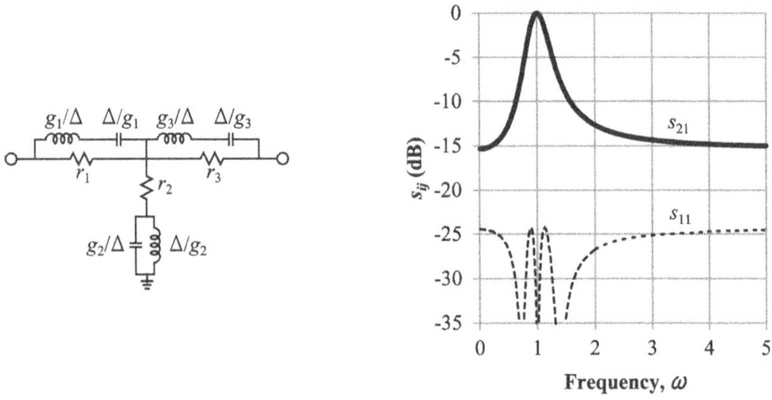

(a) (b)

Figure 2.5 (a) Third-order band-pass filter with lossy elements inserted into the resonators. Δ is the fractional bandwidth as defined in (1.41b). (b) Frequency response for $\omega_0 = 1$, $\Delta = 0.5$. The reactive elements are tuned according to a Chebyshev Type I prototype, and the normalized resistances are $r_1 = r_3 = 0.8$, and $r_2 = 0.4$.

2.4 EMPIRICAL APPROACHES

A number of empirical approaches to providing absorption of out-of-band signals in filtering structures have been tried over the years. Usually this involves the strategic addition of lossy elements to already well-established conventional filter topologies.

2.4.1 Damped Resonators

See Figure 2.5(a) for example. Based on a conventional third-order band-pass design in the Cauer topology, lossy elements have been inserted into the resonators [5]. Specifically, resistors have been placed in parallel with the series *LC* resonators and in series with the parallel *LC* resonator. This placement ensures that the resistors are inactive at the point where the *LC* pairs are in resonance (in this case, the center of the band) because the series *LC* tends toward zero impedance at the resonant frequency, shorting out the resistor that is in parallel with it, while the parallel *LC* tends toward zero admittance at the resonant frequency, disconnecting the resistor in series with it. Therefore, the minimum loss of the filter at $\omega = \omega_0$ remains 0 dB. On the other hand, when the filter is in full cutoff, the resonators effectively drop out, leaving the resistors as a tee attenuator network in their place. The best overall reflection is found when the

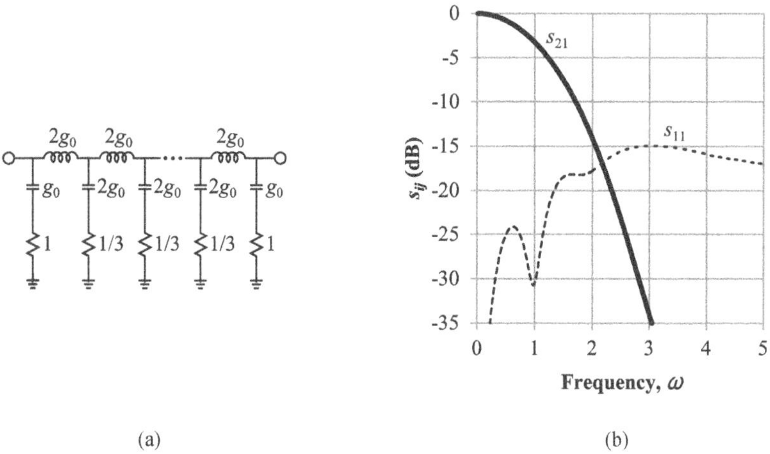

Figure 2.6 (a) Absorptive near-Gaussian filter. (b) Frequency response for $g_0 = (n\text{-}1)^{-1/2}$.

attenuator is not tuned to a perfect match, but rather a compromise — approximately 24 dB in Figure 2.5(b), though other choices are possible.

Unfortunately, while the insertion loss and reflection at three frequency points (0, ω_0, and ∞) are all prescribed by this structure, the presence of the resistors distorts the shape of the pass-band, and limits the rejection that is achieved.

2.4.2 Absorptive Gaussian Filters

Another form of absorptive filter that has found some use in practice achieves a so-called near-Gaussian filter response [6]. Filters with such a response, including another variation called a Bessel-Thompson filter, are not designed for the highest selectivity, but rather are optimized for flat group delay through the pass- and transition-bands while limiting the bandwidth of the output spectrum. As such, they find extensive use in communication and time-domain measurement systems where distortion of pulse shapes and other transient waveforms is of great importance. The absorptive version of this filter is shown in Figure 2.6(a). Note that the basic topology is the Cauer topology, except with uniform element values approximating a transmission line and lossy elements inserted periodically. The theoretical frequency response of the filter is shown in Figure 2.6(b).

2.4.3 Absorptive Transmission-Line Equalizers

One may also try incorporating resistive elements into transmission line structures to achieve low reflections across all frequencies. A number of such networks have been discovered, essentially by trial and error, some of which are shown in Figures 2.7–2.9. Design equations and performance curves are given for most of

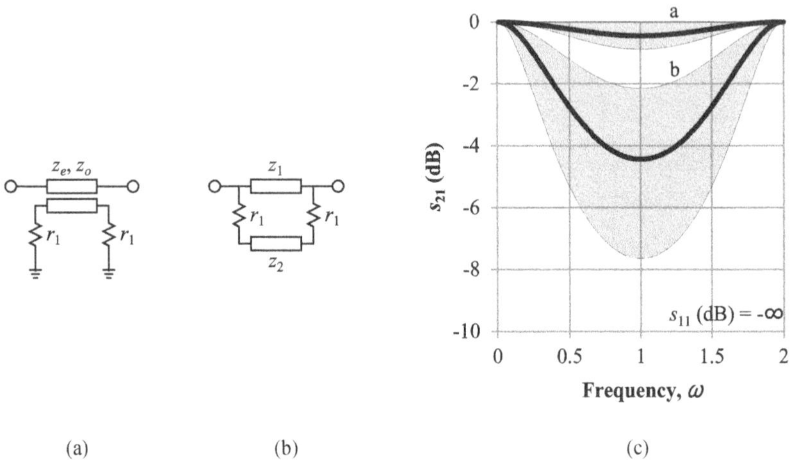

Figure 2.7 Two reflectionless equalizers using (a) coupled and (b) uncoupled transmission lines. (c) Frequency response for both circuits. The shaded areas represent the range of attenuation levels that can be achieved with realistic impedances and coupling factors.

these networks in [7].

These are all members of a class of transmission-line topologies known as commensurate-line networks [8], characterized by having transmission lines of all equal lengths (or integer multiples of that length). In this case, all transmission lines shown are a quarter-wavelength long at the center frequency. The two networks in Figure 2.7 are truly reflectionless (having identically zero reflection coefficient at all frequencies, the rest being only partially absorptive). However, these have very little out-of-band rejection (at most a few dB) given realistic constraints on characteristic impedance and coupling factor. As a consequence, these are primarily useful as absorptive slope equalizers rather than filters in the traditional sense.

Another set of transmission-line equalizers follows in Figure 2.8. While absorptive, these differ from the previous example in that their return loss is not perfect (see Figure 2.8(d) and (e)). For the range of values that represent the areas shaded in the figure, the return loss of the first circuit is better than about 17 dB, and for the second circuit, it is better than 11.5 dB. In both cases, curiously, the impedance traces out a perfect circle in the Smith chart when optimally tuned — tangent to the origin for the first network, and centered on the origin for the second, as shown in Figure 2.8(e). This feature is useful to know when deriving the element values [7].

Finally, there are many other empirically derived, absorptive transmission-line structures that have been discovered with more filter-like characteristics, such as those shown in Figure 2.9. Unlike the preceding equalizer sections, these do offer large attenuation in the stop-band (see, for example, the frequency responses

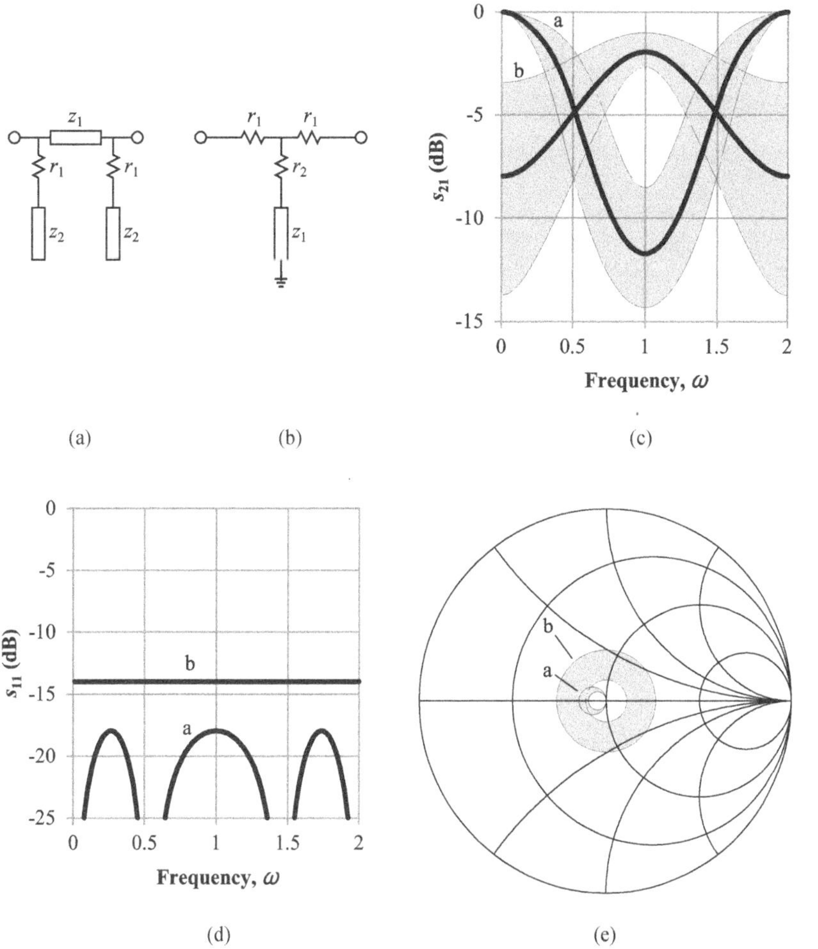

Figure 2.8 Absorptive transmission-line equalizers using (a) open-circuited and (b) short-circuited stubs. (c) Frequency response for both circuits. The shaded areas represent the range of attenuation levels that can be achieved with realistic impedances. (d) Representative reflection coefficient for both circuits. (e) Smith chart showing the extent of the impedance curves for both circuits over the same range of impedance values.

in Figure 2.9(e) and (f)), but the pass-band shape is far from optimal for the requirements of most filtering applications. Many of them also require very tightly coupled lines or extreme impedances that are difficult to realize in practice.

For example, a number of these absorptive transmission-line structures were utilized in the design of a selectable slope equalizer on the Jansky Very Large Array (JVLA) radio telescope. Data collected from the wideband front-ends of

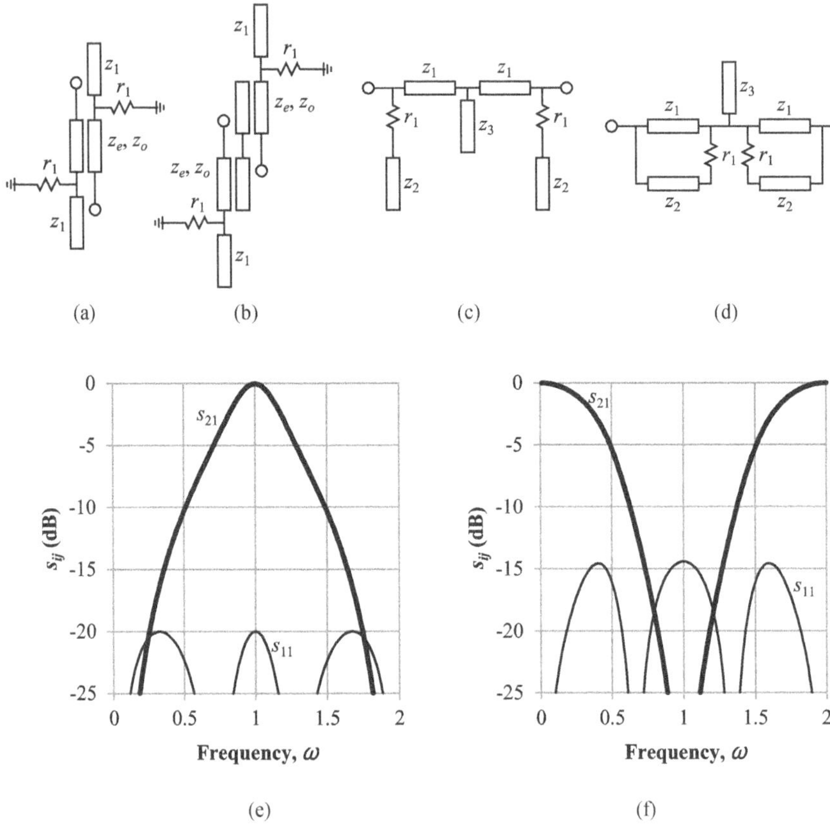

Figure 2.9 (a–d) Some empirically derived transmission-line absorptive filter structure. (e) Frequency response for the structure in (a). (f) Frequency response for the structure in (d).

this instrument covering a nearly unbroken frequency range of 1 to 50 GHz are downconverted and processed in 2 GHz bandwidths. In order to ensure that any chosen segment of the bandwidth could be processed with optimal sensitivity, a variable gain slope equalizer was needed with slope corrections ranging from -7.5 dB to +7.5 dB at the final frequency band of 2–4 GHz. The octave bandwidth configuration made the transmission-line equalizers described in this section ideal for that purpose.

Sixteen slope equalizers were fabricated monolithically in microstrip on a Polytetrafluoroethylene (PTFE) based microwave laminate substrate, along with the active switching circuitry needed to select the appropriate equalizer for each segment of processed bandwidth. Pre-amplifiers were used to compensate the midband losses of the otherwise passive equalizers so that the average gain was 0 dB in all channels. A photograph of the complete equalizer circuit board is shown

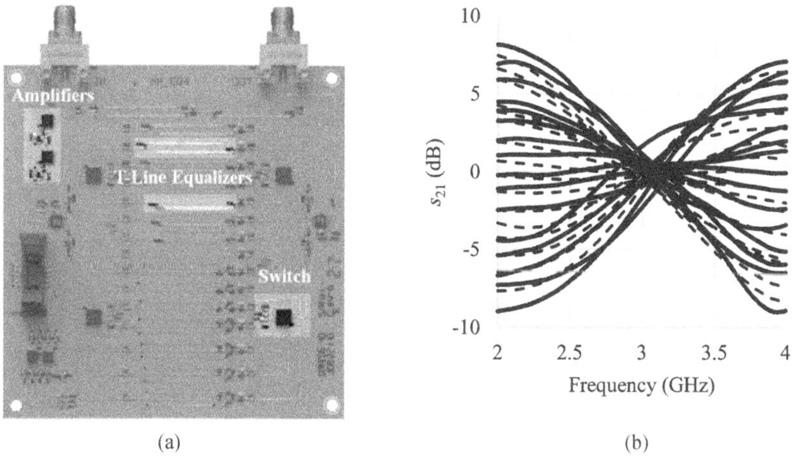

Figure 2.10 (a) Photograph of the gain slope equalizer using transmission-line absorptive filters for the JVLA radio telescope. The circuit board measures 3.5 x 3.5 inches. (b) Simulated (dashed lines) and measured (solid lines) performance of all 16 channels. (*After*: Morgan, M., Newton, T., Hayward, R., and Boyd, T., IEEE 2007 [7].)

in Figure 2.10(a). The transmission lines are printed directly on the board in the central area, while lumped-element, surface-mount resistors are soldered on to complete each equalizer section. Measured and modeled performance is shown in Figure 2.10(b). Had nonabsorptive equalizers been used, standing-waves due to mismatch effects between the equalizers and switches would have led to additional ripples in the pass-band curves [7].

2.5 CONSTANT-RESISTANCE NETWORKS

2.5.1 Topologies

A somewhat more analytical method of creating absorptive filters is to build one up from elementary constant-resistance networks [9]. A number of variants of such networks have been described, all of which fall into two general categories, or topologies. The first, shown in Figure 2.11(a), is a balanced network known as a lattice circuit. The second, Figure 2.11(b), is an unbalanced network known as a bridged-T circuit. In both cases, the series and shunt elements are dual impedances, normalized to the characteristic impedance of the system. The impedance seen looking into the input port of either circuit is constant and equal to the characteristic impedance, assuming the output port is likewise terminated. For the circuit in Figure 2.11(a), the transfer function is given by

(a) (b)

Figure 2.11 Constant-resistance networks. (a) Lattice/balanced structure. The dashed lines represent duplicate immittances of their symmetric counterparts. (b) Bridge-T/unbalanced structure. The resistors have value equal to the characteristic impedance.

$$s_{21} = \frac{1-z}{1+z} \tag{2.1}$$

whereas for the circuit in Figure 2.11(b), the transfer function is given by

$$s_{21} = \frac{1}{1+z} \tag{2.2}$$

(Lowercase z and y here and elsewhere in this book denote normalized immittance quantities.) The individual branches of these two networks can be any combination of resistors, inductors, and capacitors that one desires, so long as the duality conditions are met. Various options have been cataloged that provide arbitrary transfer functions up to the first- and second-order [9]. Higher-order topologies were deemed unnecessary because any realizable transfer function can be factored into first- and second-degree terms with real coefficients, thus allowing transfer functions of higher-order to be synthesized by cascading first- and second-order networks.

2.5.2 Realizability

However, there is a catch: the first- and second-degree factors of a realizable transfer function need not be realizable themselves. Take the transfer function in (1.15) for example. We know from the previous chapter that this function is that of a third-order Chebyshev Type II filter and that it is realizable with conventional reflective filter topologies (and we will further show in the next chapter that it has a reflectionless implementation as well). However, to realize it using constant-resistance networks, one must first factor it into first- and second-degree terms,

$$H(s) = \frac{s^2+1}{2s^3+3s^2+2s+1} = \frac{s^2+1}{(2s^2+s+1)(s+1)} = \left(\frac{s^2+1}{2s^2+s+1}\right)\left(\frac{1}{s+1}\right) \tag{2.3}$$

The last term on the right side is a realizable transfer function, but the second-degree term multiplying it is not. Most simply, the gain at the pure frequency $s = j0.5$ is given by

$$\left|\frac{s^2+1}{2s^2+s+1}\right|^2 = \left|\frac{-\frac{1}{4}+1}{-\frac{1}{2}+j\frac{1}{2}+1}\right|^2 = \left|\frac{\frac{3}{4}}{\frac{1}{2}+j\frac{1}{2}}\right|^2 = \left|\frac{3}{2+j2}\right|^2 = \frac{9}{8} > 1 \tag{2.4}$$

which is an obvious violation of passivity. Moving the numerator, s^2+1, over to the second term is not an option, as it makes that term then diverge to infinity. Ultimately, the minimum extra loss that must be added to make both terms passive is about 0.55 dB. That may not seem so serious in this case, but analysis shows that the amount of extra loss that must be added increases with the order of the filter, and is dramatically worse for other response functions.

2.5.3 Chebyshev Type I Example

Consider, for example, a seventh-order Chebyshev Type I response. To implement it using constant-resistance networks, we must factor it into second-order terms. Recall from Section 1.4.3 that the Chebyshev Type I transfer function has N finite poles and all of its zeros at infinity. The transfer function is therefore of the form,

$$H(s) = a_0\left(\prod_{k=1}^{N}(s-p_k)\right)^{-1} \tag{2.5}$$

where the poles, p_k, are given by (1.24) and the constant a_0, nominally given by (1.25), is required to ensure passivity. For this example, we have $N = 7$, so there are seven poles comprising three conjugate pairs and one real pole, all in the left half of the complex plane.

From Section 1.2.3, we know that the second-order terms, to be realizable, must have a Hurwitz polynomial in the denominator, which among other things requires the coefficients to be real and nonnegative. That means the conjugate poles must be paired together. Our transfer function will therefore consist of four terms,

$$H(s) = h_1(s)\cdot h_2(s)\cdot h_3(s)\cdot h_4(s) \tag{2.6}$$

where the first three are second-order terms and the last is a first-order term. Each term will be supplied by a constant-resistance lattice such as that shown in Figure 2.11(a), and the entire filter will comprise a cascade of all four lattice sections. The second order terms have the following form,

$$h_l(s) = \frac{a_l}{(s - p_l)(s - p_l^*)} = \frac{a_l}{s^2 - 2\,\mathrm{Re}\{p_l\}s + |p_l|^2} \tag{2.7a}$$

$$= \frac{a_l}{s^2 - 2r_l s + (r_l^2 + i_l^2)} \quad \text{for } 1 \le l \le 3 \tag{2.7b}$$

where r_l and i_l are the real and imaginary parts of the pole, respectively. To ensure passivity, we must find the maximum value of this function on the pure frequency (imaginary) axis. Thus, we substitute $s = j\omega$ and differentiate the square magnitude with respect to ω,

$$\frac{d}{d\omega}|h_l(j\omega)|^2 = \frac{d}{d\omega}\frac{a_l^2}{(r_l^2 + i_l^2 - \omega^2)^2 + 4r_l^2\omega^2} \tag{2.8a}$$

$$= \frac{d}{d\omega}\frac{a_l^2}{\omega^4 + 2(r_l^2 - i_l^2)\omega^2 + (r_l^2 + i_l^2)^2} \tag{2.8b}$$

$$= \frac{-a_l^2\left(4\omega^3 + 4(r_l^2 - i_l^2)\omega\right)}{\left(\omega^4 + 2(r_l^2 - i_l^2)\omega^2 + (r_l^2 + i_l^2)^2\right)^2} = 0 \tag{2.8c}$$

$$\therefore \omega^3 + (r_l^2 - i_l^2)\omega = \omega\left(\omega^2 - (i_l^2 - r_l^2)\right) = 0 \tag{2.8d}$$

$$\omega = 0, \pm\sqrt{i_l^2 - r_l^2} \tag{2.8e}$$

The maximum, which occurs at the second root above, has the value

$$|h_l(j\omega)|^2 = \frac{a_l^2}{(r_l^2 + i_l^2 - \omega^2)^2 + 4r_l^2\omega^2} = \frac{a_l^2}{(2r_l^2)^2 + 4r_l^2 i_l^2 - 4r_l^4} = \frac{a_l^2}{4r_l^2 i_l^2} \tag{2.9}$$

In order to make the transfer function passive, we must set $a_l = \pm 2r_l i_l$.

We must now attempt to derive a second-order constant-resistance network that has this frequency response. If we wished to use the unbalanced topology in

Figure 2.11(b), we would set s_{21} equal to $h_l(s)$ and solve (2.2) for the bridge impedance,

$$z_l = \frac{1-h_l(s)}{h_l(s)} = \frac{s^2 - 2r_l s + \left(r_l^2 + i_l^2\right) - 2r_l i_l}{2r_l i_l} \quad \text{for } 1 \le l \le 3 \tag{2.10}$$

However, this is not a positive-real immittance function because the degrees of the numerator and denominator differ by more than one. We are forced, then, to use the balanced topology in Figure 2.11(a), and instead solving (2.1) for the lattice impedance, namely

$$z_l = \frac{1-h_l(s)}{1+h_l(s)} = \frac{s^2 - 2r_l s + \left(r_l^2 + i_l^2\right) - 2r_l i_l}{s^2 - 2r_l s + \left(r_l^2 + i_l^2\right) + 2r_l i_l} = \frac{s^2 - 2r_l s + \left(i_l - r_l\right)^2}{s^2 - 2r_l s + \left(i_l + r_l\right)^2} \tag{2.11}$$

for $1 \le l \le 3$. Equation (2.11) is what is known as a biquadratic minimum function. Much literature has been devoted to the synthesis of such immittances. Seshu [10] has proven that at least seven elements are required to realize the general form above. That form is given in Figure 2.12(a). This circuit would appear in the through arm (the "z" component) of the topology in Figure 2.11(a). The diagonal arm, "y", must be the dual of this network.

Dual networks were introduced in Section 1.3, where it was stated that the dual of a given circuit may be formed by exchanging inductors and capacitors, resistors with their inverses (conductors), and series junctions with parallel junctions. Simplistic instructions such as these are easy enough to follow for small series- and parallel-groupings of elements, or even ladder topologies such as Cauer topology filters (Figure 1.10), but they are not so clear when applied to more complex networks including bridged-impedance structures such as this one. For that, a more reliable methodology is required.

There is a graphical approach to finding dual networks that one may use in these situations [11–13]. To begin, we add notional voltage generators between the terminals (if this was a multiport network, then a voltage generator would be added between each port and ground). The generator is only notional at this stage, but it is required to avoid losing track of the ports that might otherwise appear as open circuits and become shorted out (see Figure 2.12(b)). The ground node should be drawn explicitly in this step, if it isn't present already. Next, we draw the nodes of the dual network in the center of each closed loop in the original network, as in Figure 2.12(c). An additional node should be drawn surrounding the original network. In Figure 2.12(d) we have drawn in the duals of each of the original components, where the new component is connected between the two dual nodes that are separated by the original component. Note that the dual of a voltage generator is a current source. (The polarity of both the voltage generator and the current source is unimportant, as it will be removed later.) Next, in Figure

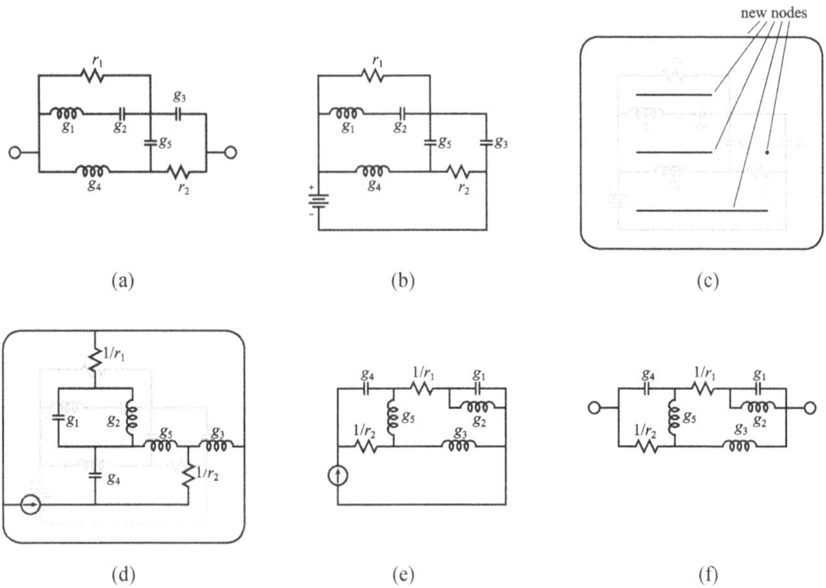

Figure 2.12 (a) Biquadratic impedance network for the lattice-section of a Chebyshev Type I constant-resistance filter. (b) First step in construction of dual; addition of port generator. (c) Additional of dual nodes. (d) Addition of dual elements. (e) Elimination of original circuit elements. (f) Final dual network.

2.12(e), the original components are removed and the dual network simplified. The final dual network, without the notional current source, is shown in Figure 2.12(f).[1]

Thus, a single second-order lattice section for the Chebyshev filter in this example requires 28 elements — 7 for each of the through arms of the lattice, and 7 for each for the dual networks in the diagonal arms — as shown in Figure 2.13(a). Three such second-order sections are needed for the complete filter.

Additionally, because the order of our example is odd ($N = 7$), we are left with one real pole that has no conjugate counterpart, and hence a single-order term of the form,

$$h_4(s) = \frac{a_4}{s - r_4} \tag{2.12}$$

The maximum of this function occurs at $s = 0$, for which we find that $a_4 = \pm r_4$ to

[1] The graphical procedure outlined here is sufficient to construct a dual of any planar network — that is, any network that can be drawn without crossings between some of its branches. More sophisticated techniques are required to form a dual of nonplanar networks [13].

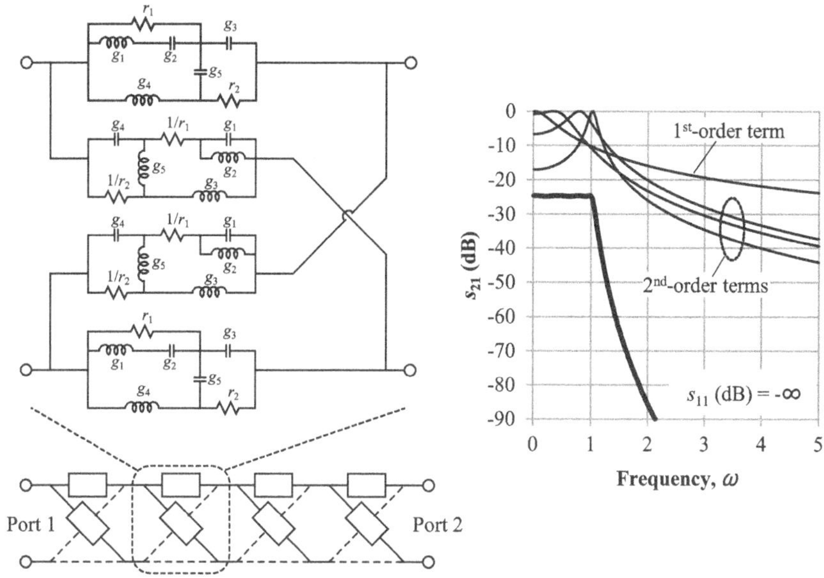

(a) (b)

Figure 2.13 (a) Second-order lattice section (top) implementing a single pair of conjugate poles. Three such sections plus a first-order section are needed (shown at bottom) to complete a seventh-order Chebyshev Type I filter. (b) Frequency response of individual lattice sections (thin lines) and complete seventh-order constant-resistance filter (thick line).

guarantee passivity. Applying this to (2.1), we find

$$z_4 = \frac{1-h_4(s)}{1+h_4(s)} = \frac{s-r_4-r_4}{s-r_4+r_4} = \frac{s-2r_4}{s} = 1-\frac{2r_4}{s} \tag{2.13}$$

which is simply a resistor in series with a capacitor (recall that $r_l < 0$ for all l).

Thus we have completed the design of a Chebyshev Type I filter using constant-resistance networks. It comprises three second-order lattice sections and one first-order lattice section, for a total of 92 elements. The frequency response (shown in Figure 2.13(b)) reveals that the scalar constants required to keep the second-order terms passive has increased the minimum attenuation to almost 25 dB!

It is possible to reduce the complexity of this circuit with a modified implementation. For example, there is an alternate biquadratic function similar to that in (2.11) that requires only 5 elements instead of 7. It matches the amplitude response exactly, but has a different phase. This would reduce the total element count from 92 to 68. However, nothing can be done about the excess 25 dB loss.

Similar analyses to those above reveal that the excess loss incurred by using a

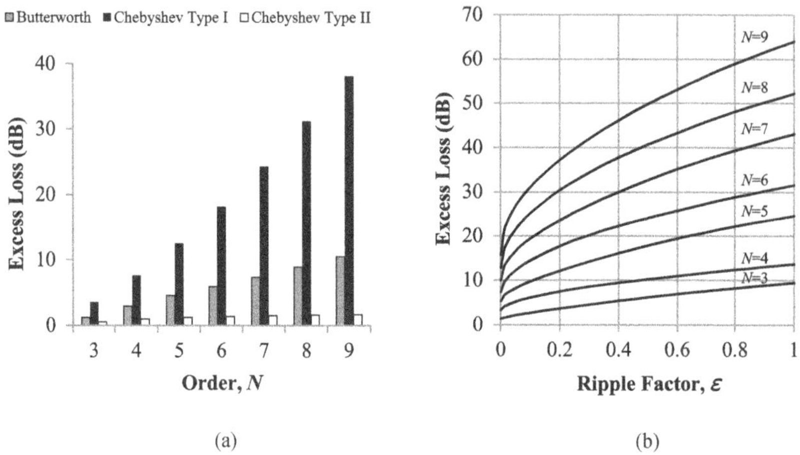

Figure 2.14 (a) Excess loss of constant-resistance Butterworth, Chebyshev Type I and Chebyshev Type II filters as a function of order. (b) Excess loss of constant-resistance Chebyshev Type I filters as a function of ripple factor.

constant-resistance network depends on the type of filter response chosen, on the order, and on the ripple factor. Figure 2.14(a) shows the excess loss as a function of filter type (Butterworth, Chebyshev Type I, and Chebyshev Type II) and as a function of filter order. Clearly, the Chebyshev Type II has a much lower need for excess loss given these topologies than other filter types. Nevertheless, Type II filters remain less popular due to their limited stop-band rejection.

In the case of the two Chebyshev filter types, the losses shown in Figure 2.14(a) were for filters having pass-band ripple (determined by the ripple factor, ε) consistent with the prototype parameters in Table 1.1. The loss depends critically on this ripple factor as well, as illustrated for Chebyshev Type I filters in Figure 2.14(b). Intuitively this makes sense, because the height of the peaks in the second-order terms shown in Figure 2.13(b) has to be taller for larger ripples and steeper transition slopes. Taller peaks lead to higher gain maxima that have to be suppressed by excess loss so that the individual terms are passive.

It is worth noting that this excess loss is a fundamental feature of the limited topologies available in the classical constant-resistance portfolio. These losses have nothing to do with component Q's or real-world electrical models. Rather, it is the signal power that is purposefully dumped into the resistor elements so that even ideal components (such as those used to generate the plots presented here) may collectively present a constant impedance to the ports while producing the desired pass-band shape.

2.5.4 Higher-Order Solutions

It bears asking whether better results could be obtained with higher-order sections, rather than cascades of only first- and second-order terms. While a thorough study of the realizability of various filter types of arbitrary order using the constant-resistance topologies in Figure 2.11 is beyond the scope of this book, a simple example suffices to show that the necessary immittance functions to use these topologies are not always realizable in higher-order sections either.

For simplicity, let us look once again at the third-order Chebyshev Type II transfer function of (1.15). Selecting the topology in Figure 2.11(b) and solving (2.2) for the bridge impedance, we find

$$z = \frac{1 - s_{21}}{s_{21}} = \frac{\left(2s^3 + 3s^2 + 2s + 1\right) - \left(s^2 + 1\right)}{s^2 + 1} = \frac{2s^3 + 2s^2 + 2s}{s^2 + 1} \tag{2.14}$$

Recall from the previous chapter that a driving point impedance must be positive-real, and that all positive-real functions have only simple poles on the $j\omega$ axis with real, positive residues. If we examine the partial fraction expansion of this expression,

$$z = \frac{2s^3 + 2s^2 + 2s}{s^2 + 1} = 2s + 2 + \frac{-2}{s^2 + 1} = 2s + 2 + \frac{-j}{s + j} + \frac{j}{s - j} \tag{2.15}$$

we find that the two poles at $\pm j$ have imaginary residues. Therefore, the requisite impedance function is not realizable, and no third-order variant of Figure 2.11(b) exists that implements the desired transfer function.

It turns out that the balanced network in Figure 2.11(a) does admit a third-order solution; however, balanced filter structures are seldom used in microwave circuit design, so this will not be pursued further.

Clearly, new topologies are needed if one hopes to make reflectionless filters of any practical value. Fortunately, topologies that are capable of producing these and other optimal filter functions without any excess loss whatsoever (beyond that which is dictated by the finite Q's of the reactive components) do exist. In the next chapter, we will take the first steps to find those topologies.

References

[1] Matthaei, G., Young, L., and Jones, E., *Microwave Filters, Impedance-Matching Networks, and Coupling Structures*, Norwood, MA: Artech House, 1980.

[2] Cohn, S. and Coale, F., "Directional channel-separation filters," *Proc. IRE*, vol. 44, no. 8, pp. 1018–1024, August 1956.

[3] Wanselow, R. and Tuttle, Jr. L., "Practical design of half-wavelength resonator directional filters," *IRE Trans. Microw. Theory Tech.*, vol. 7, no. 1, pp. 168–173, January 1959.

[4] Guyette, A., Hunter, I., and Pollard, R., "Design of absorptive microwave filters using allpass networks in a parallel-cascade configuration," *IEEE MTT-S Intl. Microw. Symp. Digest*, Boston, MA., pp. 733–736, June 7-12, 2009.

[5] Hayward, R., Morgan, M., and Saini, K., "A gain slope correction scheme for the EVLA receiver system," *EVLA Memo #80*, July 2004.

[6] Breitbarth, J. and Schmelzer, D., "Absorptive near-Gaussian low pass filter design with applications in the time and frequency domain," *IEEE MTT-S Intl. Microwave Symposium Digest*, vol. 3, pp. 1303–1306, June 2004.

[7] Morgan, M., Newton, T., Hayward, R., and Boyd, T., "Non-reflective transmission-line filters for gain slope equalization," *IEEE MTT-S Intl. Microwave Symposium*, Honolulu, HI, June 2007.

[8] Richards, P., "Resistor-transmission-line circuits," *Proc. of the IRE*, vol. 36, pp. 217–220, February 1948.

[9] Bode, H., *Network Analysis and Feedback Amplifier Design*, New York: D. Van Nostrand Company, pp. 249–275, 1945.

[10] Seshu, S., "Minimal realizations of the biquadratic minimum function," *IRE Trans. Circuit Theory*, vol. 6, no. 4, pp. 345–350, December 1959.

[11] https://en.wikipedia.org/wiki/Dual_impedance.

[12] Shenkman, A., *Circuit Analysis for Power Engineering Handbook*, Dordrecht, Netherlands: Springer Science+Business Media BV, pp. 44–48, 1998.

[13] Bloch, A., "On methods for the construction of networks dual to non-planer networks," *Proc. Phys. Society*, vol. 58, no. 6, pp. 677–694, November 1946.

Chapter 3

Foundations of Reflectionless Filters

The concepts of symmetry and duality are the fundamental principles upon which all reflectionless filter topologies presented in this book ultimately derive. The important result in this chapter is that the two even- and odd-mode equivalent half-circuits of a topological network may be both symmetric and dual at the same time [1, 2]. Though not specific enough to provide a clear and complete roadmap to all possible reflectionless filters, this does serve as a guideline to the engineer wishing to develop topologies that exhibit a loosely prescribed frequency response while having no reflections.

3.1 BASIC FILTER DERIVATION

One of the first simplifying assumptions we make about a reflectionless filter topology is that it should be symmetric, so as to have good impedance match from both ports, input and output. This is not strictly required for a network to be reflectionless, but it is desirable, it narrows down the vast field of possibilities, and it suggests a mechanism by which the topology can be analyzed, even at intermediate stages in its derivation. That mechanism is even-/odd-mode analysis [3, 4] described in Section 1.6.

3.1.1 Methodology

For a symmetric two-port network to be reflectionless, we simply assign the expression in (1.46a) for s_{11} and s_{22} to be identically zero. This requires that

$$\Gamma_{even} = -\Gamma_{odd} \tag{3.1}$$

Applying (1.5),

$$\frac{z_{even}-1}{z_{even}+1}=-\frac{z_{odd}-1}{z_{odd}+1}=\frac{1-z_{odd}}{1+z_{odd}}=\frac{y_{odd}-1}{y_{odd}+1} \tag{3.2a}$$

$$\therefore z_{even}=y_{odd} \tag{3.2b}$$

where z_{even} and y_{odd} are the even- and odd-mode equivalent input impedance and admittance, respectively. Furthermore, substituting (3.1) into (1.46b),

$$s_{21}=s_{12}=\tfrac{1}{2}\left(\Gamma_{even}-\Gamma_{odd}\right)=\Gamma_{even} \tag{3.3}$$

3.1.2 Procedure

Of particular note in (3.3) is that the pass- and stop-bands of the even-mode equivalent half-circuit become reversed in the full two-port network. In other words, frequencies that are reflective in the even-mode circuit (Γ_{even} is high) are transmissive in the final two-port circuit (s_{21} is high). Therefore, if we wish to design a low-pass filter, for example, we must ensure that the even-mode half-circuit is reflective at low frequency, but not reflective at high frequency. The most direct way to do that is to draw a high-pass filter in the even-mode equivalent circuit and then terminate its output, as shown on the left of Figure 3.1(a). For now, let us ignore the element values and consider only the topology.

The expression in (3.2b) may be recognized as the duality condition — that is, the even- and odd-mode circuit input impedances are inverses of one another. As explained in Chapter 1, the dual of a given network may be formed by substituting inductors with capacitors, series elements with shunt elements, and resistors with their complementary conductance, and vice versa. The dual of the even-mode high-pass filter is therefore drawn on the odd-mode side of Figure 3.1(a). Still a high-pass filter, it has the same transfer characteristic, but with the inverse impedance (e.g., the even-mode input impedance tends toward an open-circuit at low frequencies, while the odd-mode tends toward a short-circuit).

Of course, the resulting dual networks are not immediately recognizable as the two halves of a symmetric, two-port network. We recall here that the even- and odd-mode half-circuits are equivalent circuits only, in that they describe the behavior of the original network, but not necessarily the original network topology. In the analysis of symmetric networks, the even- and odd-mode parts are typically simplified by elimination of inactive components (such as open-ended or shorted-out elements) and by zero-sum alterations such as exchanged series components or delta-wye transformations. To recover our symmetric network, we simply need to back-out those implicit alterations. Effectively, the alterations that come to us so naturally during analysis, we must now apply in reverse, sometimes from a circuit that is relatively simple to one that is more complex.

The goal, then, is to redraw these two equivalent circuits in such a way that

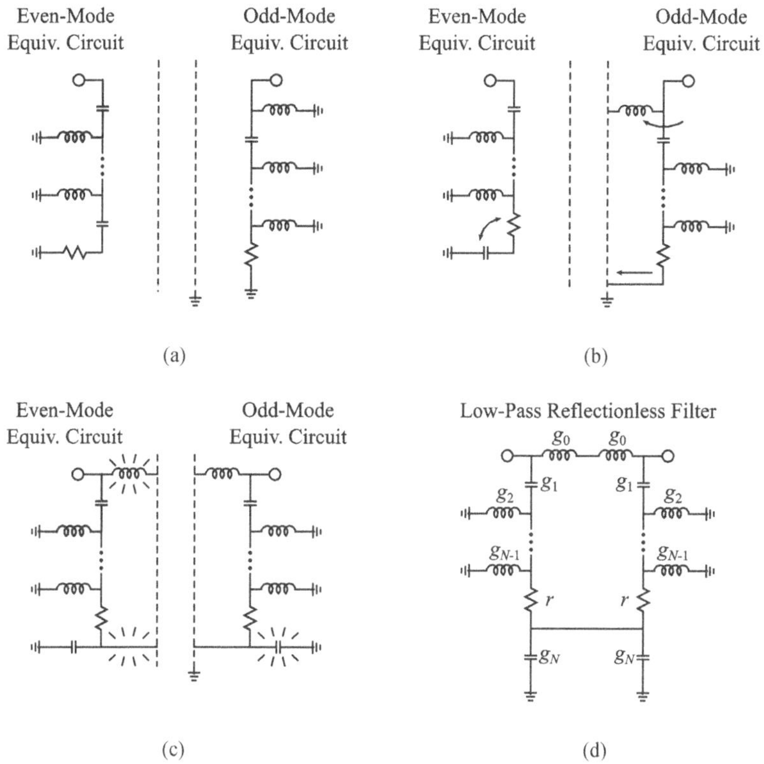

Figure 3.1 Derivation of a low-pass reflectionless filter. (a) High-pass prototypes corresponding to the even-mode equivalent circuit and its dual in the odd-mode equivalent circuit. (b) Modification by changing the ground connection of two components on the odd-mode side from absolute to virtual, and changing the order of two series elements on the even-mode side. (c) Addition of open-circuited elements on the even-mode side, and of a capacitor on the odd-mode side that is shorted at both ends. (d) Final low-pass reflectionless filter topology. (*After*: Morgan, M. and Boyd, T., IEEE, 2015 [5].)

the initial assumption of network symmetry is satisfied without changing the impedance or frequency response of either half-circuit. Fortunately, the two networks are similar, as they both consist of high-pass filters in the Cauer topology, and have series capacitors and shunt inductors. However, the circuits start and end with different kinds of elements. To resolve this, we make the following topological modifications;

1. Figure 3.1(b), even-mode side: Reverse the order of the final capacitor and termination resistor that are in series.
2. Figure 3.1(b), odd-mode side: Change the ground connections of the first inductor and the termination resistor from absolute to virtual.
3. Figure 3.1(c), even-mode side: Add an inductor between the input node and

the symmetry plane that is an open circuit. Similarly, add an open-circuited connecting line from the resistor to the symmetry plane. Since the symmetry plane is an open circuit here, these elements are inactive.

4. Figure 3.1(c), odd-mode side: Add a capacitor from the symmetry plane (a virtual short) to ground. Since the capacitor is effectively shorted at both ends, it too is inactive.

Note that none of the changes above will alter the behavior of either the even-mode circuit or the odd-mode circuit in any way. The final symmetric two-port circuit — a reflectionless low-pass filter — is illustrated in Figure 3.1(d). Although the pair of inductors at the top and the capacitors at the bottom could each be combined in series and in parallel, respectively, they have been left separate in this view for a reason that will soon become clear.

Having focused entirely on the topology in the above derivation, it is still necessary to assign the prototype element values, g_k, that satisfy the duality condition given in (3.2b). Note that the elements have been labeled with their normalized values. Starting at the top, the first inductor, labeled g_0 (which is active only in the odd-mode equivalent circuit) must be the dual of the corresponding capacitor in the even-mode circuit, labeled g_1. Therefore, based on the scaling laws that were given in Chapter 1,

$$L_0 = Z_0^2 C_1 \tag{3.4a}$$

$$\left(\frac{g_0 Z_0}{\omega_c}\right) = Z_0^2 \left(\frac{g_1 Y_0}{\omega_c}\right) \tag{3.4b}$$

$$\therefore g_0 = g_1 \tag{3.4c}$$

We have used the low-pass impedance scaling in (3.4b) because the full reflectionless filter is low-pass, even though the even- and odd-mode equivalent half-circuits were high-pass. The choice is arbitrary, however, as the same final result would be obtained if we had utilized the high-pass scaling formulae instead.

In a similar fashion, we find that the second element in the odd-mode circuit, g_1, must be the dual of the second element in the even-mode circuit, g_2. Therefore, $g_1 = g_2$, and so on. Ultimately, we find that in order to satisfy duality, we must have $g_k = g_0$ for all k. The termination resistors, r, must be their own duals, so $r = 1/r = 1$. This uniformity of element values will be a common theme in many reflectionless filter topologies derived in the early parts of this book. In their simplest normalized form, then, we may assume that all elements have unit value, unless otherwise noted. Since a value of unity is implicit, the labels can, in most cases, be omitted; otherwise, a simple numeral will suffice to specify a scaling factor. For example, were the two series inductors at the top to be drawn as one, it

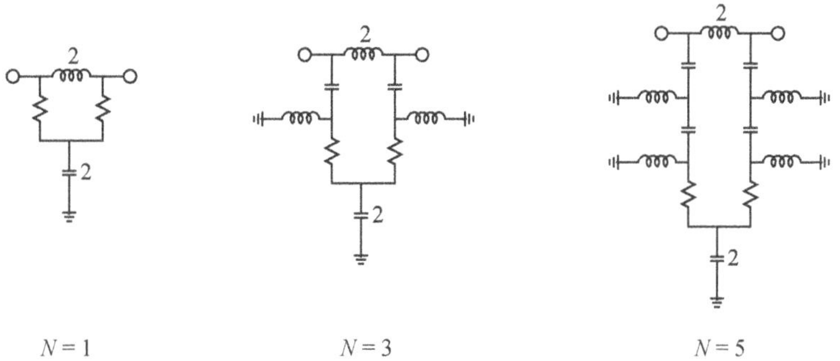

Figure 3.2 Topologies for low-pass reflectionless filters of order $N = 1$, 3, and 5. By convention here and elsewhere in this book, unlabeled elements have unit normalized value, while those labeled with a constant, e.g., "2", are scaled by that factor.

should be labeled with a single numeral, "2."

3.1.3 Performance

Topologies for the first few orders are shown in Figure 3.2. (The order must be odd for the derivation described above to work.) Recall that the order of the filter is determined by the degree of the denominator in its transfer function, which in this case, is completely determined by the even-mode equivalent circuit. The first-order topology simply replicates the bridged-T constant-resistance network mentioned in Section 2.5, but the higher-order networks differ. A plot of the transfer characteristics for these filters is shown in Figure 3.3. In all cases, the input reflection coefficient is identically zero, by construction, at all frequencies.

Note that the third-order topology is the only one in this sequence that can rightly be called an optimized filter, in this case a Chebyshev Type II, with the ripple factor fixed at $\varepsilon = 0.1925$. This is because the uniform element values of the even-mode equivalent circuit match the $N = 3$ row in Table 1.1, save for a constant scaling factor of 15.5%. This scaling factor accounts for a frequency shift between the transmission zero and the stop-band edge (defined by the ripple level), which is conventionally [6] (but not universally [7, 8]) defined as the unit frequency in Chebyshev Type II prototypes. Table 1.1 was for a Chebyshev Type I filter, but recall that according to (3.3), the transfer characteristic of the final filter is defined by the reflection coefficient of the even-mode equivalent circuit. As previously discussed, the pass-band of the even-mode circuit is the stop-band of the final network, and vice versa. So the ripples in the equivalent circuit's pass-band are inverted by this topology, appearing in the stop-band. The resulting transfer function is exactly that of a Chebyshev Type II filter. Chebyshev Type II filters are sometimes referred to as Inverse Chebyshev filters for this reason.

Importantly, the transfer functions shown here could just as well have been

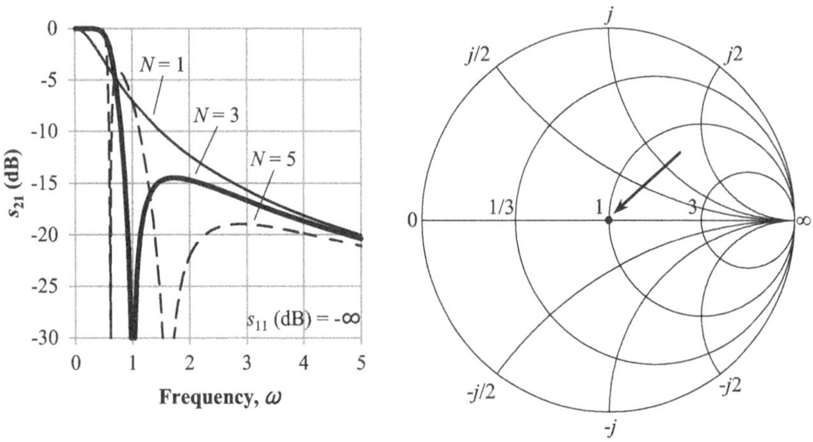

(a) (b)

Figure 3.3 (a) Plot of the transfer functions of basic reflectionless filters of first-, third-, and fifth-order cases. (b) Smith chart of input impedance for all reflectionless topologies.

implemented using other reflective topologies [9]. Reflectionless filters do not provide access to new transfer functions that were otherwise unrealizable (though they may be found easier to achieve with them); rather, they yield those same transfer functions in a reflectionless form.

3.1.4 Third-Order Case

The $N = 3$ case is in fact the very same filter plotted in Figure 1.4 with transfer function given by (1.15). To see this, consider the deconstruction of this filter illustrated in Figure 3.4. Remembering the convention that unlabeled components have unit value, the even-mode equivalent circuit has impedance

$$z_{even} = s^{-1} + \left(s^{-1} + \left(1 + s^{-1}\right)^{-1}\right)^{-1} = \frac{s^3 + 2s^2 + s + 1}{s^3 + s^2 + s} \tag{3.5}$$

whereas the odd-mode equivalent circuit has impedance

$$z_{odd} = \left(s^{-1} + \left(s^{-1} + \left(s^{-1} + 1\right)^{-1}\right)^{-1}\right)^{-1} = \frac{s^3 + s^2 + s}{s^3 + 2s^2 + s + 1} \tag{3.6}$$

Therefore, the reflection coefficients of these impedances are given by

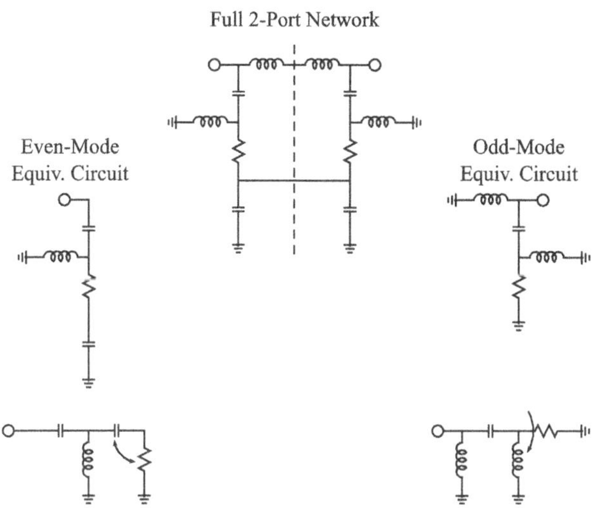

Figure 3.4 Deconstruction of third-order reflectionless filter into even- and odd-mode equivalent circuits.

$$\Gamma_{even} = \frac{z_{even} - 1}{z_{even} + 1} = \frac{s^3 + 2s^2 + s + 1 - \left(s^3 + s^2 + s\right)}{s^3 + 2s^2 + s + 1 + \left(s^3 + s^2 + s\right)} = \frac{s^2 + 1}{2s^3 + 3s^2 + 2s + 1} \tag{3.7a}$$

$$\Gamma_{odd} = \frac{z_{odd} - 1}{z_{odd} + 1} = \frac{s^3 + s^2 + s - \left(s^3 + 2s^2 + s + 1\right)}{s^3 + s^2 + s + \left(s^3 + 2s^2 + s + 1\right)} = \frac{-s^2 - 1}{2s^3 + 3s^2 + 2s + 1} \tag{3.7b}$$

Finally, by (1.46) we have

$$s_{11} = s_{22} = \tfrac{1}{2}\left(\Gamma_{even} + \Gamma_{odd}\right) = 0 \tag{3.8a}$$

$$s_{21} = s_{12} = \tfrac{1}{2}\left(\Gamma_{even} - \Gamma_{odd}\right) = \frac{s^2 + 1}{2s^3 + 3s^2 + 2s + 1} \tag{3.8b}$$

Equation (3.8a) proves that the filter is reflectionless, while (3.8b) matches the transfer function first given in (1.15).

The magnitude of the frequency response may be calculated after substituting for $s = j\omega$,

$$\left|s_{21}(j\omega)\right|^2 = \left|\frac{(j\omega)^2+1}{2(j\omega)^3+3(j\omega)^2+2(j\omega)+1}\right|^2 = \left|\frac{-\omega^2+1}{-j2\omega^3-3\omega^2+j2\omega+1}\right|^2 \tag{3.9a}$$

$$= \frac{(\omega^2-1)^2}{(2\omega^3-2\omega)^2+(3\omega^2-1)^2} = \frac{(\omega^2-1)^2}{4\omega^2(\omega^2-1)^2+(3\omega^2-1)^2} \tag{3.9b}$$

The extrema of this response are then obtained by setting its derivative equal to zero,

$$\frac{d}{d\omega}\left|s_{21}(j\omega)\right|^2 = \frac{d}{d\omega}\left(\frac{(\omega^2-1)^2}{4\omega^2(\omega^2-1)^2+(3\omega^2-1)^2}\right) = 0 \tag{3.10}$$

To simplify, let us make a substitution, $u = \omega^2\text{-}1$,

$$\frac{d}{du}\left(\frac{u^2}{4(u+1)u^2+(3u+2)^2}\right)\frac{du}{d\omega} = 0 \tag{3.11a}$$

$$\frac{\left(2u\left(4(u+1)u^2+(3u+2)^2\right)-u^2\left(4u^2+8u(u+1)+6(3u+2)\right)\right)}{\left(4(u+1)u^2+(3u+2)^2\right)^2}2\omega = 0 \tag{3.11b}$$

$$u\left(u^3-3u-2\right)\omega = 0 \tag{3.11c}$$

$$u(u-2)(u+1)^2\omega = 0 \tag{3.11d}$$

$$\omega^5\left(\omega^2-1\right)\left(\omega^2-3\right) = 0 \tag{3.11e}$$

Thus, it is found that the frequency response has extrema at $\omega = 0$, ± 1, and $\pm\sqrt{3}$. The last is the peak of the stop-band, where the transfer function reaches a value of

$$s_{21}\left(j\sqrt{3}\right) = \frac{-3+1}{-j6\sqrt{3}-9+j2\sqrt{3}+1} = \frac{1}{4+j2\sqrt{3}} \tag{3.12}$$

corresponding to a peak rejection of $10\cdot\log(28) \approx 14.47$ dB.

3.2 NORMALIZED ELEMENT SCALING

It was proven in Section 3.1.2 that all the normalized element values, g_k, have to be equal for the network to be reflectionless. That much was required. Assigning that common value to unity, however, was nothing more than a mathematical convenience. In the third-order case, described above, this leads to a Chebyshev Type II response with its sole, finite transmission zero located at $\omega_z = 1$. There are, however, other special frequencies commonly used to specify filter tuning. Some of these are shown in Figure 3.5(a). It is useful to determine the prototype parameter values, g_k, from the schematic in Figure 3.5(b) that locate each of these special points at unit frequency.

3.2.1 Stop-Band Specification ($\omega_s = 1$)

To design a Chebyshev Type II filter specified according to its stop-band frequency defined by the ripple level, it is useful to establish normalized prototype parameter values that make $\omega_s = 1$, instead of $\omega_z = 1$. We saw previously that this requires about a 15.5% scaling factor. More rigorously, we may consider the location of the transmission zero implied by (1.26). That is,

$$|H(j\omega_z)| = \frac{1}{\sqrt{1+\varepsilon^{-2}T_N^{-2}\left(\omega_z^{-1}\right)}} = \frac{\varepsilon T_N\left(\omega_z^{-1}\right)}{\sqrt{\varepsilon^2 T_N^2\left(\omega_z^{-1}\right)+1}} = 0 \tag{3.13a}$$

$$\therefore T_N\left(\omega_z^{-1}\right) = \cos\left(N\cos^{-1}\left(\omega_z^{-1}\right)\right) = 0 \tag{3.13b}$$

$$\omega_z = \frac{1}{\cos\left(\frac{\pi}{2N}\right)} \tag{3.13c}$$

$$\therefore g_k = \omega_z^{-1} = \cos\left(\tfrac{\pi}{2N}\right) = \frac{\sqrt{3}}{2} = 0.866 \tag{3.13d}$$

Thus, setting our normalized element values to $g_k \approx 0.866$ is sufficient to configure our third-order reflectionless filter so that the stop-band begins at unit frequency, or $\omega_s = 1$. (Note that 0.866 is the inverse of 1.155, the g_k value specified in Table 1.1. This is because that table was for Chebyshev Type I filters. The relationship between Type I and Type II filters will be discussed further in Chapter 6.)

This normalized prototype parameter value is then used in conjunction with the scaling laws in Table 1.2 for low-pass filters to calculate true inductance and capacitance values for a given design frequency and characteristic impedance.

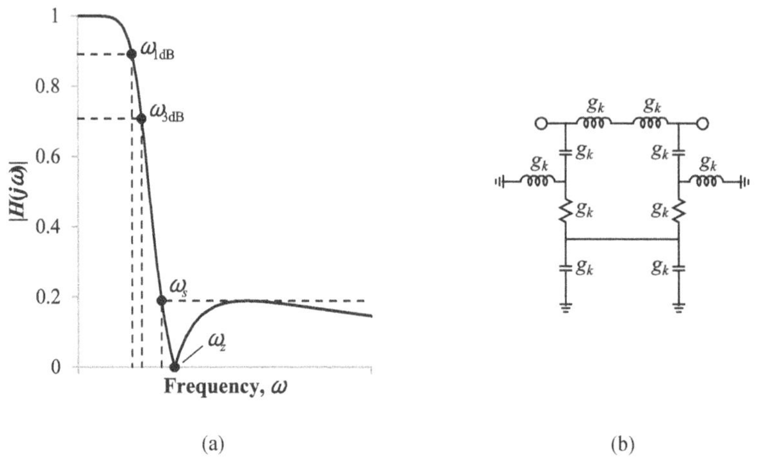

Figure 3.5 (a) Plot of third-order Chebyshev Type II response. (b) Third-order Chebyshev Type II reflectionless filter topology.

3.2.2 Pass-Band 3 dB Corner Specification ($\omega_{3dB} = 1$)

In common engineering practice, however, filters are often not specified by their transmission zeros or the beginning of their stop-bands, but by their pass-band corner, typically either 1 dB or 3 dB. To calculate these scaling factors, we once again return to (1.26). Relating the square of the transfer function to the desired insertion loss at unit frequency, we have

$$\left|H(j\omega_{IL})\right|^{-2} = 1 + \varepsilon^{-2} T_N^{-2}\left(\omega_{IL}^{-1}\right) = 10^{IL/10} \tag{3.14a}$$

$$T_N\left(\omega_{IL}^{-1}\right) = \cosh\left(N \cosh^{-1}\left(\omega_{IL}^{-1}\right)\right) = \frac{1}{\varepsilon\sqrt{10^{IL/10} - 1}} \tag{3.14b}$$

$$\omega_{IL}^{-1} = \cosh\left(\frac{1}{N}\cosh^{-1}\left(\frac{1}{\varepsilon\sqrt{10^{IL/10} - 1}}\right)\right) \tag{3.14c}$$

where IL is the specified insertion loss in dB.

To shift the frequency response so that the cutoff corner becomes unit frequency, we must multiply the prototype parameter values by the appropriate relative offset, or

$$g_k' = g_k \omega_{IL} = \frac{g_k}{\cosh\left(\frac{1}{N}\cosh^{-1}\left(\frac{1}{\varepsilon\sqrt{10^{IL/10}-1}}\right)\right)} \tag{3.15}$$

For a 3 dB cutoff specification, then, we simply have

$$g_k' = \frac{\frac{\sqrt{3}}{2}}{\cosh\left(\frac{1}{3}\cosh^{-1}\left(\frac{1}{(0.1925)\sqrt{10^{3/10}-1}}\right)\right)} \approx 0.6573 \tag{3.16}$$

3.2.3 Pass-Band 1 dB Corner Specification ($\omega_{1\text{dB}} = 1$)

If instead the pass-band is defined by its 1 dB cutoff corner, we may apply (3.15) with $IL = 1$, in which case

$$g_k' = \frac{\frac{\sqrt{3}}{2}}{\cosh\left(\frac{1}{3}\cosh^{-1}\left(\frac{1}{(0.1925)\sqrt{10^{1/10}-1}}\right)\right)} = 0.5592 \tag{3.17}$$

3.2.4 Pass-Band Ripple Factor Specification (ε)

The expression within the square root sign in (3.15) may be associated with an equivalent pass-band ripple factor. If we denote pass-band ripple as ε_I, and stop-band ripple as ε_{II}, then the appropriate scaling of prototype parameters simplifies to

$$g_k' = \frac{g_k}{\cosh\left(\frac{1}{N}\cosh^{-1}\left(\frac{1}{\varepsilon_I \varepsilon_{II}}\right)\right)} \tag{3.18}$$

This version of the element scaling is especially useful when comparing Chebyshev Type I and Type II filters having identical pass-band loss and stop-band rejection requirements.

3.3 NONUNIQUENESS

It is important to note that the topology derived in the previous section is not unique. The general principles of the derivation may be applied in different ways to arrive at alternative topologies that have the exact same scattering parameters. An example is shown in Figure 3.6. Once again, we start with a high-pass Cauer

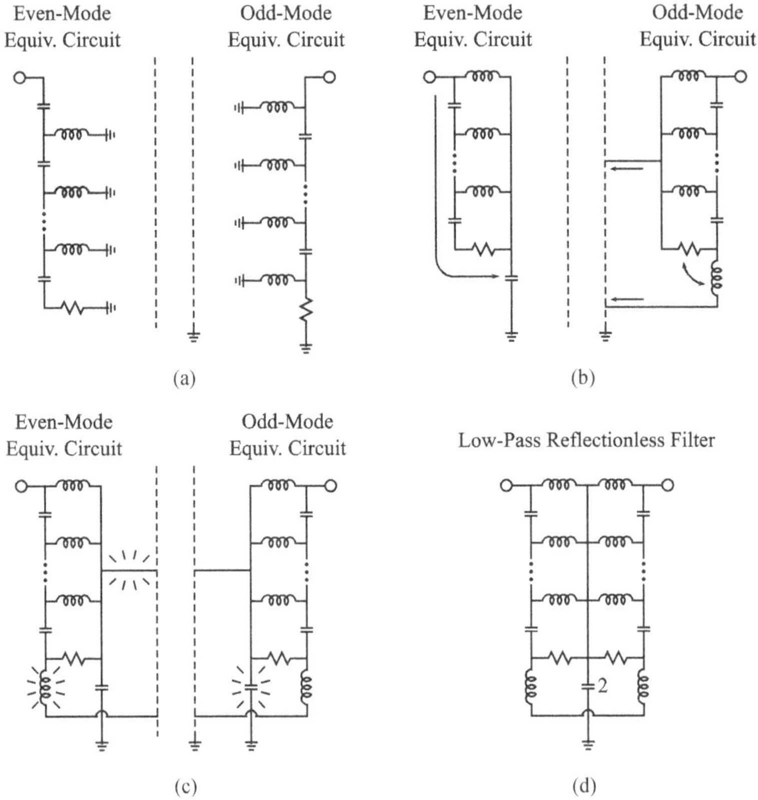

Figure 3.6. Derivation of an alternative low-pass reflectionless filter topology. (a) High-pass prototypes corresponding to the even-mode equivalent circuit and its dual in the odd-mode equivalent circuit. (b) Modification by changing the ground connection of all components on the odd-mode side from absolute to virtual, and exchanging the order of the first capacitor on the even-mode side with the rest of the equivalent circuit, with which it is in series. (c) Addition of open-circuited elements on the even-mode side, and of a capacitor on the odd-mode side that is shorted at both ends. (d) Final alternative low-pass reflectionless filter topology.

topology in the even-mode equivalent circuit, and its dual in the odd-mode equivalent circuit. Then we make the following changes to restore symmetry without altering the circuit behavior:

1. Figure 3.6(b), even-mode side: Exchange the order of the first capacitor with the rest of the circuit, with which it is in series.
2. Figure 3.6(b), odd-mode side: Change the ground connections of all components from absolute to virtual. Note that the termination resistor and the inductor in parallel with it have exchanged places as well.
3. Figure 3.6(c), even-mode side: Add an inductor between the output node

and the symmetry plane, which is an open circuit. Add a connecting line from the common node to the symmetry plane as well. Since the symmetry plane is an open circuit here, these elements are inactive.

4. Figure 3.6(c), odd-mode side: Add a capacitor between the symmetry plane node (a virtual short) and ground. Since the capacitor is effectively shorted at both ends, it too is inactive.

The final topology is shown in Figure 3.6(d). Distinct from the preceding topology, it too is constrained by duality such that all elements have equal normalized value, but for the final capacitor (which is doubled) and the frequency response is exactly the same, shown already in Figure 3.3.

3.4 AUXILIARY COMPONENTS

Except for frequency scaling, the maxima and minima calculated in Section 3.1 are fixed. Nothing can be done within the topologies presented thus far to change the element values relative to one another, alter the pass-band shape, or improve the limiting rejection in the stop-band except to cascade multiple filters together. Instead, to make progress, we must attempt to find ways of altering the topology while still meeting the requirements of symmetry and duality to ensure that it is reflectionless.

Some options do present themselves. In the original derivation when the even- and odd-mode equivalent circuits were first drawn (Figure 3.1(a)), the shunt inductors could have been split into two parallel inductors, one connected to absolute ground, and the other connected to the symmetry plane. These are labeled $g_{k,a}$ and $g_{k,b}$, respectively, in the modified topology of Figure 3.7(a). This would have no effect on the equivalent circuit topology in either mode, but would allow the inductor values in the odd-mode circuit to be smaller than those in the even-mode circuit. Similarly, additional parallel elements connected between the two resistors, which would appear only at the end of the odd-mode equivalent circuit, may be compensated by their dual elements in series with the final capacitor, thus appearing only at the end of the even-mode equivalent circuit. See, for example, g_x, g_y, r_x, and r_y in Figure 3.7(a).

Duality again constrains the values of the elements, yielding in this case the following relationships,

$$g_1 = g_0 \tag{3.19a}$$

$$g_{k,a} = g_{k-1} \text{ for } k \text{ even between 2 and } N\text{-}1 \tag{3.19b}$$

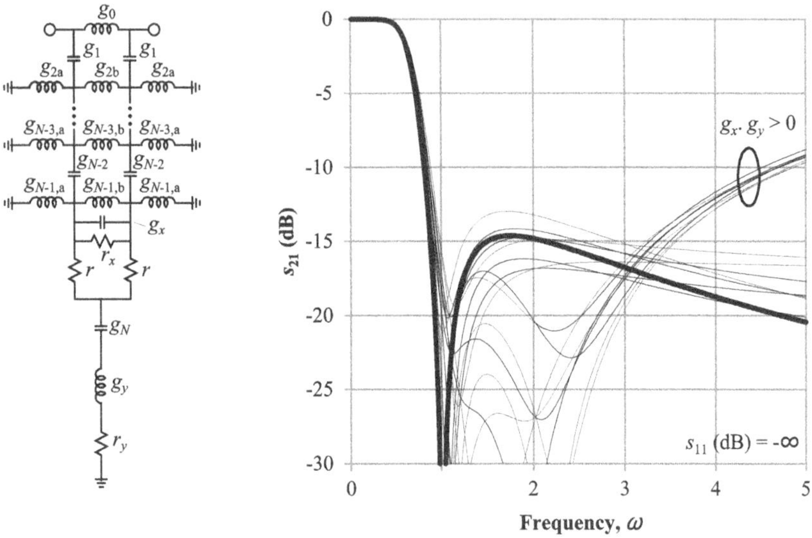

(a) (b)

Figure 3.7 (a) Modified low-pass reflectionless filter topology with (b) a sampling of possible pass-band characteristics for $N = 3$, $\omega_0 = 1$. In all cases, $s_{11} = 0$. The thick line corresponds to the nominal case matching Figure 3.4.

$$g_k = \frac{g_{k-1,a}\, g_{k-1,b}}{2g_{k-1,a} + g_{k-1,b}} \quad \text{for } k \text{ odd between 3 and } N\text{-}2 \tag{3.19c}$$

$$g_N = \frac{2g_{N-1,a}\, g_{N-1,b}}{2g_{N-1,a} + g_{N-1,b}} \tag{3.19d}$$

$$g_y = g_x \tag{3.19e}$$

$$r_y = \tfrac{1}{2}\left(r^{-1} - r\right) + r_x^{-1} \tag{3.19f}$$

Unlike the previous topologies, these duality constraints leave some parameters undetermined. In the case where $N = 3$, there are five degrees of freedom — four, if we discount simple frequency scaling. A sampling of possible pass-band curves for different selections of elements, all meeting the conditions for reflectionlessness, is shown Figure 3.7(b). Generally, none of these extra elements permit a sharper cutoff (note that the thick line corresponding to the previously derived topology is still steepest). Some do, however, permit slightly more rejection in the

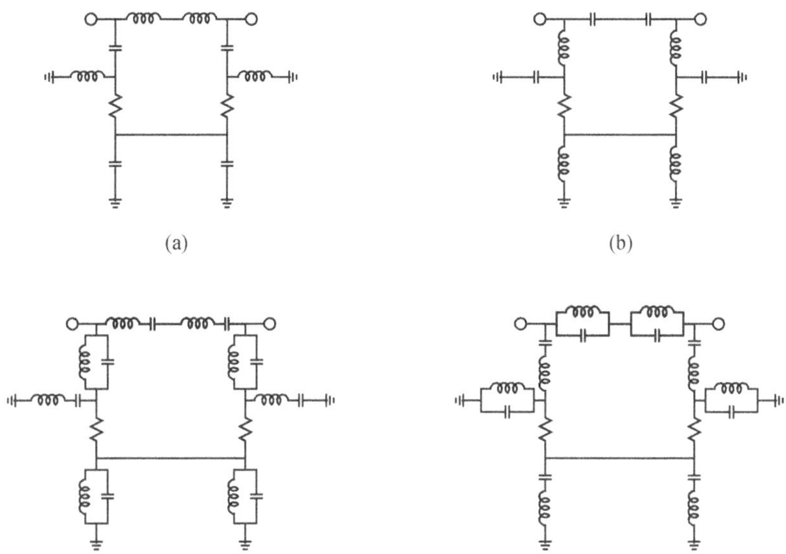

Figure 3.8 Third-order reflectionless filter topologies. (a) Low-pass. (b) High-pass. (c) Band-pass. (d) Band-stop. Networks are drawn so that all prototype parameters, g_k, have unit value. Scaled element values are given by Table 1.2.

stop-band, specifically those for which r_x or r_y are finite and nonzero. We will encounter this result again later in Chapter 4 in connection with a more sophisticated method of creating new filter topologies.

It is noteworthy that when g_x and g_y are greater than zero, the filter is no longer strictly low-pass, since it will have a transmission maximum at infinity. Strictly speaking, this makes it more like a band-stop filter that has a weak upper-frequency cutoff. Such a response may not seem that useful at present, but it will turn out to have interesting properties when used in combination with some other techniques to follow in later chapters.

3.5 FREQUENCY TRANSFORMATIONS

The derivations above targeted a low-pass transfer function. From these beginnings, Table 1.2 provides the necessary transformations to create high-pass, band-pass, and band-stop implementations. The resulting topologies appear in Figure 3.8. The networks are drawn such that all prototype parameters may be considered as having unit value (i.e., $g_k = 1$ for all k). This allows the actual element values to be determined most simply from Table 1.2. For example, the

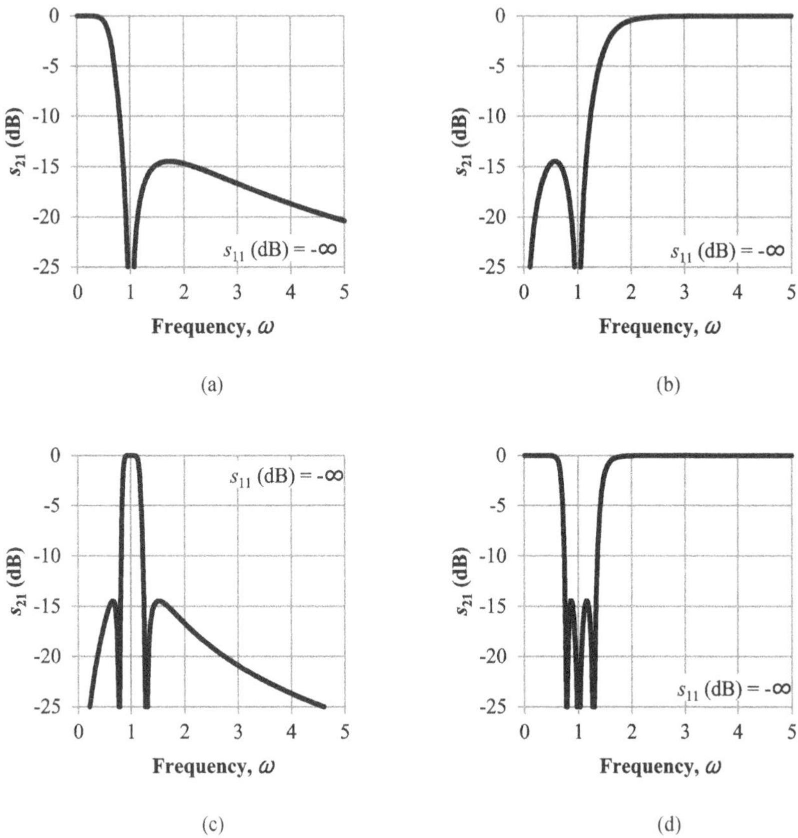

Figure 3.9 Third-order reflectionless filter responses. (a) Low-pass. (b) High-pass. (c) Band-pass. (d) Band-stop. For these illustrations, $\Delta = 0.5$.

series resonators in the band-pass circuit, Figure 3.8(c), have elements given by,

$$L_{series} = \frac{Z_0}{\omega_0 \Delta} \tag{3.20a}$$

$$C_{series} = \frac{Y_0 \Delta}{\omega_0} \tag{3.20b}$$

whereas the parallel resonators contain the following elements,

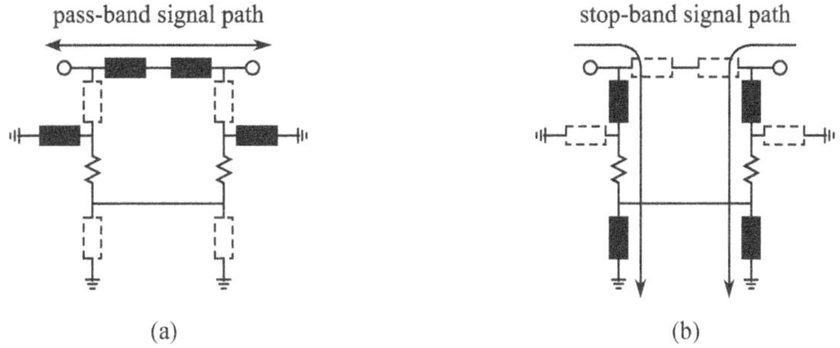

Figure 3.10 Signal paths for reflectionless filters in (a) the pass-band, and (b) the stop-band. The dashed, hollow boxes represent high-impedance elements, while the dark, solid boxes represent low-impedance elements.

$$L_{parallel} = \frac{Z_0 \Delta}{\omega_0} \tag{3.21a}$$

$$C_{parallel} = \frac{Y_0}{\omega_0 \Delta} \tag{3.21b}$$

where ω_0 is the center frequency of the pass-band and Δ is the fractional bandwidth (as measured between the two transmission zeros).

Plots of the transfer functions for all four basic third-order filter types are shown in Figure 3.9. Note that the peak stop-band rejection, 14.47 dB, is the same in each case.

Regardless of which frequency response is implemented, the operation of the circuit can be understood in terms of the limiting impedance of the connective elements or resonators in the pass-band and stop-band, as shown Figure 3.10. At the limiting pass-band frequency (e.g., DC for low-pass filters, or at the resonant frequency of the resonators for band-pass filters) the elements shaded black in Figure 3.10 become electrical shorts, while the hollow dashed elements become open circuits. The circuit thus simplifies to a direct connection between input and output for the pass-band. At the other extreme (e.g., infinite frequency for low-pass, or DC and infinity for band-pass), the impedance elements trend in the opposite direction, routing the stop-band energy directly through the resistor terminations, as indicated in Figure 3.10(b).

3.6 DESIGN EXAMPLE

The equality of element values makes the topologies presented thus far especially easy to implement monolithically. One needs only to optimize a single spiral-

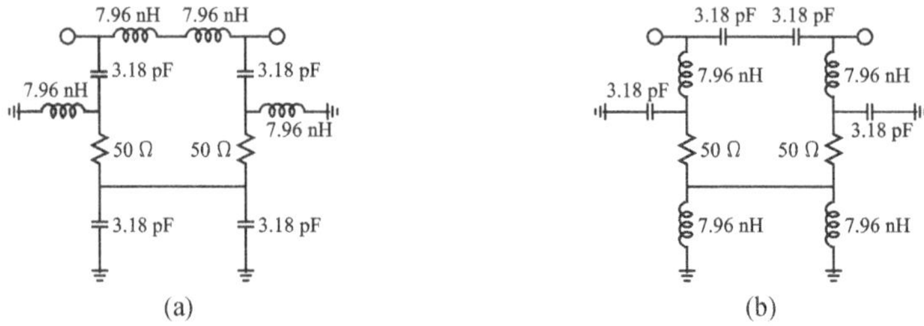

Figure 3.11 Frequency- and impedance-scaled element values for a third-order (a) low-pass and (b) high-pass reflectionless filter.

inductor, for example, and then copy and paste it as necessary to complete the filter. In fact, both low- and high-pass filters may be designed simultaneously using the exact same set of layout elements, given a common frequency of the transmission zero.

Prior to designing the layouts, however, we must first determine the element values from our circuit prototypes, given in Figure 3.8(a)–(b). Recall that unlabeled components in these figures have unit normalized value (that is, $g_k = 1$). As explained in Section 3.2, this will give us a filter with its design frequency defined by the location of the transmission zero. Had we instead wanted a filter tuned according to its 1 dB cutoff, we would use the prototype element value given in (3.17).

We therefore calculate the required inductor and capacitor values for a low-pass filter using the formulae in the left-most column of Table 1.2,

$$L = \frac{g_k Z_0}{\omega_0} = \frac{(1)(50\,\Omega)}{(2\pi)(1\,\text{GHz})} = 7.96\,\text{nH} \tag{3.22a}$$

$$C = \frac{g_k Y_0}{\omega_0} = \frac{(1)(50\,\Omega)^{-1}}{(2\pi)(1\,\text{GHz})} = 3.18\,\text{pF} \tag{3.22b}$$

where the characteristic impedance is to be 50Ω and the single transmission zero is to be located at 1 GHz.

For the high-pass filter, we should use the formulae in the second column of Table 1.2. However, since $g_k = 1$ for these filters, the results are exactly the same. Thus, the final schematics for the low- and high-pass reflectionless filters after frequency and impedance scaling are shown in Figure 3.11.

3.6.1 Monolithic Implementation

Consider the basic layout elements shown in Figure 3.12(a). These are common to

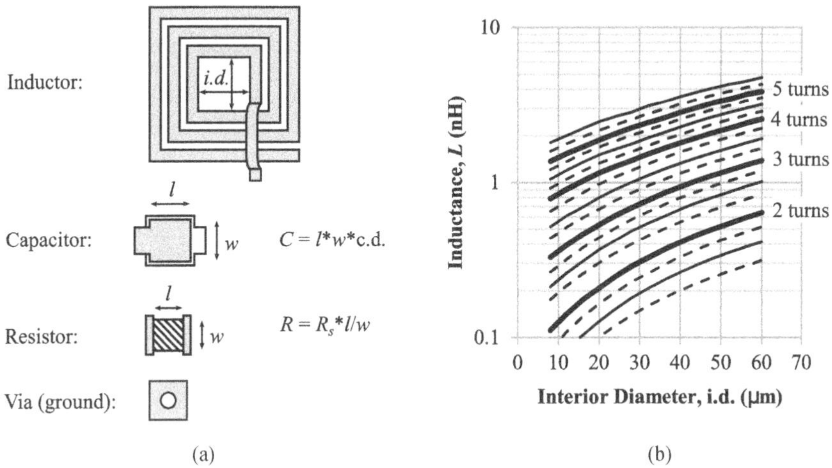

Figure 3.12 (a) Basic layout elements of a reflectionless filter in an integrated passive device (IPD) process. (b) Representative tuning curves for planar spiral inductors. Thick lines represent whole numbers of turns, thin lines for half-turns (e.g., 3.5 or 4.5 turns), and dashed lines for quarter-turns.

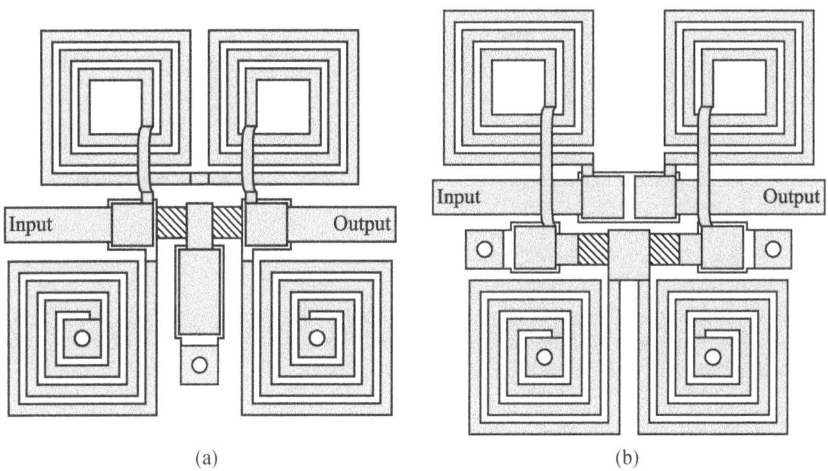

Figure 3.13 Physical layouts for third-order, monolithic, (a) low-pass and (b) high-pass reflectionless filters.

most integrated passive device (IPD) fabrication processes. The capacitors are available as metal-insulator-metal (MIM) stack-ups. The dielectric material and its thickness are usually fixed, which determines a process nominal capacitance density (c.d.) leaving the circuit designer to choose only the length and the width of the metal plates to realize the desired capacitance. The resistors are likewise

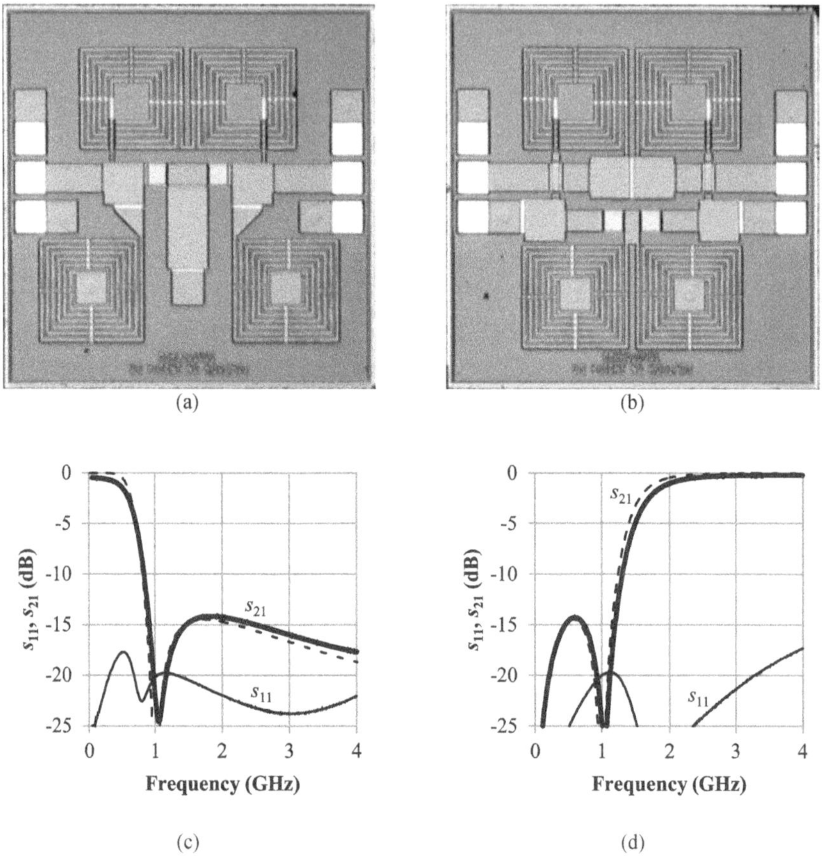

Figure 3.14 (a) Low-pass and (b) high-pass reflectionless filters implemented as integrated passive devices (IPDs) on GaAs substrates with dimensions 1.0 x 1.0 x 0.1 mm. (c) Measured results for low-pass filter [10]. (d) Measured results for high-pass filter [11]. Dashed lines indicate modeled performance assuming ideal lumped-elements.

provided as a resistive film with predetermined sheet resistance, so that the designer may select the desired value by controlling the lateral aspect ratio.

The planar spiral inductors are somewhat more tricky to design, typically requiring electromagnetic simulations to generate the appropriate tuning curves, such as that shown in Figure 3.12(b).

Having thus determined the geometry for resistors, inductors, and capacitors with the proper values, one needs only to connect them together in a layout matching the schematics from Figure 3.11. The four inductors dominate the overall dimensions, so the chip area is optimized if they are situated in roughly the four quadrants of the die, as indicated in Figure 3.13.

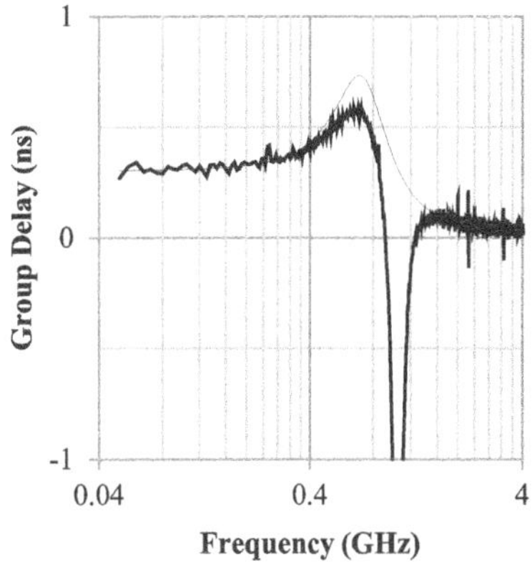

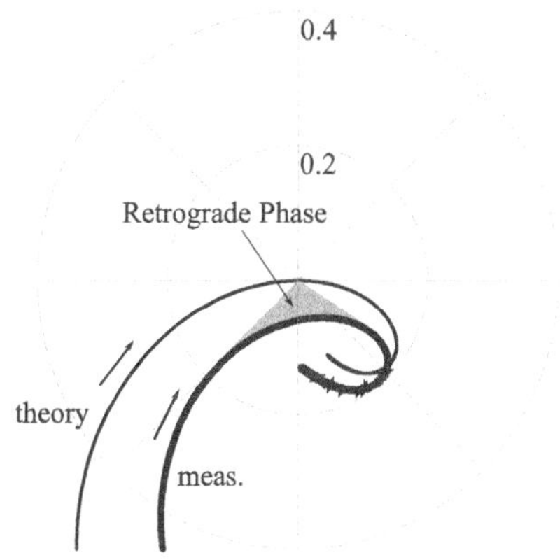

Figure 3.15 (a) Measured versus modeled group delay for low-pass reflectionless filter in Figure 3.14(a). (b) Polar plot of measured (thick line) and modeled (thin line) transfer coefficient (s_{21}) showing retrograde phase that causes negative group delay.

3.6.2 Experimental Results

Microphotographs of two chips made to these specifications are shown in Figure 3.14. These were fabricated on GaAs substrates in an IPD process. One is low-pass, the other high-pass, but both use the same inductor and capacitor geometries, simply rearranged. Their measured performance is given in Figure 3.14(c) and Figure 3.14(d). Although the ideal element model predicts infinite input return loss at all frequencies, the parasitics of real elements lead to finite reflection coefficients, in this case, near 20 dB for both designs in the pass-band and stop-band. The total chip size in both cases is 1 mm square.

The measured versus modeled group delay for the low-pass filter is shown in Figure 3.15(a). The measured performance is exactly as predicted throughout the pass-band; however, a large negative delay suckout has shown up in the transition-band that was not predicted by the idealized lumped-element model. Figure 3.15(b) reveals why. This polar plot of s_{21} shows that, theoretically, the curve should have passed through the origin at this frequency. The nonidealities of the physical components, however, cause the curve to miss the origin by a small factor. This should have been a transmission zero, but in practice, the signal cancellation at the output is imperfect. Throughout most of the frequency range, the trajectory of s_{21} progresses clockwise around the origin. However, by chance, the curve here passes slightly below the origin, temporarily reversing the direction in which the phase changes (the region highlighted in grey on the plot), thereby causing the group delay to become negative. Had the curve passed slightly above the origin instead, the group delay would have spiked positive. This illustrates why care must be taken in the interpretation of group delay in regions of high attenuation. The velocity associated with group delay does not always represent the propagation velocity of energy or information, but it can and does represent the propagation of nearly undistorted pulse waveforms while taking on values that are negative, superluminal, and even infinite [12].

3.7 THE PROBLEM OF DISCOVERING NEW TOPOLOGIES

This chapter has presented a fundamental technique for generating filter topologies that are reflectionless. Two illustrations of this methodology have been provided (Figures 3.1 and 3.6), yielding reflectionless low-pass filters of arbitrary order that may then be transformed into high-pass, band-pass, and band-stop equivalents. These two general topologies yield the same responses, however, which in turn are very limited by the constraints imposed by the need for even- and odd-mode duality.

One might ask if this is a fundamental limitation of the reflectionless condition on symmetric circuits, or are there other as-yet undiscovered topologies that are also possible? The answer is definitely yes, as the later chapters will show. However, these topologies can be very difficult to discover using this technique alone. The first step, generating the desired even-mode equivalent circuit, is easy enough. Finding a dual network to represent the odd-mode equivalent circuit is also straightforward, thanks to the techniques given in the previous chapters — but dual networks need not be unique, and their equivalents are not always easy to recognize as such. Identifying dual network pairs that admit symmetric realizations, given the appropriate boundary conditions, thus often presents quite a challenge. The inherent difficulty of this task is emphasized by the following examples.

3.7.1 Nonuniqueness of Dual Networks

The first thing we must realize is that the dual of a particular network may not be unique. Consider the simple network in Figure 3.16(a). It has a normalized admittance given by

$$y_a = \left(3s + s^{-1}\right)^{-1} + \left(s^{-1} + \left(s^{-1} + \left(\left(2s\right)^{-1} + 1\right)^{-1}\right)^{-1}\right)^{-1} \tag{3.23a}$$

$$= \frac{6s^5 + 8s^4 + 8s^3 + 4s^2 + 2s}{6s^5 + 9s^4 + 8s^3 + 6s^2 + 2s + 1} \tag{3.23b}$$

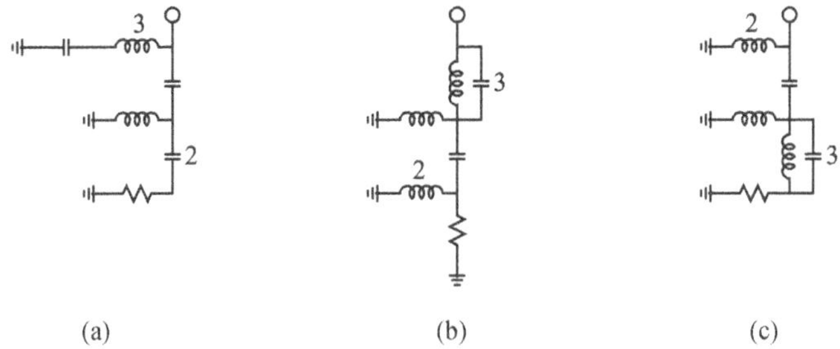

Figure 3.16 (a) An immittance network with at least two duals. (b) First dual network. (c) Second dual network. Recall that unlabeled components have unit normalized value.

The dual network in Figure 3.16(b) was formed by substitution of series elements with parallel elements, inductors with capacitors, etc., as taught in Chapter 1. We know from construction, then, that its normalized impedance, z_b, will be the same as the expression above for y_a. What about the last network, however? Does $z_c = z_b = y_a$? At a glance, we can see that Figure 3.16(c) is essentially the same network as Figure 3.16(b) in reverse, but with the port and termination resistor fixed in their original positions. That alone, however, is not sufficient to guarantee that two networks will be equivalent; almost any random example is sufficient to prove that this is not, in general, the case. However, it is the case here, as one may verify by calculating the normalized impedance of the last network in Figure 3.16(c),

$$z_c = \left((2s)^{-1} + \left(s^{-1} + \left(s^{-1} + \left((3s + s^{-1})^{-1} + 1 \right)^{-1} \right)^{-1} \right)^{-1} \right)^{-1} \tag{3.24a}$$

$$= \frac{6s^5 + 8s^4 + 8s^3 + 4s^2 + 2s}{6s^5 + 9s^4 + 8s^3 + 6s^2 + 2s + 1} = y_a = z_b \tag{3.24b}$$

Thus it is shown that Figure 3.16(b) and Figure 3.16(c) are equivalent.

3.7.2 Nontrivial Equivalent Circuits

Having established that dual network pairs are not necessarily unique, a simpler question might be: given a dual circuit that we have formed using the previously described techniques, how might we find other duals that are equivalent to it, but more amenable to symmetric realization? In other words, how can we write down a circuit that is equivalent to one we already have? This sounds like it should be

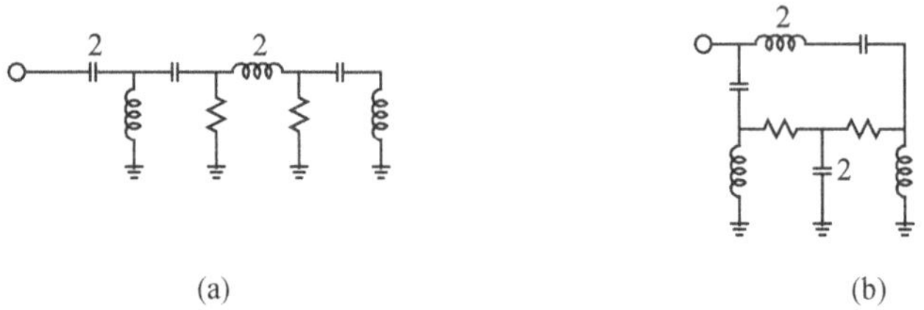

Figure 3.17 Two networks which are equivalent.

Figure 3.18 The circuit in Figure 3.17(b) after (a) first delta-wye transformation, and (b) second delta-wye transformation. The remaining shunt element is nonphysical (not positive-real).

trivial, but it is not. Figure 3.16(b) and Figure 3.16(c) are equivalent, but there is no simple way to transform one into the other by exchanging groups of elements in series or in parallel, or by applying delta-wye transformations. The only proof of their equivalence is to go through the math, which is tedious and time-consuming.

Consider another example, shown in Figure 3.17(a) and (b). These two networks are equivalent, but this is very difficult to see. The first network, Figure 3.17(a), may be analyzed much as above, by using series and parallel combinations of elements,

$$z = (2s)^{-1} + \left(s^{-1} + \left(s^{-1} + \left(1 + \left(2s + \left(1 + \left(s^{-1} + s\right)^{-1}\right)^{-1} \right)^{-1} \right)^{-1} \right)^{-1} \right)^{-1} \tag{3.25a}$$

$$= \frac{2s^6 + 6s^5 + 9s^4 + 8s^3 + 6s^2 + 2s + 1}{2\left(s^6 + 3s^5 + 4s^4 + 4s^3 + 2s^2 + s\right)} \tag{3.25b}$$

The second network in Figure 3.17(b) is much more difficult to analyze. It requires the application of a sequence of delta-wye transformations. The first such transformation is applied to the pair of resistors and the shunt capacitor between them, the result of which is shown in Figure 3.18(a). This set of components was chosen for the initial transformation because it is one of the very few involving mixed elements that yields physically realizable components (see Appendix A for a more extensive listing). The next transformation shown in Figure 3.18(b), however, does produce a nonphysical — that is, not positive-real — impedance element.

The input impedance of the network may then be found as usual,

$$z = \frac{s^2}{2\left(s^2+s+1\right)\left(s+1\right)} + \left(\left(2s + s^{-1} + \left(s^{-1}+1\right)^{-1}\right)^{-1} + \left(s^{-1} + \left(s^{-1}+1\right)^{-1}\right)^{-1}\right)^{-1} \tag{3.26a}$$

$$= \frac{s^2}{2\left(s^2+s+1\right)\left(s+1\right)} + \left(\frac{s^2+s}{2s^3+3s^2+s+1} + \frac{s^2+s}{s^2+s+1}\right)^{-1} \tag{3.26b}$$

$$= \frac{2s^6+6s^5+9s^4+8s^3+6s^2+2s+1}{2\left(s^6+3s^5+4s^4+4s^3+2s^2+s\right)} \tag{3.26c}$$

Through much labor, then, we have shown that the normalized impedance of both networks in Figure 3.17 is the same, hence they are equivalent.

Keep in mind, however, that this is actually the reverse of the problem we wish to solve. Assuming that the circuit in Figure 3.17(a) is the dual of our even-mode equivalent circuit, we would like to be able to come up with the circuit in Figure 3.17(b) as an alternate, having not seen it before. Considering the relative difficulty we have encountered just convincing ourselves that these two circuits are equivalent, to infer the second having only seen the first would be virtually impossible.

This inability to instantly deduce and recognize circuits that are equivalent to each other (much less duals of each other) is a serious impediment to developing new and more sophisticated reflectionless filter topologies. Simply writing down circuits at random and then tediously solving for their impedance in hopes of finding dual pairs is not at all practical. We need a more intuitive method to sift these topologies out from the enormous parameter space that is available to us. In the next chapter, we will begin to explore some simple ways of doing that.

References

[1] Morgan, M., and Boyd, T., "Theoretical and experimental study of a new class of reflectionless filter," *IEEE Trans. Microw. Theory Tech.*, vol. 59, no. 5, pp. 1214–1221, May 2011.

[2] Morgan, M., *Reflectionless Filters*, U.S. Patent No. 8,392,495, March 5, 2013. People's Republic of China Patent No. 201080014266.1, July 30, 2014.

[3] Guillemin, E., *Synthesis of Passive Networks*, New York: Wiley, pp. 196–198, 1957.

[4] Pozar, D., *Microwave Engineering*, 4th ed., New York: Wiley, pp. 344–345, 2011.

[5] Morgan, M., and Boyd, T., "Reflectionless filter structures," *IEEE Trans. Microw. Theory Tech.*, vol. 63, no. 4, pp. 1263–1271, April 2015.

[6] https://en.wikipedia.org/wiki/Chebyshev_filter.

[7] Ahn, J., Choi, S., Yun, C., and Kim, D., "Some prominent characteristics of the modified inverse Chebyshev function," *Proc. 40th Midwest Symp. on Circuits and Systems*, pp. 328–331, August 1997.

[8] Kim, H., Song, S., and Kim, D., "Some prominent aspects of the inverse Chebyshev functions," *IEEE Trans. Circuits and Systems*, vol. 38, no.3, March 1991.

[9] Roy, S. C. D, "Comments on 'Theoretical and experimental study of a new class of reflectionless filter'," *IEEE Trans. Microw. Theory Tech.*, vol. 60, no. 3, pp. 632–633, March 2012.

[10] *Mini-Circuits datasheet*, XLF-421+, available at http://www.minicircuits.com/pdfs/XLF-421+.pdf, October 2015.

[11] *Mini-Circuits datasheet*, XHF-23+, available at http://www.minicircuits.com/pdfsXHF-23+.pdf, October 2015.

[12] Milonni, P., *Fast Light, Slow Light and Left-Handed Light*, New York: Taylor & Francis, pp. 26–58, 2005.

Chapter 4

Subnetwork Expansion

A reflectionless filter — indeed, any kind of absorptive filter, no matter what its basic design — may be considered a kind of diplexer, where the absorptive elements terminate the unwanted outputs. These terminations do not always stand out as ports in the usual sense. The lumped resistors within the reflectionless filter topologies described in the previous chapter, for example, are embedded internally to the network, without either terminal sharing a common ground with the main ports. Nonetheless, it is possible to connect internal subnetworks in place of these terminations. Under the right conditions, this can be used to modify the frequency response while maintaining the reflectionless condition. This will turn out to be a prolific method of creating new and more sophisticated reflectionless filter topologies [1, 2].

The frequency responses achieved with this technique may not be ideal by themselves, but the inherent cascadability of reflectionless filters provides a means of combining them to leverage the best features of multiple topologies. A number of useful cascade combinations of the filters developed thus far and in Sections 4.1–4.4 will therefore be presented in Section 4.5.

4.1 THE REFLECTIONLESS FILTER AS A DIPLEXER

In Chapter 3, stop-band energy from the input ports of a reflectionless filter is absorbed by terminations consisting of lumped-element resistors. In principle, however, any matched network may serve as a termination if connected to the same terminals as the original resistor. This viewpoint of the reflectionless filter structure is shown in Figure 4.1(a). The internal ports have been labeled ports 3 and 4.

The full scattering-parameter matrix of this four-port structure may be derived using even-/odd-mode analysis as described in Section 1.6. The even- and odd-mode equivalent circuits (illustrated in Figure 4.1(b)) are now two-port networks consisting of simple tee- and pi-network high-pass filters, respectively (the even-mode circuit has been simplified by reversing the order of the internal

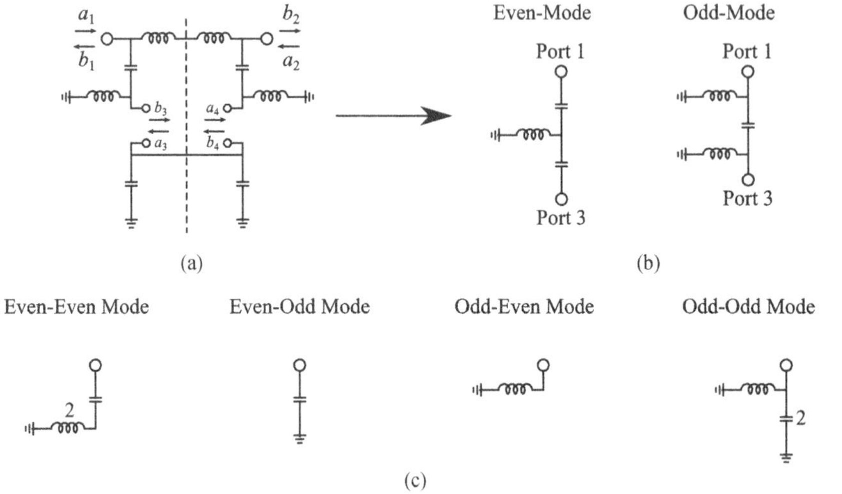

Figure 4.1 (a) Low-pass reflectionless filter with explicit internal ports in place of the terminations. (b) Even- and odd-mode equivalent circuits. (c) Even-even, even-odd, odd-even, and odd-odd equivalent circuits. As before, all unlabeled elements are assumed to have unit normalized value, and those labeled with a numeral are scaled accordingly.

port 3 and the capacitor with which it is in series). The full four-port scattering parameters of the original filter can now be obtained [3],

$$s_{11} = s_{22} = \tfrac{1}{2}\left(s_{11,even} + s_{11,odd}\right) \tag{4.1a}$$

$$s_{21} = s_{12} = \tfrac{1}{2}\left(s_{11,even} - s_{11,odd}\right) \tag{4.1b}$$

$$s_{31} = s_{13} = s_{42} = s_{24} = \tfrac{1}{2}\left(s_{31,even} + s_{31,odd}\right) \tag{4.1c}$$

$$s_{41} = s_{14} = s_{32} = s_{23} = \tfrac{1}{2}\left(s_{31,even} - s_{31,odd}\right) \tag{4.1d}$$

$$s_{33} = s_{44} = \tfrac{1}{2}\left(s_{33,even} + s_{33,odd}\right) \tag{4.1e}$$

$$s_{43} = s_{34} = \tfrac{1}{2}\left(s_{33,even} - s_{33,odd}\right) \tag{4.1f}$$

As before, we assume that the unlabeled elements have unit value. The results of the first two expressions (4.1a,b) were already given in (3.8). Due to the symmetry of the even- and odd-mode equivalent circuits, the same results apply to the last two expressions, (4.1e,f). One may evaluate (4.1c) and (4.1d) by employing a second level of even-/odd-mode analysis. The tee- and pi-network high-pass equivalent circuits are thus divided in half again into even-even, even-odd, odd-even, and odd-odd equivalent circuits (Figure 4.1(c)). The resulting scattering parameters are thus found,

$$s_{31,even} = \tfrac{1}{2}\left(s_{11,even,even} - s_{11,even,odd}\right) = \tfrac{1}{2}\left(\frac{z_{even,even}-1}{z_{even,even}+1} - \frac{z_{even,odd}-1}{z_{even,odd}+1}\right) \tag{4.2a}$$

$$= \tfrac{1}{2}\left(\frac{s^{-1}+2s-1}{s^{-1}+2s+1} - \frac{s^{-1}-1}{s^{-1}+1}\right) = \tfrac{1}{2}\left(\frac{2s^2-s+1}{2s^2+s+1} + \frac{s-1}{s+1}\right) = \frac{2s^3}{2s^3+3s^2+2s+1} \tag{4.2b}$$

$$s_{31,odd} = \tfrac{1}{2}\left(s_{11,odd,even} - s_{11,odd,odd}\right) = \tfrac{1}{2}\left(\frac{z_{odd,even}-1}{z_{odd,even}+1} - \frac{z_{odd,odd}-1}{z_{odd,odd}+1}\right) \tag{4.2c}$$

$$= \tfrac{1}{2}\left(\frac{s-1}{s+1} - \frac{\left(s^{-1}+2s\right)^{-1}-1}{\left(s^{-1}+2s\right)^{-1}+1}\right) = \tfrac{1}{2}\left(\frac{s-1}{s+1} + \frac{2s^2-s+1}{2s^2+s+1}\right) = \frac{2s^3}{2s^3+3s^2+2s+1} \tag{4.2d}$$

$$\therefore s_{31} = s_{13} = s_{42} = s_{24} = \frac{2s^3}{2s^3+3s^2+2s+1} \tag{4.2e}$$

$$\therefore s_{41} = s_{14} = s_{32} = s_{23} = 0 \tag{4.2f}$$

Therefore, the full scattering-parameter matrix for the four-port reflectionless filter is

$$\mathbf{S} = \frac{1}{2s^3+3s^2+2s+1}\begin{pmatrix} 0 & s^2+1 & 2s^3 & 0 \\ s^2+1 & 0 & 0 & 2s^3 \\ 2s^3 & 0 & 0 & s^2+1 \\ 0 & 2s^3 & s^2+1 & 0 \end{pmatrix} \tag{4.3a}$$

$$= \begin{pmatrix} 0 & H(s) & P(s) & 0 \\ H(s) & 0 & 0 & P(s) \\ P(s) & 0 & 0 & H(s) \\ 0 & P(s) & H(s) & 0 \end{pmatrix} \tag{4.3b}$$

Thus it is found that all of the stop-band energy incident upon port 1 (minus the 14.47 dB maximum leakage from port 1 to 2) is routed to port 3, while all of the

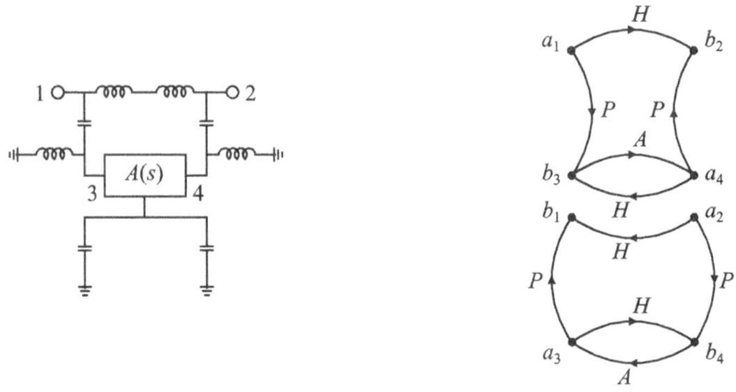

(a) (b)

Figure 4.2 (a) Low-pass reflectionless filter with matched, internal subnetwork having transfer function $A(s)$. (b) Signal flow diagram. $H(s)$ is the intrinsic transfer function of the filter, given by (3.8b), and $P(s)$ is the subnetwork coupling function given by (4.2e).

absorbed stop-band energy from port 2 terminates in port 4. That is, port 4 is perfectly isolated from port 1, and port 3 from port 2. The filter may be thought of as a bidirectional diplexer, with ports 3 and 4 (the stop-band ports) nominally terminated. The expression in (4.2e) which appears in all terms coupling the main ports to the internal ports will be referred to as the *subnetwork coupling function*, and will be labeled $P(s)$.

Note that because the network is now considered a lossless four-port, its scattering-parameter matrix is unitary. It immediately follows then from (1.7) that

$$|H|^2 + |P|^2 = 1 \tag{4.4a}$$

$$H^* P + P^* H = 0 \tag{4.4b}$$

4.2 SUBNETWORKS

4.2.1 Internal Subnetwork Analysis

It was stated in the previous section that any matched network may serve as a termination if attached to the internal ports. As long as the subnetwork is matched, the filter as a whole will still be reflectionless. Consider a two-port subnetwork connecting the two internal ports, as shown in Figure 4.2(a). We assume that it is matched, and that is has a transfer function given by $A(s)$. A signal flow diagram for this case is drawn in Figure 4.2(b). Note that two independent signal flow graphs have been obtained, corresponding to the forward and reverse directions of the signal path, respectively. This highlights the fact that the reflectionless condition has not been altered because there is no path in the signal flow diagram connecting the input wave, a_1, to the output wave at the same port, b_1 (nor any connection from a_2 to b_2). On the other hand, a loop has been created in the through-paths, altering the frequency response. By Mason's rule [4], the modified transfer function is

$$H'(s) = H(s) + \frac{A(s)P^2(s)}{1 - A(s)H(s)} \tag{4.5}$$

4.2.2 Subnetwork Attenuator

Intuitively, the value of the subnetwork is to judiciously add leakage from input to output that interferes constructively or destructively with the primary signal path so as to alter the frequency response. Ideally, we'd like the subnetwork pass-band to interfere constructively at the start of the transition, and destructively at the end, thus steepening the cutoff, or to add destructively at the peak of the stop-band to deepen its rejection. The latter may be achieved, if only partially, by using a simple attenuator as the subnetwork. See, for example, Figure 4.3(a). This offers only a modest improvement, since a simple attenuator itself does not produce the phase shift required in combination with the subnetwork coupling function for perfect cancellation. Applying (4.5) with the known transfer function of the basic reflectionless filter cell given in (4.3), we obtain

$$H'(s) = H(s) + \frac{A \cdot P^2(s)}{1 - A \cdot H(s)} = \frac{H(s) - A \cdot \left(H^2(s) - P^2(s)\right)}{1 - A \cdot H(s)} \tag{4.6a}$$

$$= \frac{\left(s^2+1\right)\left(2s^3+3s^2+2s+1\right) - A\left(\left(s^2+1\right)^2 - \left(2s^3\right)^2\right)}{\left(2s^3+3s^2+2s+1\right)^2 - A\left(s^2+1\right)} \tag{4.6b}$$

$$= \frac{\left(2s^5+3s^4+4s^3+4s^2+2s+1\right) + A\left(4s^6 - s^4 - 2s^2 - 1\right)}{\left(4s^6+12s^5+17s^4+16s^3+10s^2+4s+1\right) - A\left(s^2+1\right)} \tag{4.6c}$$

$$= \frac{4As^6 + 2s^5 + (3-A)s^4 + 4s^3 + 2(2-A)s^2 + 2s + (1-A)}{4s^6 + 12s^5 + 17s^4 + 16s^3 + (10-A)s^2 + 4s + (1-A)} \tag{4.6d}$$

The optimum value of attenuation has been found to be about 19.43 dB, yielding a worst-case stop-band rejection just over 17 dB. Better rejection overall would require more carefully tuning the subnetwork response according to (4.5).

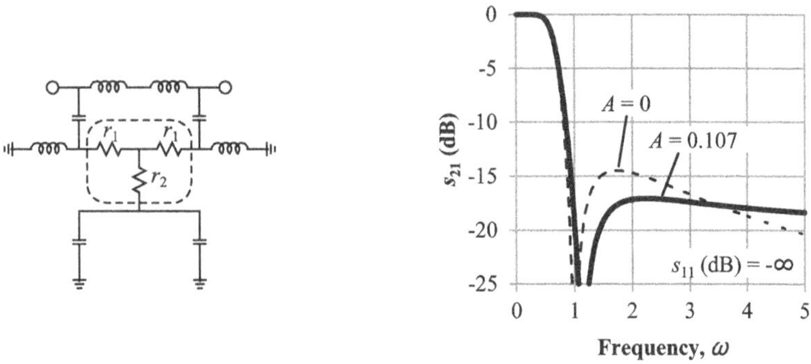

(a) (b)

Figure 4.3 (a) Reflectionless low-pass filter with attenuator subnetwork. (b) Frequency response with and without attenuator at $A = 0.107$ (~19.43 dB).

4.2.3 Cross-Connected Subnetwork Analysis

It is also permissible to cross-connect the subnetwork terminations of two cascaded reflectionless filters, as in Figure 4.4. The signal flow diagram in Figure 4.4(b) once again reveals two independent graphs for the forward and reverse directed signal paths, and there is no reflection. Once again by Mason's rule, the modified frequency response is

$$H'(s) = \frac{H^2(s)}{1 - P^2(s)Q(s)} \tag{4.7}$$

where $Q(s)$ is the coupling network transfer function.

Since there is no common ground for the two internal ports, they have been connected in this example with a 1:1 transformer to avoid common-mode loading of the network. In this case, the coupling transfer function is simply $Q(s) = 1$. Therefore,

$$H'(s) = \frac{H^2(s)}{1 - P^2(s)} = \frac{\left(s^2 + 1\right)^2}{\left(2s^3 + 3s^2 + 2s + 1\right)^2 - \left(2s^3\right)^2} \tag{4.8a}$$

$$= \frac{\left(s^2 + 1\right)^2}{12s^5 + 17s^4 + 16s^3 + 10s^2 + 4s + 1} \tag{4.8b}$$

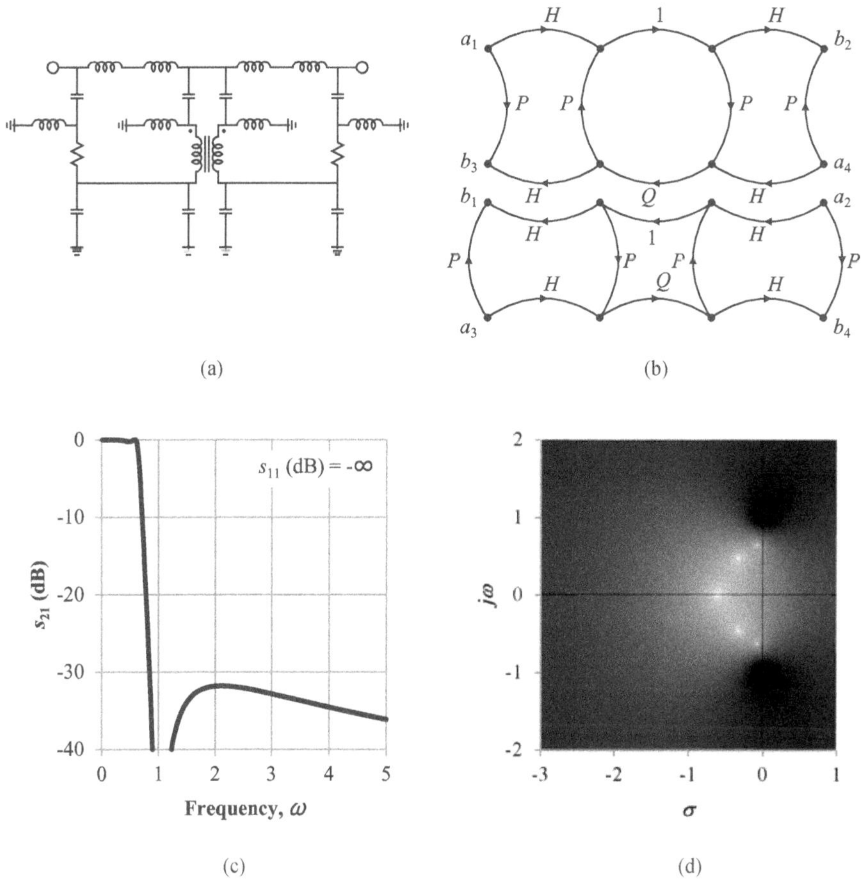

Figure 4.4 Two cascade reflectionless low-pass filters with cross-connected subnetwork comprising a 1:1 transformer. (a) Circuit diagram. (b) Signal flow diagram. (c) Frequency response. (d) Complex frequency space plot.

$$= \frac{\left(s^2+1\right)^2}{\left(3s^2+2s+1\right)\left(4s^3+3s^2+2s+1\right)} \tag{4.8c}$$

This frequency response is shown Figure 4.4(c). The overall rejection, a bit more than 30 dB, is comparable to what would be achieved with a simple cascade of two reflectionless filters, but the extra loop in the signal path has created a much sharper transition. The log magnitude plot, Figure 4.4(d) shows unevenly spaced poles in a semicircular arc around the origin with two double-zeros at $\pm j$. This does not correspond precisely to any optimized pole-zero configuration, but

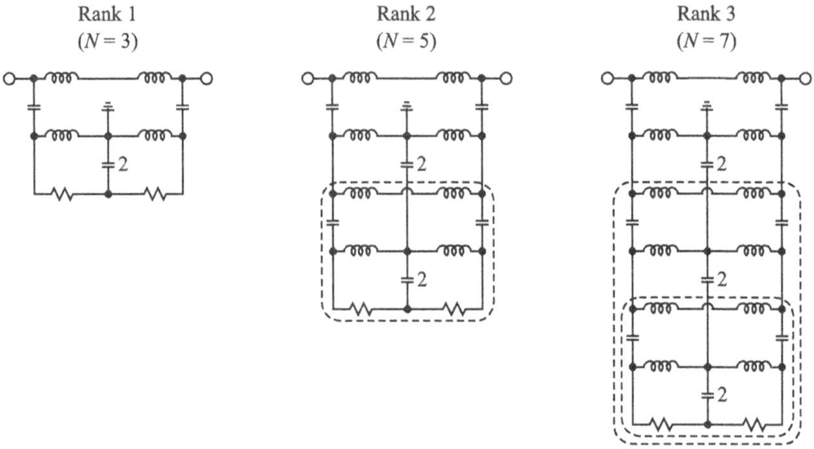

Figure 4.5 Compound low-pass reflectionless filters of first-, second-, and third-rank. The dashed lines enclose the subnetworks.

it is qualitatively similar to an elliptic response, having ripples in both the pass-band and the stop-band.

In practice, unfortunately, the transformer element is difficult to realize with losses low enough to produce a response close to the idealized version presented here. Its value as an example is to show the range of frequency responses that can be achieved with a reflectionless structure if the proper topology is found. Many of the examples in this book will be Chebyshev Type II, but as this one shows the variety of transfer functions that can be implemented in a reflectionless configuration is fundamentally much broader.

As an aside, this network is also the basis of the example given earlier in Figure 3.16. If one draws a symmetry plane down the middle between the two cross-connected filter cells in Figure 4.4(a), the even- and odd-mode equivalent circuits will be those shown in Figure 3.16(a) and Figure 3.16(c). Since the combination is known to be reflectionless, the two equivalent half-circuits must be duals, but one could find it most difficult to arrive at this conclusion given the two half-circuits only. Similarly, the subsequent example from Figure 3.17 is based on the even- and odd-mode decomposition of the same two cascaded filters without the subnetwork transformer.

4.3 COMPOUND REFLECTIONLESS FILTERS

Any matched network whatsoever may form a subnetwork within a reflectionless filter — including another reflectionless filter. These subfilters in turn may embed additional filters within them, and so on. In general, the number of reflectionless

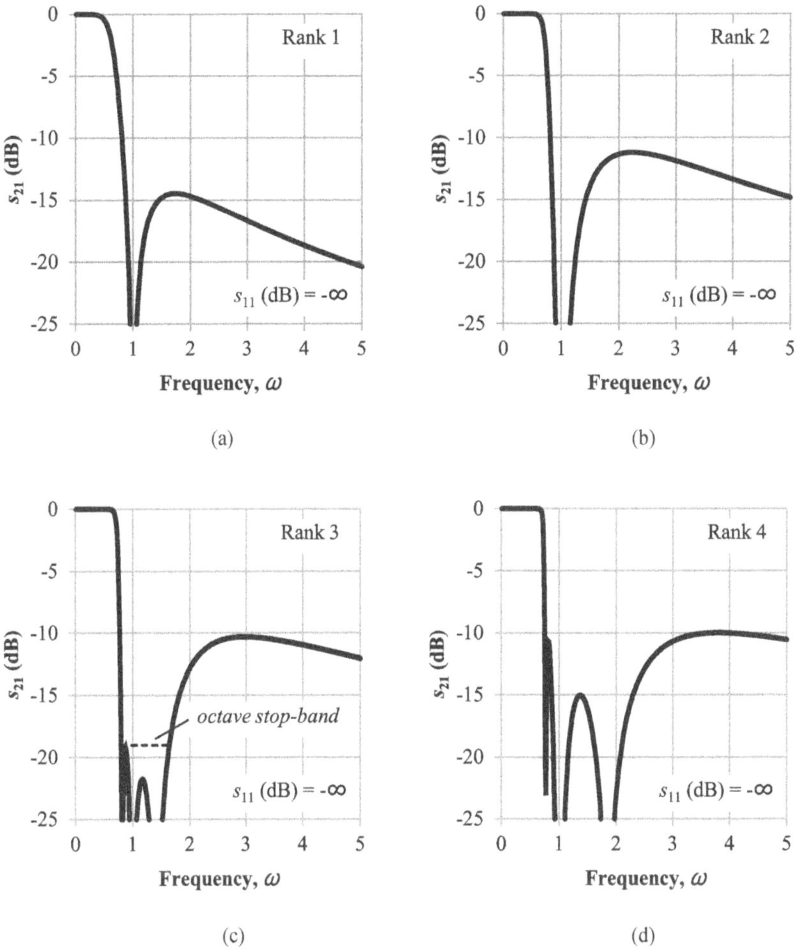

Figure 4.6 Frequency responses for reflectionless filters of (a) first-, (b) second-, (c) third-, and (d) fourth-rank.

filters embedded in this way will be referred to as the *rank* of the filter. Filters of first, second, and third rank are thus shown in Figure 4.5. The first-rank filter may be recognized as the original third-order reflectionless filter derived in Chapter 3, Figure 3.8(a), only it has been redrawn to make the extension to higher rank more clear. Dashed lines in the figure enclose the subnetwork filters. At this point, we know in principle that the individual filter stages do not need to be tuned to the same frequency as their subnetworks; the only requirement was that the subnetwork itself be matched. However, in practice, little advantage has been found to doing otherwise. Therefore, in keeping with the previous convention, the elements in Figure 4.5 are all assumed to have unit normalized value unless otherwise marked with a scaling factor.

The frequency response for a compound filter of this kind may be calculated using (4.5) where $A(s) = H(s)$, and is plotted for several ranks in Figure 4.6. The order of the filter response is one more than twice the rank, but it does not conform to any optimized filter function. In general, higher rank leads to sharper transitions, but often with higher stop-band peaks. The third-rank filter is of interest, however, because its rejection is close to 20 dB for the first octave of the stop-band (from $\omega = 0.8$ to 1.6), making it useful as a relatively simple anti-aliasing filter — simple compared to conventional seventh-order filters, in that all of its inductors are equal-valued.

4.3.1 Third-Rank Compound Filter

Since the third-rank filter will be of considerable use later on, it is worth deriving its transfer function explicitly. Rather than apply (4.5) twice, let us write the equation for the normalized input impedance of the even-mode equivalent circuit directly, and then apply (3.3). The even-mode equivalent circuit can be simplified to that of a simple Cauer topology, as illustrated in Figure 4.7. Its input impedance is given by

$$z_e = s^{-1} + \left(s^{-1} + \left(2s^{-1} + \left(s^{-1} + \left(2s^{-1} + \left(s^{-1} + \left(s^{-1} + 1 \right)^{-1} \right)^{-1} \right)^{-1} \right)^{-1} \right)^{-1} \right)^{-1} \tag{4.9a}$$

$$= \frac{s^7 + 6s^6 + 9s^5 + 19s^4 + 12s^3 + 16s^2 + 4s + 4}{s^7 + 3s^6 + 9s^5 + 8s^4 + 12s^3 + 4s^2 + 4s + 4} \tag{4.9b}$$

The frequency response of the filter is then given by the even mode reflection coefficient, which is

$$s_{21} = \Gamma_{even} = \frac{z_e - 1}{z_e + 1} = \frac{3s^6 + 11s^4 + 12s^2 + 4}{2s^7 + 9s^6 + 18s^5 + 27s^4 + 24s^3 + 20s^2 + 8s + 4} \tag{4.10a}$$

$$= \frac{\left(s^2+1\right)\left(s^2+2\right)\left(3s^2+2\right)}{\left(s^3+3s^2+2s+2\right)\left(2s^4+3s^3+5s^2+2s+2\right)} \tag{4.10b}$$

Equation (4.10b) is useful for studying the pole-zero configuration of this filter. By examination, we see that it has zeros at $\pm j$, $\pm j\sqrt{2}$, and $\pm j\sqrt{2/3}$. The final factorization of the denominator is too complex to carry out by hand, but the roots

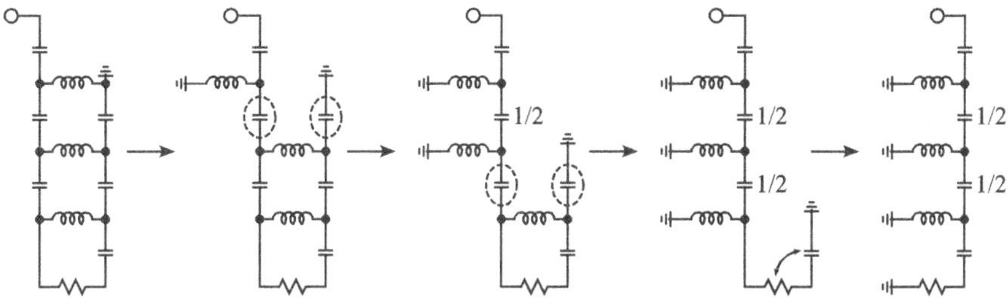

Figure 4.7 Simplification of the even-mode equivalent circuit for a third-rank reflectionless filter. The elements enclosed in dashed circles at each step are in series.

of the polynomials may be found using computer software. These tools reveal poles at the following locations: -2.52138, -0.23931 ± j0.857874, -0.0621417 ± j0.750946, and -0.687858 ± j1.13494. The stop-band peaks reach approximately 19.3 dB, 21.8 dB, and 10.3 dB, at $\omega \approx 0.88$, 1.17, and 2.98, respectively.

As a practical matter, the higher stop-band peak of the third-rank filter at $\omega \approx 2.98$ is of somewhat lesser consequence for low-pass filters than it might at first seem, because in actual implementations the parasitics of the elements lead to degradation in stop-band performance at high frequencies anyway — a phenomenon sometimes referred to as *reentry* or, if the insertion loss comes close to 0 dB again, a parasitic pass-band. This practical consideration would not apply to high-pass, band-pass, or band-stop versions of these filters, however, since the high stop-band peaks would get transformed to frequencies that are not typically degraded by parasitics. This will be discussed in greater detail in Chapter 6. Better optimized reflectionless filter responses will ultimately be needed.

4.3.2 Design Example

Let us work through an example for a third-rank (N = 7) reflectionless filter having its 1 dB cutoff corner at 1.25 GHz. First, considering the plot in Figure 4.6(c), we find that the 1 dB corner is at a normalized frequency of approximately $\omega \approx 0.71$. To scale in frequency, then, such that the 1 dB corner corresponds to the design frequency, we may consider $g_k = 0.71$. Therefore, we calculate the nominal inductor and capacitor values using the left-most column of Table 1.2 as follows,

$$L = \frac{g_k Z_0}{\omega_c} = \frac{(0.71)(50\Omega)}{2\pi(1.25\,\text{GHz})} = 4.52\,\text{nH} \tag{4.11a}$$

$$C = \frac{g_k Y_0}{\omega_c} = \frac{(0.71)(50\Omega)^{-1}}{2\pi(1.25\,\text{GHz})} = 1.81\,\text{pF} \tag{4.11b}$$

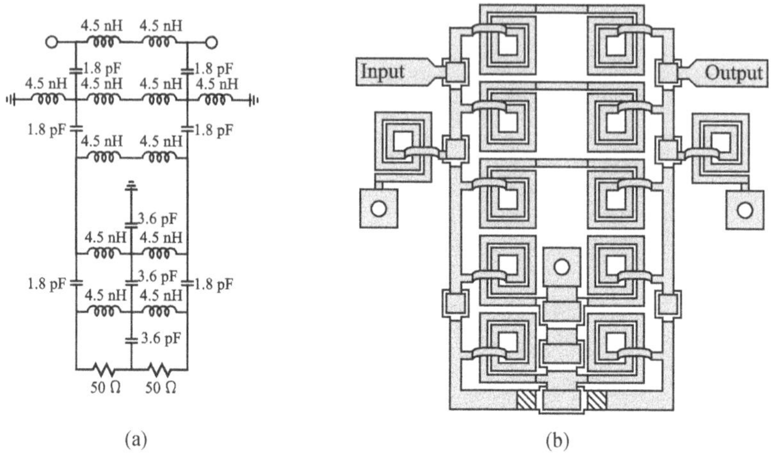

Figure 4.8 (a) Schematic for third-rank ($N = 7$) reflectionless filter with 1 dB corner at 1.25 GHz. (b) Sample layout for monolithic fabrication.

The final schematic with these element values is shown in Figure 4.8(a). A possible layout for monolithic fabrication is shown in Figure 4.8(b). Note that since all the inductors have the same value, the same spiral layout may be copied 12 times in this circuit, considerably simplifying the effort required to design such a filter monolithically. The mutual coupling between inductors in a layout such as this will slightly affect the frequency response, but can be accounted for in practice with a simple offset bias.

4.4 COMBINATION FILTERS

Thus far, we have seen how a reflectionless filter (having, in principle, identically zero reflection at all frequencies) may be derived from first principles to arbitrary order (Section 3.1), modified by the addition of complementary pairs of auxiliary elements (3.4), enhanced by embedding matched subnetworks (4.2), and built up hierarchically as filters within filters to achieve ever more complex reflectionless topologies (4.3). By now, it should be clear that each of these techniques may be applied in tandem, and the variety of possible reflectionless structures that we know how to create is growing exponentially.

This section will explore some of the ways these techniques may be combined to achieve a desired effect.

4.4.1 Second-Rank Filter with Subnetwork Attenuator

Take, for example, the structure shown in Figure 4.9. Here we have a second-rank

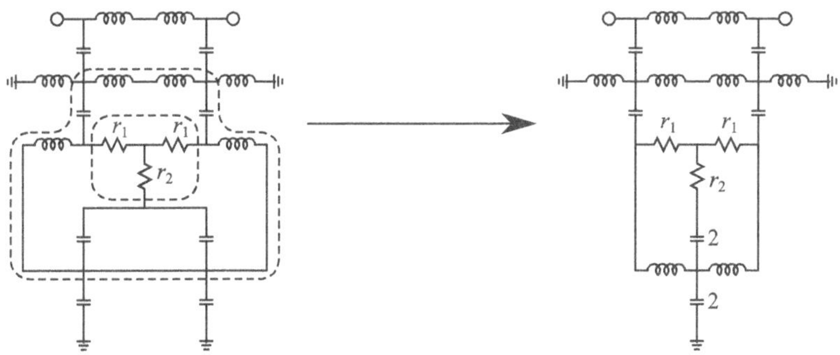

Figure 4.9 Second-rank filter having an attenuator subnetwork at the innermost level.

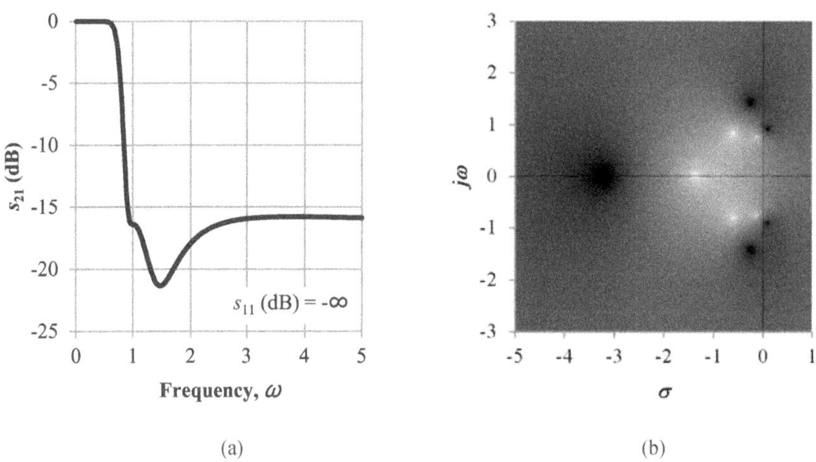

Figure 4.10 (a) Frequency response and (b) complex frequency space plot of the second-rank filter with internal attenuator set to A = 16.39 dB.

filter where the innermost level has been augmented with a subnetwork attenuator. Its frequency response is given in Figure 4.10(a). Its cutoff is similar to that of a standard second-rank filter, but with superior stop-band rejection. The transmission zeros have been lost, however.

The complex frequency space plot in Figure 4.10(b) reveals that the order of the filter has not been diminished, rather the transmission zeros have been moved off of the pure-frequency (imaginary) axis, a result typical for filters with attenuators in them. In fact, note that two of the zeros have moved slightly into the right-half plane. Recall that right-half plane poles are forbidden for passive networks, but right-half plane zeros are permitted.

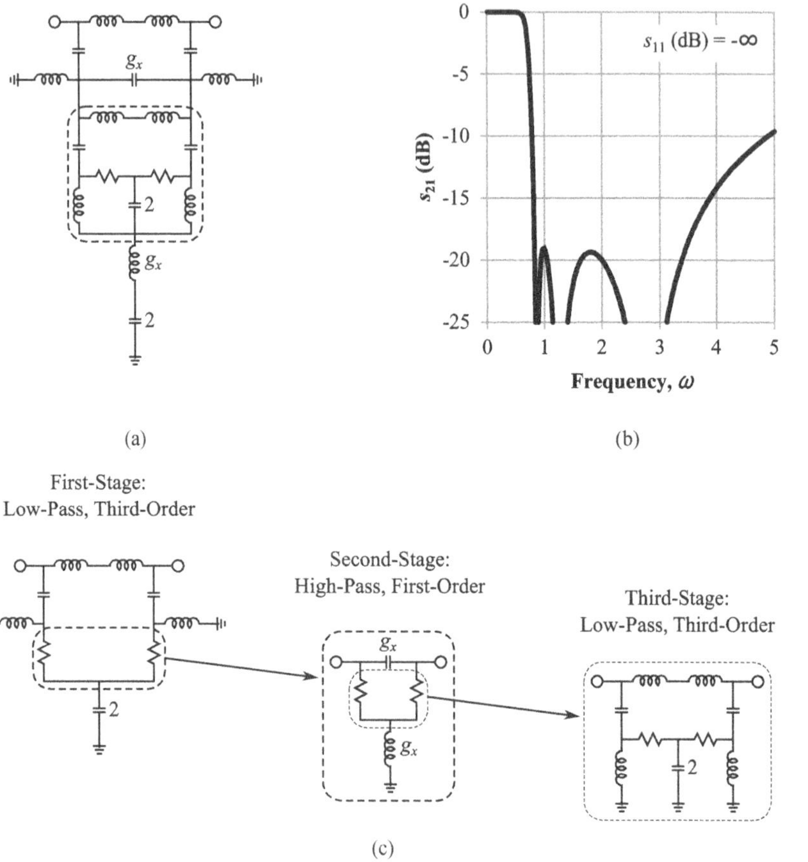

Figure 4.11 (a) Second-rank filter where the outermost filter has been modified with a pair of auxiliary components. (b) Frequency response for $g_x = 1/9$. (c) Alternate construction of the same filter involving high-pass, first-order middle-stage.

4.4.2 Second-Rank Filter with Auxiliary Elements

Alternatively, consider the filter in Figure 4.11. This also is a second-rank filter, but in this case, the outermost filter has been modified with auxiliary components using the techniques described in Section 3.4. The optimal value of the auxiliary element, g_x, has been found to be approximately $1/9^{th}$. This gives limited stop-band ripple close to 20 dB, similar to the third-rank filter in Figure 4.6(c), but with much broader stop-band. Note that the filter is no longer strictly low-pass, as the transmission response asymptotically approaches 0 dB at high frequency. As previously described, however, this is not a significant practical disadvantage, at least for low-pass filters, as the parasitics usually result in some reentry at high frequencies anyway.

In the course of discovering new reflectionless filter topologies, one may observe that the same topology can often be arrived at by different routes. In Section 3.4, the auxiliary elements used in this design were thought of as a modification to the primary filter stage in this two-stage filter. Another way to view this topology is that it has three stages, with the standard third-order, low-pass filters in the first and third stage, and a first-order high-pass filter in the second stage (that is, stage two is a high-pass transformed version of the filter shown for $N = 1$ in Figure 3.2). This viewpoint is illustrated in Figure 4.11(c). To avoid confusion, the *rank* of the filter will refer only to the number third-order stages embedded in a compound filter, making this still a *second-rank* filter, despite having an alternate derivation that involves embedding three distinct stages.

4.4.3 Reflectionless Notch Filters

Notch filters are a particular class of band-stop filters where the rejection band is designed to be very narrow, typically in order to reject a spurious tone or narrow-band interferer. They are also among the most commonly attempted types of absorptive filters [5, 7], suggesting that there is a compelling need for such components in many applications. Most of these solutions, however, are either largely empirical with poor reflection coefficients in the pass-band, or require unrealistic components such as ideal transformers and quadrature hybrids, limiting the bandwidth over which low reflections can be achieved.

Fortunately, absorptive notch filters are especially easy to design using the reflectionless filter techniques described here. Perhaps the simplest approach is to apply the band-stop transformation in Table 1.2 to a simple first-order ($N = 1$) reflectionless filter, known earlier as a constant-resistance network, given in Figure 3.2. However, to obtain very narrow stop-bands in this way requires resonators with extreme reactive element values.

The subnetwork approach offers a better solution. Starting with a third-order low-pass filter in the first stage, we embed a third-order high-pass filter as its subnetwork, Figure 4.12(a). This has the effect of passing the high-frequency signals through the subnetwork and back out through the output port of the filter. Intuitively, then, we know this is going to have some kind of band-stop response (i.e., passing at very low and very high-frequencies while rejecting somewhere in between).

One might expect that if the low-pass filter has a much higher cutoff than the high-pass subnetwork, then any stop-band energy routed through the subnetwork would simply pass through it to re-emerge at the output, and hence the overall response would be all-pass, however this is not the case. The internal loop closed by the subnetwork in the signal flow diagram (Figure 4.2(b)) acts in such a way to interfere with the primary leakage path, creating a transmission zero at midband.

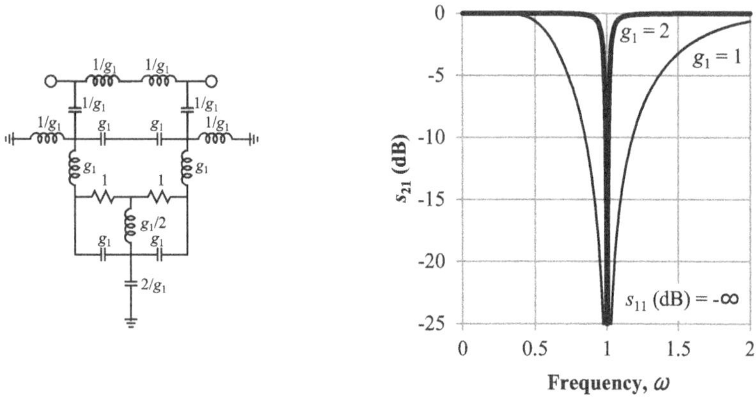

(a) (b)

Figure 4.12 (a) Reflectionless notch filter comprising a third-order low-pass outer stage with a third-order high-pass subnetwork. (b) Frequency response for $g_1 = 1$ (thin line) and $g_1 = 2$ (thick line).

To see this, consider that the scaling of the outer low-pass filter and the internal high-pass filter indicated in Figure 4.12(a) guarantee that their transmission responses are conjugate at midband. That is, $A = H^*$ at $s = j1$ where $H(s)$ is the frequency response of the outer low-pass filter and $A(s)$ is the frequency response of the subnetwork high-pass filter. Combining this relationship with the conditions of unitarity in (4.4) and substituting into (4.5), we find

$$H' = H + \frac{AP^2}{1 - AH} = H + \frac{H^* P^2}{1 - |H|^2} = H + \frac{H^* P}{P^*} = H - H = 0 \tag{4.12}$$

The nature of this midband cancellation allows the stop-band to be very narrow even with moderately valued components, while still maintaining the circuit's reflectionless properties. Figure 4.12(b) shows the frequency response for this topology with $g_1 = 1$ (thin line) and $g_1 = 2$ (thick line). To achieve a comparably narrow notch using the first-order reflectionless filter of Figure 3.2 with a band-stop transformation would require resonators with reactive elements 16 times larger and others 16 times smaller!

4.5 CASCADE COMBINATIONS

It has been observed that most of the higher-order filter responses derived so far fall short of the optimal curves derived for classical filters in Section 1.4. Often this manifests as a degradation of stop-band rejection at high-frequency (for low-

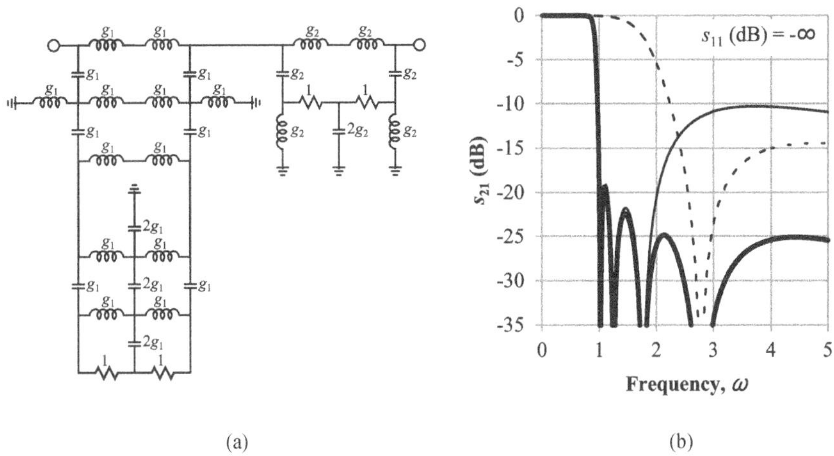

Figure 4.13 Combination of high-order and low-order low-pass filters for sharp transition with extended stop-band. (a) Schematic. (b) Frequency response for individual filters (thin/dashed lines) and cascade combination (thick solid line) where $g_1 = 0.8$, and $g_2 = 0.36$.

pass filters), a phenomenon previously referred to as *reentry*. This situation may be ameliorated by considering reflectionless filters in cascade. Unlike conventional filters where the interaction of reactive stop-bands prevents cascades from behaving as predicted, the combination of two or more reflectionless filters is truly the sum of its parts.

4.5.1 Cascaded High-/Low-Order Low-Pass Filters

This is especially beneficial for low-pass designs, as one may cascade a sharp-cutoff high-order design (such as the third-rank filter) with a lower-order first-rank filter tuned to a higher frequency. The latter in practice will have much lower parasitics, allowing the well-matched stop-band to extend to very high frequencies.

Consider the cascade combination of a high-order and a low-order low-pass filter shown in Figure 4.13(a). The element values are fully labeled in this case because the two cascaded filters are not tuned to the same frequency, so unity normalized value can no longer be assumed. The first filter, a third-rank ($N = 7$) design, is used to realize the sharp cutoff, while a first-rank ($N = 3$) design tuned to a higher frequency suppresses the reentry in the higher-order filter, giving a more extended stop-band.

4.5.2 Cascaded High-/Low-Frequency Low-Pass Filters

In the previous theoretical example, the reentry at high frequency is a fundamental

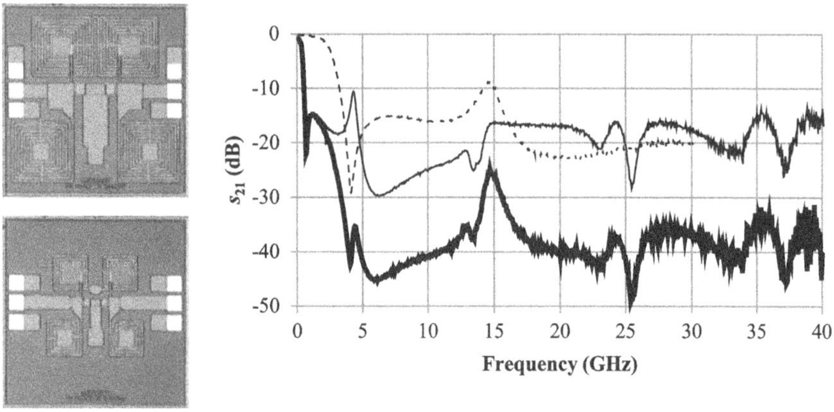

(a) (b)

Figure 4.14 Combination of high-frequency and low-frequency low-pass filters for greatly extended stop-band. (a) Micro-photographs of individual chips having cutoffs of about 400 MHz (top) and 2 GHz (bottom). Both chips measure 1.0 x 1.0 x 0.1 mm in size. (b) Frequency response for individual filters (thin/dashed lines) and cascade combination (thick solid line) showing good stop-band rejection out to 100 times the cutoff frequency [8, 9].

feature of the third-rank topology. The first-rank design does not have this feature, so in principle, the stop-band extent of the cascaded combination is infinite. Nevertheless, in practice, parasitics of the physical elements will always lead to some degradation at high frequencies. As previously stated, however, even this can be addressed by cascading higher-frequency designs with lower-frequency designs, similar to the way that large bypass capacitors are often combined with lower-valued bypass capacitors to account for their differing self-resonant frequencies.

An example of this is shown in Figure 4.14. Here, we have combined two first-rank filters, one with a cutoff of about 400 MHz, and another at about 2 GHz. Actual measurements are used in this case, as the effects of real parasitics are at the heart of this example. Note that the lower-frequency filter, which provides the dominant cutoff for the combination, would have had a resonant peak at around 4 GHz, but with the second filter to help reject it, this pairing has a good stop-band up to at least 40 GHz, roughly 100 times higher than the pass-band edge.

4.5.3 Cascaded High-/Low-Pass Ultrawideband Filters

Another way in which combinations may be used is illustrated in Figure 4.15, where a low-pass and a high-pass design have been combined to create an ultrawideband filter. Filters such as these are notoriously difficult to design using conventional techniques, but become almost trivial with reflectionless filters where the upper and lower cutoffs can be designed and verified independently,

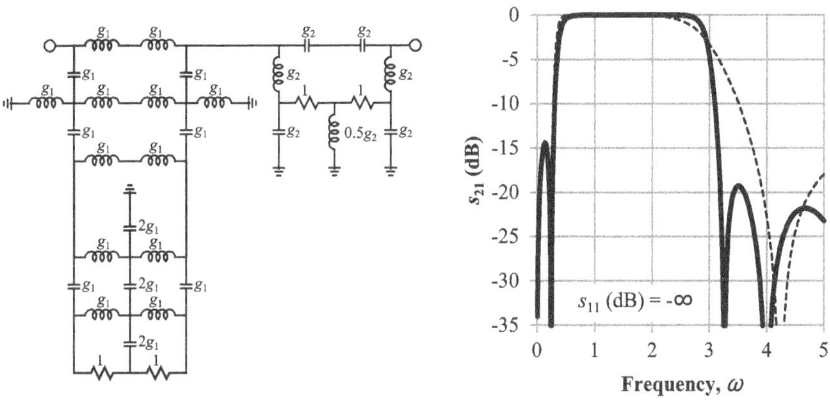

(a) (b)

Figure 4.15 Combination of low-pass and high-pass filters for ultrawide pass-band. (a) Schematic. (b) Frequency response where $g_1 = 0.25$, and $g_2 = 4$. The dashed line shows the frequency response for the same range of element values ($\Delta = 4$) using instead the topology in Figure 3.8(c).

and the good impedance match from both parts ensure that the cascade combination will behave as expected.

This approach has the added benefit of allowing the high-pass and low-pass response to be of different orders. Because of the large scale factor between the high- and low-frequency cutoffs, conventional band-pass responses exhibit a remarkable degree of asymmetry. This is illustrated with the dashed line in Figure 4.15(b), which is the response for the direct band-pass topology shown earlier in Figure 3.8(c). The relative bandwidth, Δ, is selected such that the range of element values is exactly the same as that in Figure 4.15(a). The low-frequency, high-pass cutoff is very similar in both designs; however, the high-frequency, low-pass cutoff of the direct band-pass design is unacceptably gradual. By using a cascade approach, the high-frequency, low-pass cutoff may be independently designed for sharper characteristics without altering the performance of the lower band edge. The reflectionless property of the two cascaded cells guarantees that the combined result is well-behaved to a fine degree, in contrast with conventional filters where even slight in-band mismatches may cause standing-wave interactions that degrade performance.

4.5.4 Predistortion Filters

In cases where the fabrication technology for a filter is relatively lossy, one well-known technique for improving the performance is to apply predistortion [10]. In conventional filters, this involves perturbing the locations of the poles to compensate for signal loss (e.g., moving the poles closer to the $j\omega$ axis) peaking

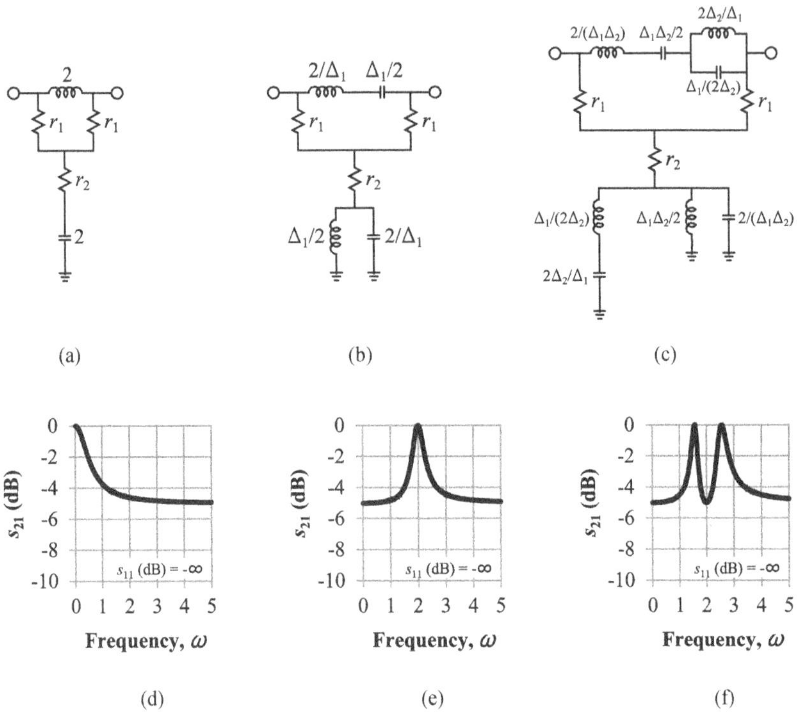

Figure 4.16 (a) First-order reflectionless filter prototype with attenuator subnetwork. (b) Predistortion filter configuration for low-pass and high-pass filters. (c) Predistortion filter configuration for band-pass and band-stop filters. Δ_1 controls the sharpness of the predistortion peaks, and Δ_2 controls the spacing between them. (d) Response of the low-pass filter in (a). (e) Response of the predistortion filter in (b). (f) Response of the predistortion filter in (c).

up the ideal response near cutoff so that the losses merely restore the desired flat pass-band rather than rounding it off. Invariably, this means an overall increase in the minimum loss of the pass-band, but this is accepted in favor of a more abrupt transition into the stop-band.

Precisely tailoring the pole-zero configuration of reflectionless filters is problematic using the topologies presented thus far — the need to maintain symmetry and duality does not permit such flexibility. However, the excellent cascadability of such filters offers another way of achieving essentially the same effect.

To accomplish this, we propose cascading one of the previously demonstrated reflectionless filters with a low-order predistortion filter that has peaks near the desired cutoff frequencies. We start with a first-order (N = 1) reflectionless filter prototype from Figure 3.2, and embed within it an attenuator as its subnetwork. This is shown in Figure 4.16(a). The attenuator limits the rejection of the

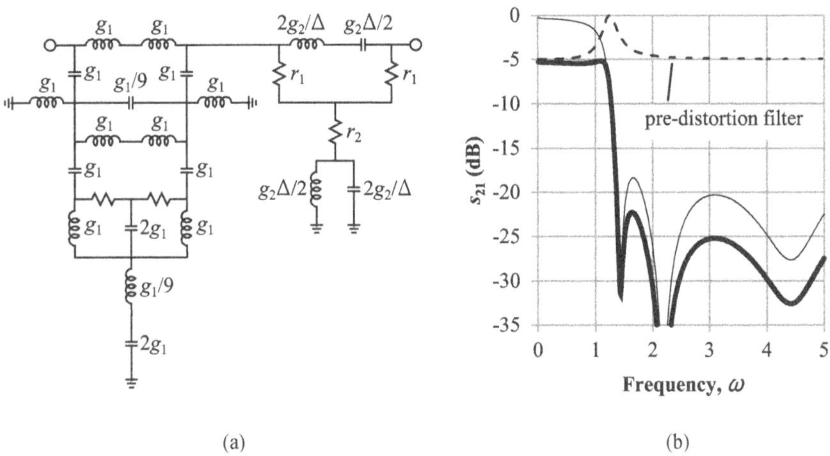

Figure 4.17 (a) Application of predistortion to a modified, second-rank, low-pass reflectionless filter. (b) Simulated results of original filter with significant losses included (thin solid line), of predistortion filter (dashed line), and of the combination (thick solid line).

predistortion filter (and thereby determines the height of the corrective peaking) by an amount, A, in decibels. The normalized resistor values are given by

$$r_1 = \frac{1-T}{1+T} \tag{4.13a}$$

$$r_2 = \frac{2T}{1-T^2} \tag{4.13b}$$

$$T = 10^{-A/20} \tag{4.13c}$$

To apply this correction to low-pass or high-pass filters, the peak of the predistortion filter must be translated to the cutoff frequency. This is accomplished with a band-pass transformation as given in Table 1.2. The resulting predistortion filter is shown in Figure 4.16(b). Its use is illustrated in Figure 4.17. In this example, a modified, second-rank, low-pass filter is simulated with significant losses, resulting in severe rounding of the transition, and then combined with a predistortion filter having A = 5 dB and Δ = 1. The resulting transition corner, though relatively high in attenuation due to the losses of the physical components, is much sharper.

To apply predistortion to band-pass or band-stop filters, we must have a predistortion filter with two peaks. This requires that a second band-pass transformation be applied to the predistortion filter, as illustrated in Figure

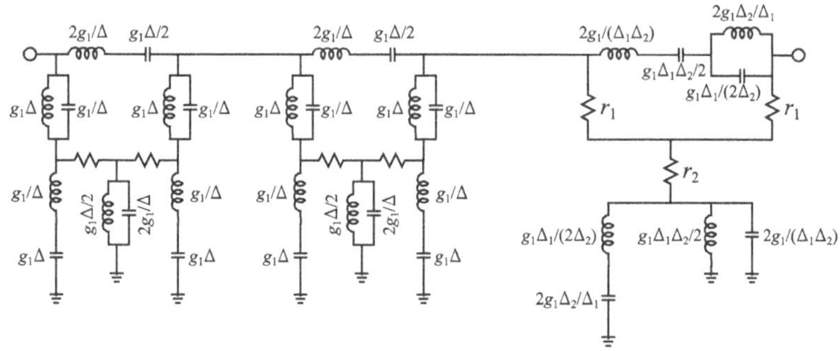

(a)

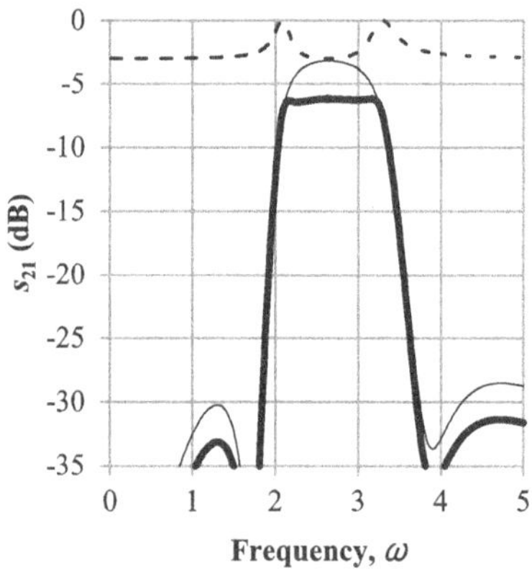

(b)

Figure 4.18 (a) Application of predistortion to a cascaded band-pass reflectionless filter. (b) Simulated results of original filter with significant losses included (thin solid line), of predistortion filter (dashed line), and of the combination (thick solid line).

4.16(c). Now there are two bandwidth parameters, Δ_1 and Δ_2. The first, associated with the initial transformation, controls the sharpness of the corrective peaks, whereas the second, associated with the follow-up transformation, controls the spacing between them. This is illustrated by the example in Figure 4.18. Here, a cascaded band-pass filter has been simulated with significant losses, resulting in fairly poor transition corners. A double-peak predistortion filter is applied, in this case with $A = 3$ dB, $\Delta_1 = 3$, and $\Delta_2 = 0.46$.

In these examples, a fairly high level of loss is simulated in order to better illustrate the use of predistortion in reflectionless filters. Many fabrication technologies are capable of realizing components of much higher quality than that illustrated here, in which case the degree of corrective peaking needed will be

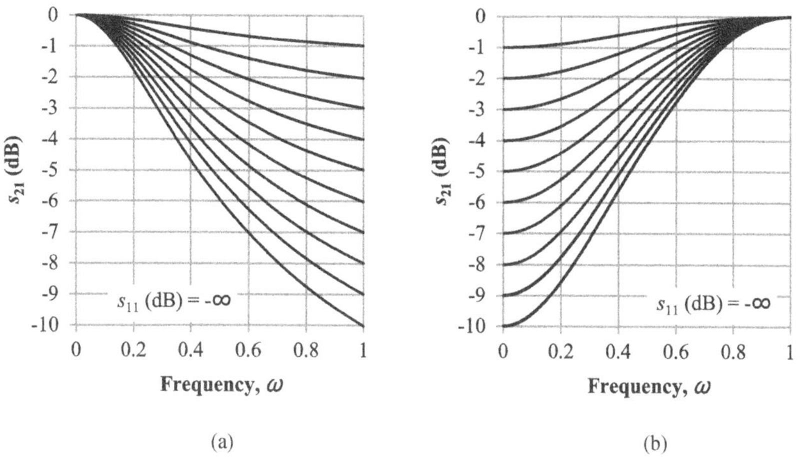

Figure 4.19 (a) Simulated response of the negative-sloping equalizer in Figure 4.16(a) for slopes ranging from 1 dB to 10 dB. (b) Simulated response of the positive-sloping equalizer in Figure 4.16(b) for the same range of slopes, with $\Delta_1 = 30/A$.

much more subtle, but the basic principles remain the same.

4.5.5 Slope Equalizers

Subnetworked reflectionless filters also provide a straightforward solution for matched slope equalizers. Such circuits are often needed to compensate for the loss slopes associated with other components, including broadband gain stages or cable runs.

In fact, one simple form of a negative-sloping equalizer has just been seen in another context in Figure 4.16(a), and its response is shown in Figure 4.16(d). There is no universal rule for the desired attenuation curve of an equalizer, but something close to a linear slope in decibels is often chosen for simplicity and is good enough for most applications. The curve in Figure 4.16(d) is sufficiently close to linear from DC to $\omega = 1$ to meet this loose requirement.

On the other hand, positive-sloping equalizers are more often required due to the normal frequency dependence of Ohmic losses in passive interconnects and the negatively trending available gain of active devices. A simple high-pass transformation from Table 1.2 applied to the aforementioned topology yields a positive-sloping equalizer, but since the high-pass version only reaches its theoretically minimum insertion loss of 0 dB at infinite frequency, it is difficult to use this topology to achieve near-linear slopes at baseband without incurring substantial excess losses. A possibly better solution is to use the band-pass form in Figure 4.16(b). A good rule of thumb to achieve linearity from DC to $\omega = 1$ with this topology is to let $\Delta_1 = 30/A$, where A is the peak attenuation (slope) in decibels, and the normalized resistor values are given by (4.13).

The simulated responses of near-linear slope equalizers using the topologies in Figure 4.16(a) and Figure 4.16(b) are shown for slopes ranging from 1 dB to 10 dB in Figure 4.19.

These relatively simple structures lend themselves well to monolithic fabrication; selectivity is not a dominant concern for equalizers, so a monolithic implementation allows them to be made very small without any meaningful loss of performance. Say, for example, that we need a positively sloping equalizer for DC to 6 GHz with 8 dB of slope. Utilizing the topology in Figure 4.16(b), we may calculate the element values as follows,

$$\Delta_1 = \frac{30}{A} = \frac{30}{(8\text{dB})} = 3.75 \tag{4.14a}$$

$$L_1 = \frac{2Z_0}{\Delta_1 \omega_c} = \frac{2(50\Omega)}{(3.75)(2\pi \cdot 6\text{GHz})} = 0.71 \text{ nH} \tag{4.14b}$$

$$L_2 = \frac{\Delta_1 Z_0}{2\omega_c} = \frac{(3.75)(50\Omega)}{2(2\pi \cdot 6\text{GHz})} = 2.49 \text{ nH} \tag{4.14c}$$

$$C_1 = \frac{\Delta_1 Y_0}{2\omega_c} = \frac{(3.75)(50\Omega)^{-1}}{2(2\pi \cdot 6\text{GHz})} = 0.99 \text{ pF} \tag{4.14d}$$

$$C_2 = \frac{2Y_0}{\Delta_1 \omega_c} = \frac{2(50\Omega)^{-1}}{(3.75)(2\pi \cdot 6\text{GHz})} = 0.28 \text{ pF} \tag{4.14e}$$

$$T = 10^{-A/20} = 10^{-(8\text{dB})/20} = 0.3981 \tag{4.14f}$$

$$R_1 = \frac{1-T}{1+T} Z_0 = \frac{1-0.3981}{1+0.3981} 50\Omega = 21.5\Omega \tag{4.14g}$$

$$R_2 = \frac{2T}{1-T^2} Z_0 = \frac{2(0.3981)}{1-0.3981^2} 50\Omega = 47.3\Omega \tag{4.14h}$$

The final schematic for the desired equalizer is thus shown in Figure 4.20(a). The layout can be made very small, as illustrated in Figure 4.20(b), estimated at roughly 0.5 mm in size for the 6 GHz frequency range specified. Electromagnetically simulated performance for this equalizer is shown in Figure 4.20(c). It maintains better than 20 dB return loss while providing a smooth, positive 8-dB slope from DC to 6 GHz.

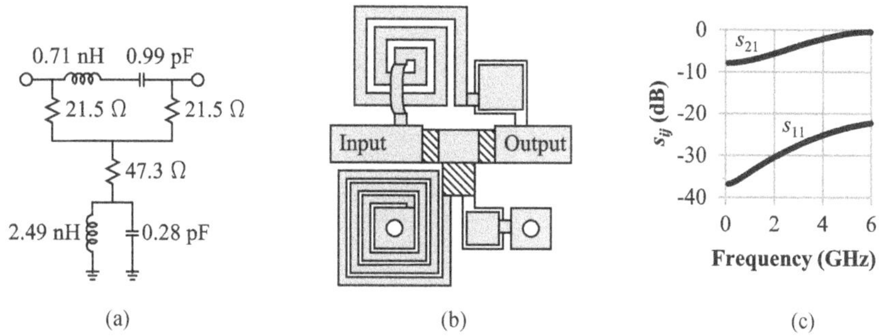

Figure 4.20 (a) Final schematic of the proposed +8-dB slope equalizer. (b) Monolithic layout. (c) EM simulated performance with losses and parasitics included.

References

[1] Morgan, M., and Boyd, T., "Reflectionless filter structures," *IEEE Trans. Microw. Theory Tech.*, vol. 63, no. 4, pp. 1263–1271, April 2015.

[2] Morgan, M., *Sub-Network Enhanced Reflectionless Filter Topologies*, U.S. Patent Application No. 14/724,976, May 29, 2015, PCT Application No. PCT/US15/33118, May 29, 2015, and Taiwan Patent Application No. 104120606, June 25, 2015.

[3] Pozar, D., *Microwave Engineering*, 4th ed., New York: Wiley, pp. 344–345, November 2011.

[4] Mason, S., "Feedback theory – further properties of signal flow graphs," *Proc. IRE*, vol. 44, pp. 920–926, July 1956.

[5] Chieh, J. C., and Rowland, J., "Quasi-lumped element bridged-T absorptive bandstop filter," *Microw. & Wireless Comp. Lett.*, vol. 26, no. 4, pp. 264–266, April 2016.

[6] Guyette, A., Hunter, I., and Pollard, R., "Design of absorptive microwave filters using allpass networks in a parallel-cascade configuration," *IEEE MTT-S Intl. Microw. Symp. Digest*, Boston, pp. 733–736, June 7-12, 2009.

[7] Lee, J., Lee, T., and Chappell, W., "Lumped-element realization of absorptive bandstop filter with anomalously high spectral isolation," *IEEE Trans. Microw. Theory Techn.*, vol. 60, no. 8, pp. 2424–2430, August 2012.

[8] *Mini-Circuits datasheet*, XLF-192+, available at http://www.minicircuits.com/pdfs/XLF-192+.pdf, October 2015.

[9] *Mini-Circuits datasheet*, XLF-221+, available at http://www.minicircuits.com/pdfs/XLF-221+.pdf, October 2015.

[10] Cameron, R., Kudsia, C., and Mansour, R., *Microwave Filters for Communications Systems*, New Jersey: John Wiley & Sons, pp. 230–238, 2007.

Chapter 5

Duality Reduction

In Chapter 4, it was shown that reflectionless filter topologies may be expanded to higher order by embedding subnetworks within them, including, potentially, other reflectionless filters, to any rank that is desired. Now we exploit the concept of duality, first introduced in Section 1.3, as a means of deriving circuit identities for the core reflectionless filter topologies, and then utilizing them to simplify compound reflectionless filters, achieving the same circuit response with fewer elements.

5.1 DUAL REFLECTIONLESS FILTER NETWORK

5.1.1 Properties of Dual Multiport Networks

In Section 1.3, it was shown that every circuit has a dual obtained (for lumped elements) by replacing inductors with capacitors, capacitors with inductors, series connections with parallel connections, and parallel connections with series connections. It was further shown that the rational polynomial immittance functions of a lumped element network and its dual are multiplicative inverses. But what of the transfer functions of dual multiport networks? To answer that question, let us first consider an arbitrary, linear two-port network. Such a network may be fully described by its admittance matrix,

$$\mathbf{Y}(s) = \begin{pmatrix} a_{11} & a_{12} \\ a_{21} & a_{22} \end{pmatrix} \tag{5.1}$$

For a passive, reciprocal network, $a_{12} = a_{21}$. Although the actual topology may differ, an equivalent circuit for this network has port 1 and port 2 connected by a series admittance element, $-a_{12}$, corresponding to the negative of the off-diagonal terms, and shunt admittance elements at both ports equal to the sum of the terms in the corresponding row (or column) [1]. This is shown in Figure 5.1(a). Note

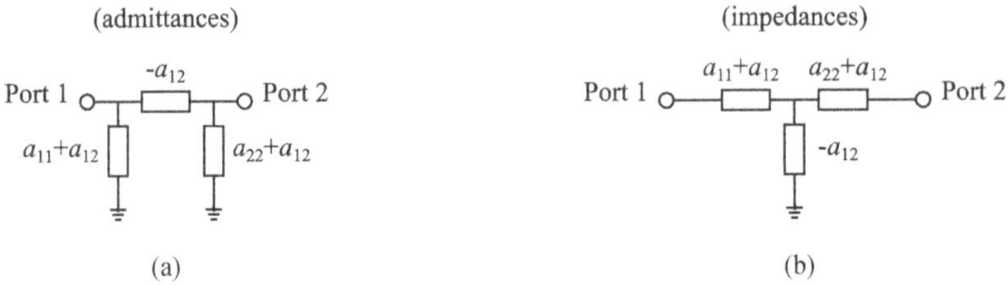

Figure 5.1 (a) Two-port network equivalent circuit. (b) Dual of the equivalent circuit.

that each term, a_{ij}, may be an arbitrary admittance function in s, and the circuit elements need not be individual resistors, inductors, or capacitors, but rather comprise full admittance networks.

The dual of this equivalent circuit is a tee-network where the series element (now an impedance) at port 1 is $z = a_{11} + a_{12}$, the series element at port 2 is $z = a_{22} + a_{12}$, and the shunt element between them is $z = -a_{12}$, shown in Figure 5.1(b). Note that we have assumed normalized element values. The impedance matrix, $\mathbf{Z}^d$, for this dual network may be found by considering the voltages developed on each side if a current is injected into one port while the other is left open. For example, if a current, i, is injected into port 1, the voltage developed at port 1 is $v = i\cdot(a_{11} + a_{12} - a_{12}) = i\cdot a_{11}$, so the impedance at port 1 is a_{11}. Filling in the rest of the dual network impedance matrix, we have

$$\mathbf{Z}^d(s) = \begin{pmatrix} z_{11}^d & z_{12}^d \\ z_{12}^d & z_{22}^d \end{pmatrix} = \begin{pmatrix} a_{11} & -a_{12} \\ -a_{21} & a_{22} \end{pmatrix} \tag{5.2}$$

We can solve for the scattering parameters of the original two-port network and its dual using (1.6),

$$\mathbf{S} = (\mathbf{I}+\mathbf{Y})^{-1}(\mathbf{I}-\mathbf{Y}) = \begin{pmatrix} 1+a_{11} & a_{12} \\ a_{12} & 1+a_{22} \end{pmatrix}^{-1} \begin{pmatrix} 1-a_{11} & -a_{12} \\ -a_{12} & 1-a_{22} \end{pmatrix} \tag{5.3a}$$

$$= \frac{1}{\det(\mathbf{Y})} \begin{pmatrix} (1-a_{11})(1+a_{22})+a_{12}^2 & -2a_{12} \\ -2a_{12} & (1+a_{11})(1-a_{22})+a_{12}^2 \end{pmatrix} \tag{5.3b}$$

$$\mathbf{S}^d = (\mathbf{Z}^d+\mathbf{I})^{-1}(\mathbf{Z}^d-\mathbf{I}) = \begin{pmatrix} a_{11}+1 & -a_{12} \\ -a_{12} & a_{22}+1 \end{pmatrix}^{-1} \begin{pmatrix} a_{11}-1 & -a_{12} \\ -a_{12} & a_{22}-1 \end{pmatrix} \tag{5.3c}$$

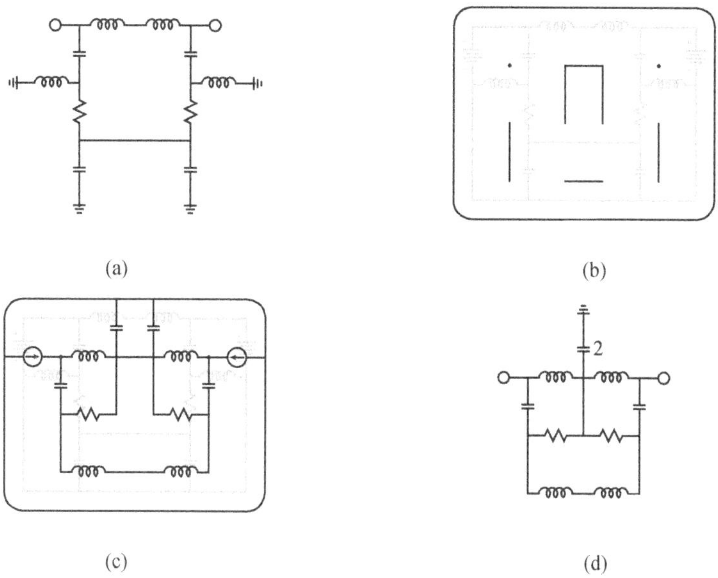

Figure 5.2 Derivation of dual low-pass reflectionless filter topology. (a) Original third-order low-pass reflectionless filter. (b) Redrawn filter with dual nodes added. (c) Dual-filter after element-by-element substitution. (d) Final dual, low-pass reflectionless filter. In keeping with the established convention of the previous chapters, all unlabeled elements are assumed to have unit normalized value. Those labeled with a constant are scaled accordingly.

$$= \frac{1}{\det\left(\mathbf{Z}^d\right)}\begin{pmatrix} (a_{11}-1)(a_{22}+1)-a_{12}^2 & -2a_{12} \\ -2a_{12} & (a_{11}+1)(a_{22}-1)-a_{12}^2 \end{pmatrix} \tag{5.3d}$$

where

$$\det(\mathbf{Y}) = \det\left(\mathbf{Z}^d\right) = (a_{11}+1)(a_{22}+1)-a_{12}^2 \tag{5.3e}$$

More simply,

$$\mathbf{S} = \begin{pmatrix} s_{11} & s_{12} \\ s_{21} & s_{22} \end{pmatrix} = \begin{pmatrix} -s_{11}^d & s_{12}^d \\ s_{21}^d & -s_{22}^d \end{pmatrix} \tag{5.4}$$

This shows that the reflection coefficients of a network and its dual are additive inverses, whereas the transmission coefficients of the two networks are identical. For a reflectionless filter then (since its reflection coefficients are identically zero)

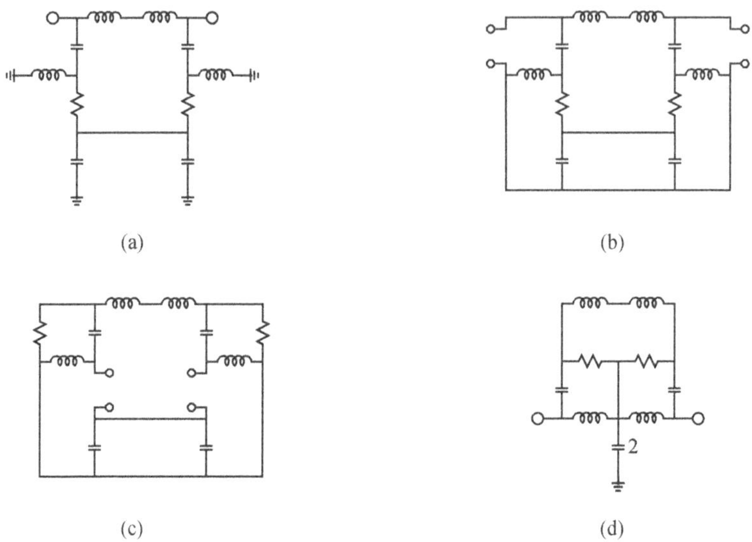

Figure 5.3 Extraction of dual low-pass reflectionless filter topology from internal ports. (a) Original reflectionless filter. (b) Redrawn, two-wire representation of reflectionless filter. (c) Filter with terminations exchanged between internal and external ports. (d) Top-wire representation of internal filter network.

we find that all of its scattering parameters are identical to those of its dual.[1]

5.1.2 Derivation

This might at first seem to imply that reflectionless filters are their own duals — that the dual topology of a given reflectionless filter is the same as the original. However, this is not the case.

Take the now-familiar third-order reflectionless low-pass filter topology in Figure 5.2(a). We will follow the same graphical procedure to derive its dual that was introduced in Section 2.5.3. First, the circuit has been redrawn in Figure 5.2(b) with an explicit ground node and the nodes of the dual network identified. Next, we substitute each element and source with their dual counterparts. The result of this substitution is shown in Figure 5.2(c). The new circuit is then simplified, and its final representation drawn in Figure 5.2(d). Not the same network as the original, but having the same port parameters, this network is a circuit identity for that in Figure 5.2(a) [2]. In fact, all reflectionless filters have duals which are circuit identities for them.

---

[1] Though illustrated here for two-port networks only, the same general statement applies to arbitrarily large multiport networks. One may prove this by following the same procedure above for every pair of ports with the remaining ports in each step terminated by load resistors.

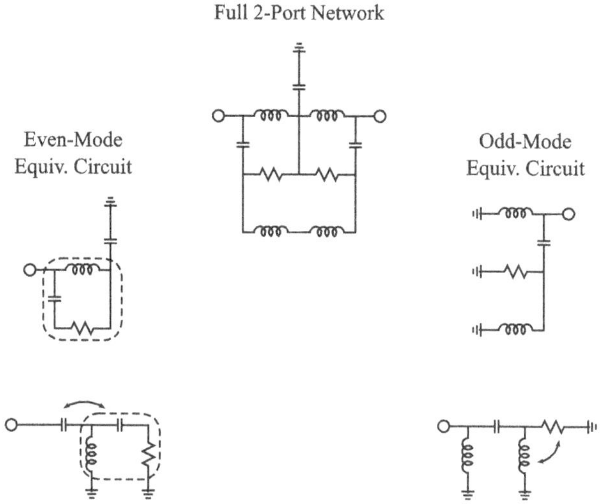

Figure 5.4 Deconstruction of dual reflectionless filter into even- and odd-mode equivalent circuits.

5.1.3 Alternative Derivation

Interestingly, the dual topology in Figure 5.2(d) appears in another place that we have already seen, though we may not have recognized it as such. Recall in Section 4.1 when we derived the scattering parameters of the full reflectionless filter including its internal ports and found that the transfer function between ports 3 and 4 was the same as that between ports 1 and 2. More generally, the scattering matrix in (4.3b) shows that if port 1 is swapped with port 3, and port 2 with port 4, the identical scattering matrix is obtained. Topologically, this corresponds to moving the terminations from ports 3 and 4 to ports 1 and 2, as illustrated in Figure 5.3. Note that the common-node between the two ports in Figure 5.3(c) was redefined as the new ground in order to convert to a top-wire representation in the last step. The final result, somewhat surprisingly, is the same as the dual obtained in Figure 5.2 [3].

Let us look at the even- and odd-mode equivalent circuits of this alternative topology, shown in Figure 5.4. In this case, on the even-mode side, simplification requires us to exchange the order of the series capacitor with the resistor-inductor-capacitor loop with which it is in series. The end result is a tee-network high-pass filter of third order (compare with Figure 3.4). Thus, we find that the even(/odd)-mode equivalent circuit of the dual reflectionless filter is not the dual of, but rather identical to, the even(/odd)-mode equivalent circuit of the original reflectionless filter. Put more simply, the equivalent half-circuits of the dual are *not* the duals of the equivalent half-circuits.

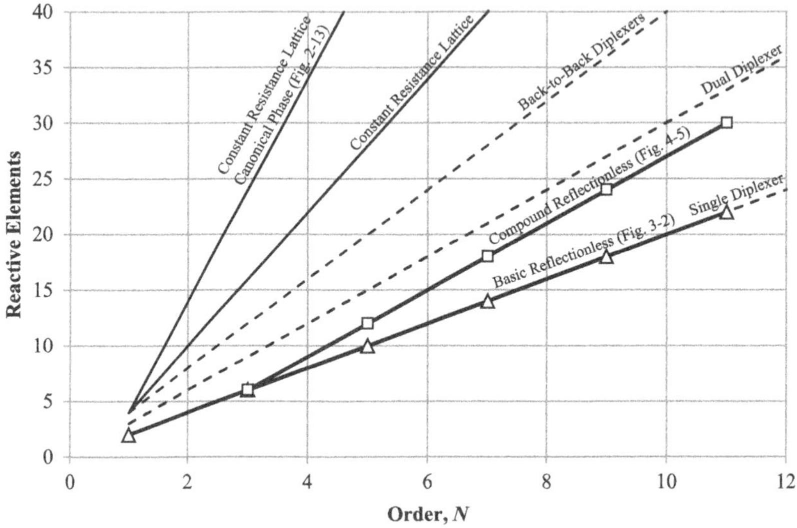

Figure 5.5 Number of reactive elements required to realize (low-pass) absorptive and reflectionless filters of order N.

5.2 COMPOUND FILTER REDUCTION

We saw in Chapter 4 that compound filters may be formed by embedding reflectionless filters as subnetworks within other reflectionless filters. The rank, or total number of reflectionless filters thus nested, determines the order of the filter response that is achieved. However, the number of elements required by these topologies for a given order has increased compared to the basic reflectionless filters initially derived (consider Figure 4.5 vs. Figure 3.2). Figure 5.5 shows the number of elements required by these two topologies to realize filters having a frequency response of a given order, N, not including terminations (these are considered free from a standpoint of cost and complexity). For comparison, the elements required for a number of other previously described techniques are shown as well — dashed lines for diplexers of various forms [4] and solid lines for constant-resistant networks [5].

Note that the basic reflectionless filter of Figure 3.2 has complexity similar to that of a unidirectional diplexer, while still being matched from both sides. The compound reflectionless filter of Figure 4.5, on the other hand, is similar in complexity to a bidirectional diplexer. Complexity, in this case, is measured by the total number of elements, but that could be misleading. We have assumed in the plot above that identical series and parallel elements are combined into one to

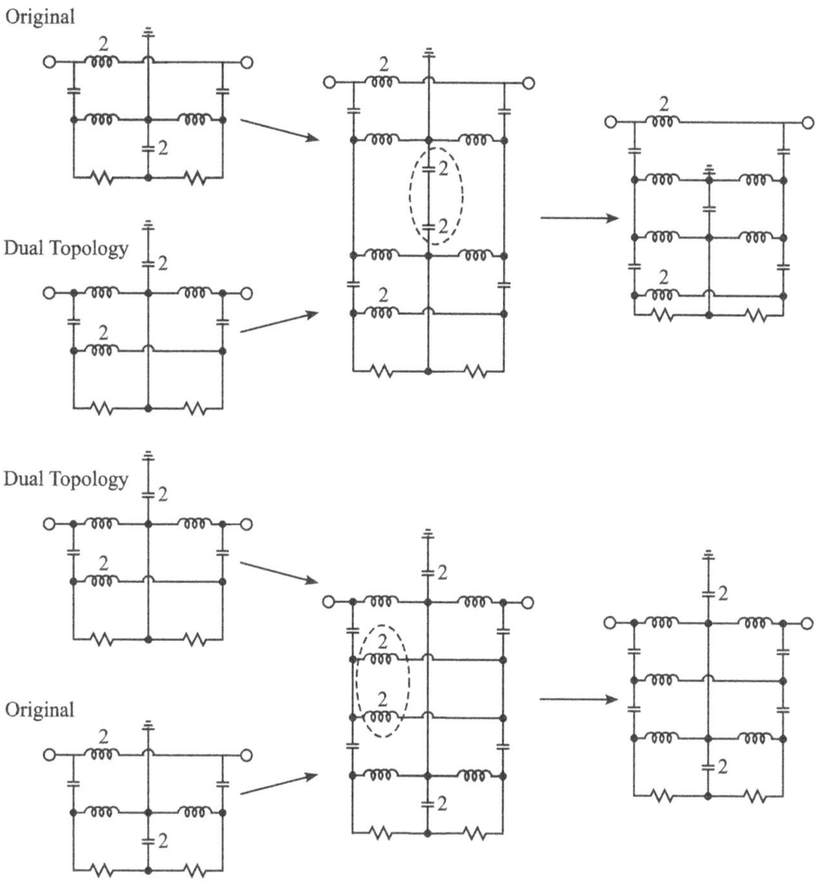

Figure 5.6 Derivation of reduced compound reflectionless filters of second rank by embedding dual topologies. The encircled components combine in series/parallel.

minimize this number, however the number of unique (i.e., nonrepeating) elements may be of greater interest in some applications (e.g., monolithic fabrication) where replication of elements is easy, but individually tuning them is hard. For most of the reflectionless filters described so far, the total number of unique reactive element values required at minimum is only 2 (one inductor and one capacitor) regardless of the order.

The increase in the number of elements for a compound filter was the price of a more desirable filter response, but it suggests the possibility that higher-order filters, even reflectionless ones, should not need so many elements. The challenge, as always, is to find the right topology.

Some progress can be made by taking advantage of the circuit identity

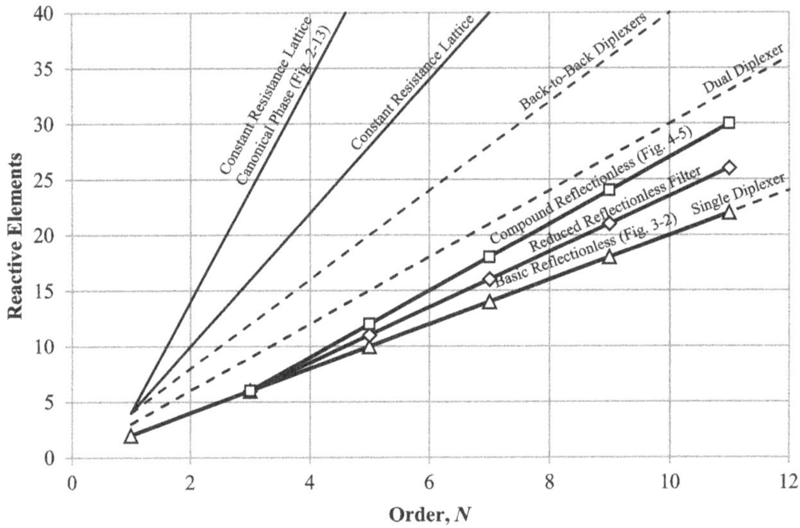

Figure 5.7 Number of reactive elements required to implement (low-pass) absorptive and reflectionless filters of order N.

discovered Section 5.1. Consider the second-rank ($N = 5$) filter shown in the middle column of Figure 4.5. This was produced by embedding one standard first-rank reflectionless filter inside another. If the dual topology was used instead for both stages of the filter, a new compound topology would be obtained with the same number of elements and having the exact same frequency response. However, if the outer and inner stages are of mixed types (one original and one dual), then some cancellation of elements occurs, as shown in Figure 5.6. Once again, we have used the folded schematic arrangement in order to make the subnetwork embedding more clear. In the first case, where the dual topology was inserted as the subnetwork inside the original topology, a capacitor from each side connects in series and may be combined. In the second case, the original topology is inserted as the subnetwork within the dual topology, and the two parallel inductors may be combined. For filters of higher rank, simply alternating between the original and dual topologies guarantees that one element will be eliminated with each added stage. This means that all of the compound filters we have seen thus far could be implemented with fewer elements, as shown in Figure 5.7.

In principle, any combination of original and dual reflectionless networks in a high-rank filter is a valid one, and each yields a different compound topology. While only the pattern of alternating original and dual networks minimizes the number of elements, this flexibility could be used in special cases to select a topology that, for one reason or another, is easier to lay out or minimizes a particular type of component (such as inductors). More importantly, this

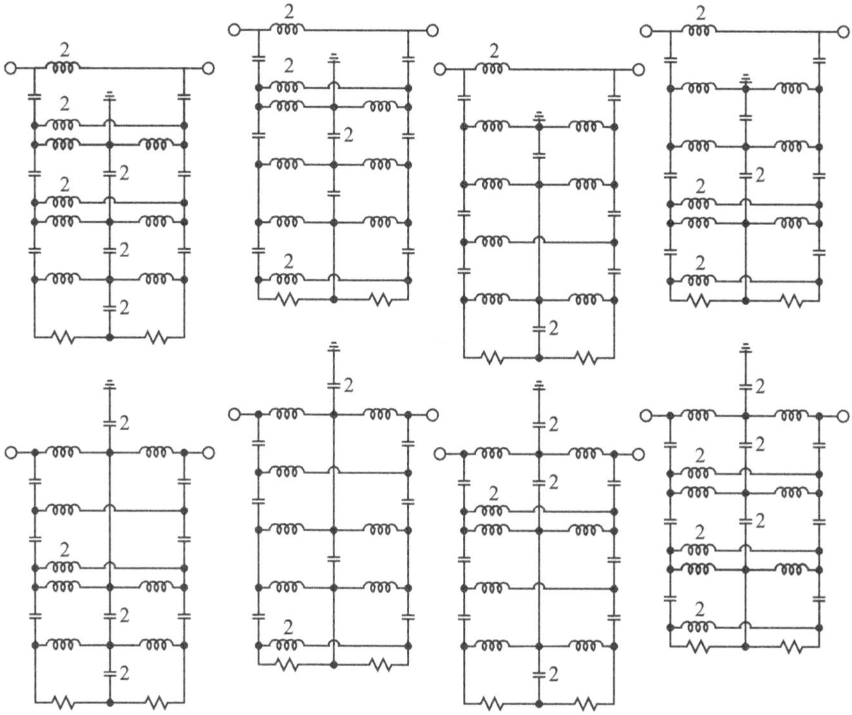

Figure 5.8 Eight possible topologies for the third-rank low-pass reflectionless filter. Frequency response is shown in Figure 4.6(c).

highlights an important fact about reflectionless filter topologies, namely that they are not unique. For rank R, the number of distinct topologies that can be obtained in this way is 2^R, all of which have identical frequency response. To underscore the point, the eight variants of the third-rank reflectionless filter are shown in Figure 5.8. Each of these distinct topologies has the same frequency response, given earlier in Figure 4.6(c). Of course, for each of these there is also a high-pass, band-pass, and band-stop equivalent, all distinct, deriving from the frequency transformations provided in Table 1.2.

5.3 LAYOUT CONSIDERATIONS

5.3.1 Monolithic Versions

Often, the value of using reduced topologies is to optimize the layout for a particular fabrication technology. In the case of planar integrated circuits, where

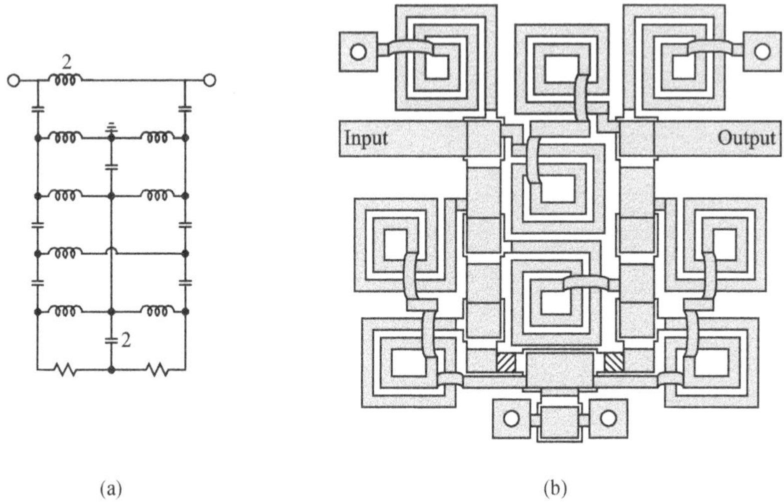

Figure 5.9 (a) Reduced third-rank filter topology. (b) Layout for monolithic fabrication showing nine equal-valued inductors instead of twelve.

the capacitors are usually metal-insulator-metal (MIM) stacks, and the inductors are planar spirals, it is usually the inductors that dominate the layout. For the low- and high-pass chips in Figure 3.14, for example, the layout was optimized such that each of the four spiral inductors occupied roughly one quadrant of the die area, with the rest of the components fitting in between. For a standard third-rank filter, however, such as that illustrated in Figure 4.8(b), twelve inductors of equal value would be required (or six of the same plus three of twice the size if series inductors were combined). Such a layout would have required a much larger chip for the same or similar cutoff frequency as those shown earlier for the low-order filters. By using the reduced topology in Figure 5.9(a), however, only nine equal-valued inductors are needed, and can be approximately arranged in a 3x3 grid. See, for example, the sample layout in Figure 5.9(b) [3].

Similarly, a reduced version of the second-rank filter modified with auxiliary components, first introduced in Figure 4.11, could likewise be optimized for fabrication on a small integrated chip. In this case, the dual topology was used in the primary stage of the filter and the original topology in the second stage so that the inductors would combine in parallel, as shown in Figure 5.10(a). The monolithic layout for this filter is shown in Figure 5.10(b). Once again, the reduced topology permits the filter to fit on a small chip, 1 mm square for 1.125 GHz cutoff. Simulated performance for this frequency is given in Figure 5.10(c) using the ideal lumped-element model (dashed line) and a full-wave electromagnetic simulation including losses (solid line) [3].

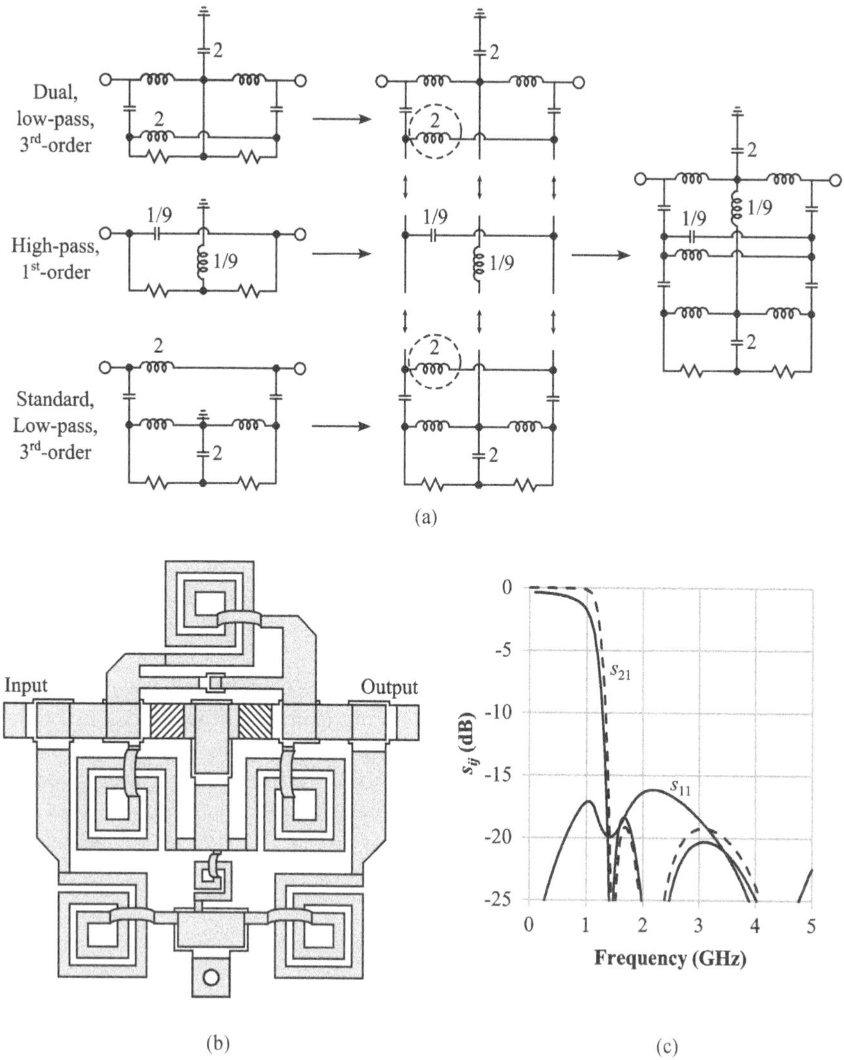

Figure 5.10 (a) Construction of a reduced version of the filter shown in Figure 4.11. The two inductors circled in the intermediate step combine in parallel. (b) Layout of a 1.125 GHz filter on a 1.0 x 1.0 x 0.1 mm chip. (c) EM-simulated performance (solid line) compared to ideal lumped-element model (dashed).

5.3.2 Discrete-Element Versions

If, instead of a monolithic fabrication approach, the design is to be implemented

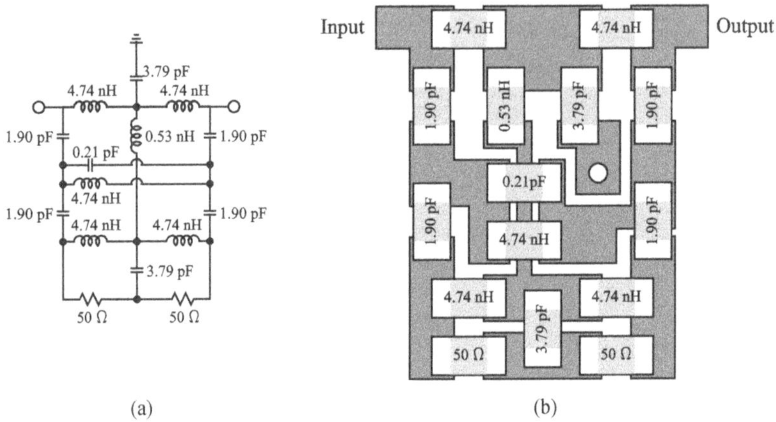

Figure 5.11 (a) Schematic of a modified second-rank reflectionless filter tuned for 1 GHz cutoff frequency. (b) Proposed circuit board layout.

with discrete, surface-mount elements, the total number of components is of greater importance than the variety of values. The inductors no longer dominate the dimensions, and higher-value elements are usually the same size as those with lower values.

To work an example, let us consider once again the second-rank filter with auxiliary elements from Figure 5.10(a). We find that a total of two inductor values and three capacitor values are needed to complete the circuit. Examination of the original, ideal-element frequency response for this filter, which was given in Figure 4.11, shows that the 1 dB corner occurs at a normalized frequency of $\omega \approx 0.67$. Using this as the base prototype parameter in the reduced topology and again assuming a 1.125 GHz cutoff, we calculate

$$L_1 = \frac{g_1 Z_0}{\omega_c} = \frac{(0.67)(50\Omega)}{2\pi(1.125\,\text{GHz})} = 4.74\text{ nH} \tag{5.5a}$$

$$L_2 = \frac{g_2 Z_0}{\omega_c} = \frac{\left(\frac{1}{9}g_1\right)Z_0}{\omega_c} = \frac{(0.67/9)(50\Omega)}{2\pi(1.125\,\text{GHz})} = 0.53\text{ nH} \tag{5.5b}$$

$$C_1 = \frac{g_1 Y_0}{\omega_c} = \frac{(0.67)(50\Omega)^{-1}}{2\pi(1.125\,\text{GHz})} = 1.90\text{ pF} \tag{5.5c}$$

$$C_2 = \frac{g_2 Y_0}{\omega_c} = \frac{(2g_1)Y_0}{\omega_c} = \frac{(2\cdot 0.67)(50\Omega)^{-1}}{2\pi(1.125\,\text{GHz})} = 3.79\text{ pF} \tag{5.5d}$$

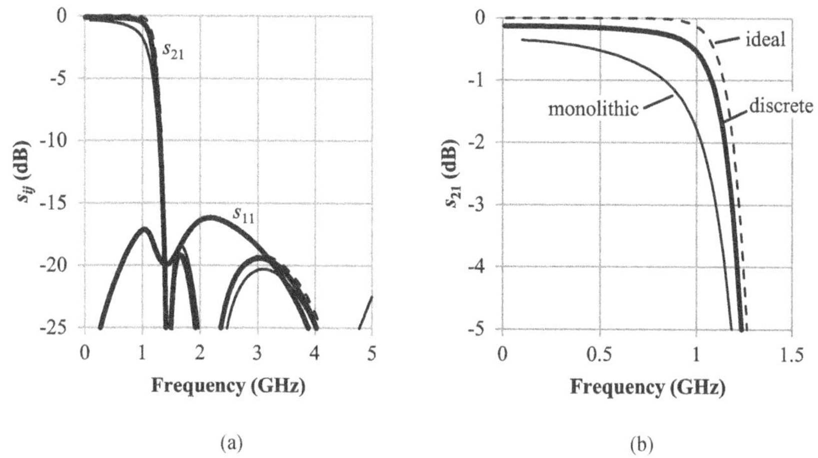

Figure 5.12 Comparison of second-rank reflectionless filter with auxiliary components using ideal elements (dashed line), monolithic fabrication (thin solid line), and discrete elements (thick solid line). (a) Overall performance. (b) Pass-band detail. (The s_{11} curve shown is for the monolithically fabricated circuit only. The s_{11} for the ideal case is infinite, and for the discrete elements is too far down to see on this scale.)

$$C_3 = \frac{g_3 Y_0}{\omega_c} = \frac{\left(\frac{1}{9} g_1\right) Y_0}{\omega_c} = \frac{(0.67/9)(50\Omega)^{-1}}{2\pi(1.125\,\text{GHz})} = 0.21\text{ pF} \tag{5.5e}$$

These values are shown schematically in Figure 5.11(a), and a proposed circuit-board layout for discrete components is given in Figure 5.11(b).

The advantage of a discrete-element, circuit board approach as opposed to monolithic fabrication is that components with much lower losses are available. A comparison of the simulated performance for these two approaches is shown in Figure 5.12. For the discrete-element version, the inductor quality factor (Q) was assumed to be 100 at 2 GHz, and 400 for the capacitors.

References

[1] Ramo, S., Whinnery, J., and Van Duzer, T., *Fields and Waves in Communications Electronics*, 3rd ed., New York: Wiley, pp. 536–539, 1993.

[2] Morgan, M., and Boyd, T., "Reflectionless filter structures," *IEEE Trans. Microw. Theory Tech.*, vol. 63, no. 4, pp. 1263–1271, April 2015.

[3] Morgan, M., *Sub-Network Enhanced Reflectionless Filter Topologies*, U.S. Patent Application No. 14/724,976, May 29, 2015, PCT Application No. PCT/US15/33118, May 29, 2015, and Taiwan Patent Application No. 104120606, June 25, 2015.

[4] Matthaei, G., Young, L., and Jones, E., *Microwave Filters, Impedance-Matching Networks, and*

Coupling Structures, Norwood, MA: Artech House, pp. 991–999, 1980.

[5] Bode, H., *Network Analysis and Feedback Amplifier Design*, New York: D. Van Nostrand Company, pp. 249–275, 1945.

Chapter 6

Generalized Reflectionless Filters

In Chapter 3, we demonstrated a method utilizing even- and odd-mode analysis to derive a general topology that could, in principle, be reflectionless, given an appropriate selection of element values. We then calculated what those element values had to be within a frequency scaling factor in order to meet the reflectionless condition. This led to a third-order, Chebyshev Type II response. In Chapters 4 and 5, new rules were developed that yielded more sophisticated topologies with higher-order responses by building upon that basic structure in combination with others. The validity of those techniques rested upon the assumption that the basic building blocks themselves were individually reflectionless; however, this need not be the case. In this chapter, we generalize those earlier techniques as a method of creating topologies that can, in principle, be made reflectionless by an appropriate selection of element values, but without prescribing those element values prior to assembling the complete, high-order network. Only after the final topology is fixed do we derive the minimum conditions upon the element values that lead to reflectionless behavior. We shall find in many cases that the degrees of freedom afforded by the forgoing topologies is greater than it was first thought.

6.1 REEVALUATION OF PRIOR TOPOLOGIES

6.1.1 Compound High-Pass Filters

For convenience, almost all of the filter examples shown thus far have been of the low-pass configuration, as is customary for filter prototypes since the transformation to other configurations (e.g., high-pass or band-pass) is well-understood and universally applicable. However, the reader may recall that many of the more sophisticated topologies developed in the preceding chapters have had undesirable characteristics at frequencies far away from the pass-band. These features were easily dismissed for practical low-pass filters because the parasitics of physical components would usually result in some high-frequency

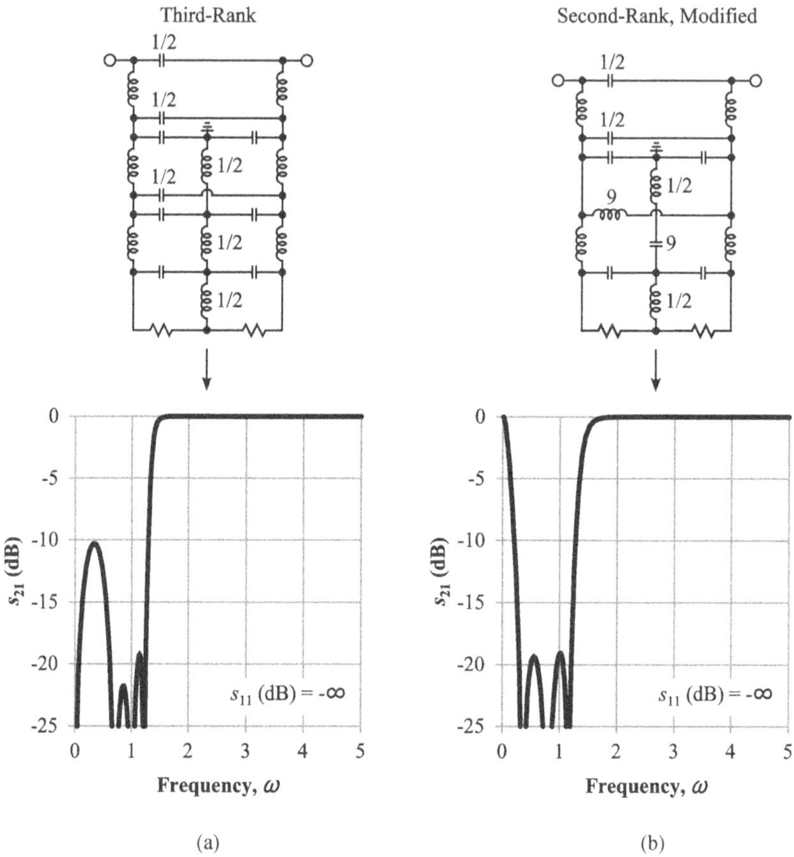

Figure 6.1 Topologies and frequency responses of high-pass reflectionless filters. (a) Third-rank, and (b) second-rank with auxiliary components. As in the previous chapters, unlabeled components have unit normalized value, while those with labels are appropriately scaled.

abnormalities anyway, and they could always be addressed with reflectionless filters by cascading with another low-order filter having higher cutoff frequency (and, consequently, lower parasitics near the primary pass-band). These rationalizations fail to be convincing, however, when configurations other than low-pass are considered.

Referring to Figure 6.1, two different high-pass topologies are shown using techniques that we have seen so far. The first is a third-rank filter. As with the low-pass version (shown previously in Figure 4.6(c)), the two stop-band peaks nearest cutoff are low, approximately -19.3 dB and -21.8 dB, respectively, but the third peak is much higher at -10.3 dB. In the low-pass version, this was accepted because the two lower peaks occupied the first octave after cutoff, making such a filter still useful in some applications, and the -10.3 dB peak was much further

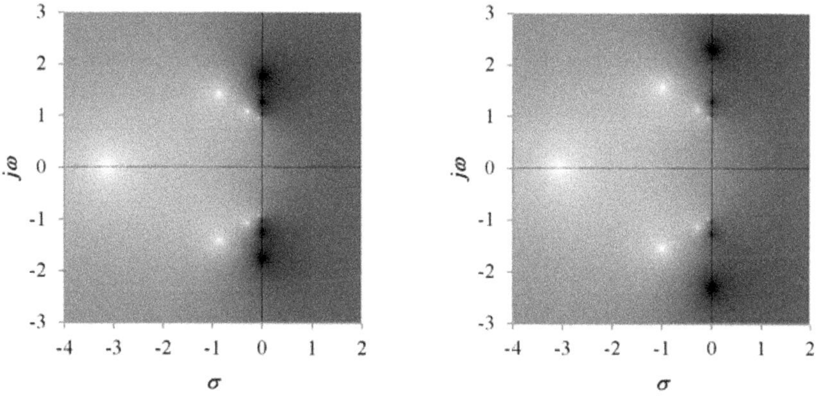

(a) (b)

Figure 6.2 Comparison of pole-zero configurations for (a) third-rank low-pass reflectionless filter, and (b) seventh-order Chebyshev Type II filter. The responses have been scaled such that the first zero is coincident.

away. However, the presence of the higher stop-band peak is more problematic for the high-pass filter in which the stop-band is more localized.

The second high-pass filter in Figure 6.1 — a second-rank topology with auxiliary components — exhibits even worse behavior. Recall the low-pass form of this filter shown first in Figure 4.11, where it was derived by embedding a high-pass filter in the intermediate subnetwork stage between two low-pass filters. Logically, when the outermost filter begins to cut off, it directs the out-of-band signal energy to its subnetwork. If that subnetwork is itself a high-pass filter, as it was in Figure 4.11, then at some frequency the subnetwork will bypass that signal energy through the filter and back out through the output port. In other words, this was not actually a low-pass filter at all, but instead was a type of asymmetrical band-stop filter. This feature was not that obvious in the earlier form because the higher-frequency cutoff where the filter starts passing again was far away from the primary band of interest and only reached 0 dB again (theoretically) at infinity. The high-pass transformation on that network, however, makes this feature more obvious, because that infinite frequency is now mapped to DC.

6.1.2 Clues in the Pole-Zero Configuration

One of the very first reflectionless filters developed in this book (Figure 3.2, $N = 3$) achieved a third-order Chebyshev Type II response. Regrettably, this was also the last that realized a classically optimized filter response of any kind. Later responses were similar, such as the compound filters in Figure 4.6, but the stop-band peaks were no longer of equal height. Let us examine the pole-zero

Table 6.1
Poles and Zeros of Third-Rank Reflectionless vs. Chebyshev Type II Filters

	Third-Rank Reflectionless	*Chebyshev Type II*
poles, p_k	-3.167	-3.094
	$-0.078 \pm j0.943$	$-.068 \pm j0.971$
	$-0.301 \pm j1.078$	$-0.282 \pm j1.148$
	$-0.864 \pm j1.426$	$-0.995 \pm j1.558$
zeros, z_k	$\pm j1.026$	$\pm j1.026$
	$\pm j1.256$	$\pm j1.279$
	$\pm j1.777$	$\pm j2.305$
	$j\infty$	$j\infty$

configuration of that filter in Figure 6.2(a) to see why. The configuration for an optimized Chebyshev Type II response of the same order ($N = 7$) is shown in Figure 6.2(b) for comparison. The two maps have been scaled in frequency such that the first zero is coincident just above $\omega = 1$. The pole-zero configurations are clearly very similar, except that in the case of the reflectionless filter, the zeros are too close together on the $j\omega$ axis. The poles are also very slightly out of position. In fact, the locations of the poles and zeros are tabulated in Table 6.1.

The striking similarity in the pole-zero configuration of these two filters suggests that, topologically at least, they are equivalent (a somewhat unrigorous and perhaps confusing statement, considering that no actual topology for the Chebyshev Type II filter has yet been given). What we mean is that because the poles and zeros of the two filters differ by position only (there are none missing, and no extras, nor any located at infinity that are not so for both) their rational polynomial transfer functions should have the same form, but with slightly different coefficients. Since the transfer function is an algebraic expression of the topology, and its coefficients directly calculable from the values of the inductors, capacitors, and resistors, we believe that any topology that is capable of producing one of these two frequency responses should (in theory) be capable of producing the other, with a suitable modification of the element values. The schematic, however, namely the arrangement of those inductors, capacitors, and resistors, should not need to change in any other way.

In other words, we should be able to implement a Chebyshev Type II filter response using our third-rank reflectionless filter topology with modified element values.

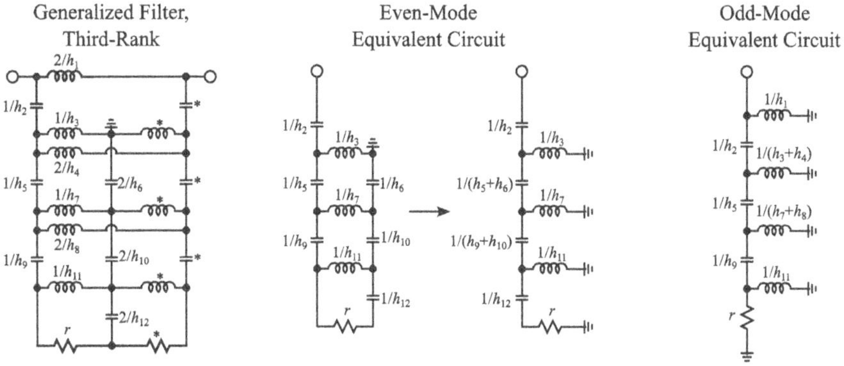

Figure 6.3 Generalized third-rank reflectionless filter with its even- and odd-mode equivalent circuits. Elements labeled with an asterisk are equal to their symmetric counterparts.

6.2 GENERALIZED COMPOUND TOPOLOGIES

What is not guaranteed is that the modified element values will permit the network to remain reflectionless (a feature that is not captured by the pole-zero configuration of the transfer function alone) or that the required element values will be feasible (e.g., finite, real, and nonnegative). There is hope, nonetheless, since the poles and zeros need only a small translation (roughly 7% on average, in Table 6.1), that the resultant element values will be realistic, and that the resulting reflection coefficient will not be "too bad."

6.2.1 Seventh-Order Derivation

Let us try to make this more rigorous. The third-rank topology was derived by embedding three stages of reflectionless filters within one another as subnetworks. However, those subnetworks were each of third order, and the net response of the triplet is only seventh order, not ninth order as one might instinctively expect. Is it possible that the rich complexity and flexibility of the topology has not yet been fully tapped? The ability to reduce the number of elements without altering the frequency response in Chapter 5 provides further evidence that this is probably the case.

These arguments make it clear that a reexamination of the duality constraints for the third-rank topology is in order. To be certain that we are not inadvertently denying ourselves a useful degree of freedom in assigning element values, we will use the original third-rank network with the full complement of reactive components from Figure 4.5, rather than the reduced versions of Chapter 5. This is shown in Figure 6.3. The elements are labeled with parameters representing their normalized values. Symmetric elements are labeled with an asterisk. This guarantees that the symmetry condition is met, but we have yet to ensure that the duality conditions are satisfied. To do so, let us examine the even- and odd-mode equivalent circuits given in Figure 6.3 and equate the normalized values of the dual counterparts in each circuit,

$$h_2 = h_1 \tag{6.1a}$$

$$h_3 = h_2 \tag{6.1b}$$

$$h_5 + h_6 = h_3 + h_4 \tag{6.1c}$$

$$h_7 = h_5 \tag{6.1d}$$

$$h_9 + h_{10} = h_7 + h_8 \tag{6.1e}$$

$$h_{11} = h_9 \tag{6.1f}$$

$$h_{12} = h_{11} \tag{6.1g}$$

$$r = r^{-1} = 1 \tag{6.1h}$$

Note that we have thirteen unknowns, but only eight equations. This reveals that there are five degrees of freedom in assigning element values to this structure, while still retaining the reflectionless property. Of the five degrees of freedom we now have at our disposal, one may be allocated to simple frequency scaling, leaving four that can be utilized for tailoring the pass-band shape.

Contrast that with the original construction in which there were only two degrees of freedom for pass-band shaping (namely, the tuning frequencies of the two independent subnetwork filters relative to the root filter). The subnetwork design technique only promised a reflectionless result if the subnetworks themselves were individually matched. That turns out to be overly restrictive. We have now shown that individual matching of the subnetworks, though sufficient, is not a necessary condition for the filter as a whole to be reflectionless. The root and subnetworks may in fact be intentionally mismatched in a compensatory way such that the overall duality constraints are still met [1]. (Recall that the *duality constraints* refer to the substantial equivalence of the normalized even- and odd-mode equivalent circuit impedance and admittance, respectively.)

In the more generalized context presented in this section, the reflectionless filter comprises the topology of a simpler reflectionless filter with a subnetwork inside a subnetwork of additional reflectionless filters (that is, it comprises the

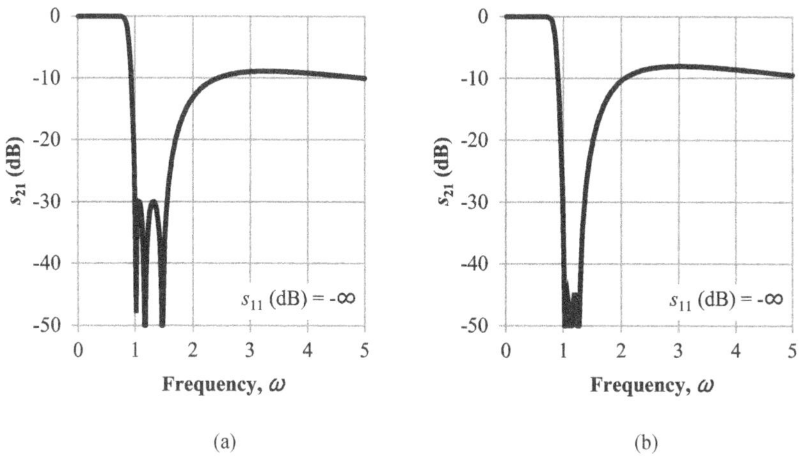

Figure 6.4 Examples of possible customized frequency responses for a generalized, seventh-order reflectionless low-pass filter with close-in rejection bands of (a) 30 dB and (b) 45 dB.

topology *only*, but without the usual assignment of element values so that each subnetwork meets the duality constraints on its own). Instead, the network as a whole is constrained to meet the duality condition, while having more degrees of freedom to do so than if the subnetworks were to be made reflectionless independently.

6.2.2 Customized Responses

The degrees of freedom afforded by the generalized structure allow customizable frequency responses to be achieved. Say, for example, that there was a need to reject an interferer close in to the pass-band with deep attenuation, while frequencies further out were of less consequence. One may optimize the element values, subject to the constraints in (6.1), to achieve deeper rejection close in to the pass-band cutoff, while still having a reflectionless filter. Two examples of such optimized filter functions are shown in Figure 6.4. These, having one large, outermost stop-band ripple where the rest are equal and smaller in amplitude, are what may be referred to as *Inverse Zolotarev* or *Zolotarev Type II* filter functions. Zolotarev filters will be discussed in more detail in Section 6.8.

6.3 CHEBYSHEV TYPE II REFLECTIONLESS FILTERS

However, is it possible for the pass-band to take on the form of a Chebyshev Type II filter, as the similarity of the pole-zero configurations would seem to imply? The answer is both yes and no. By (3.3) we know that the transfer function is

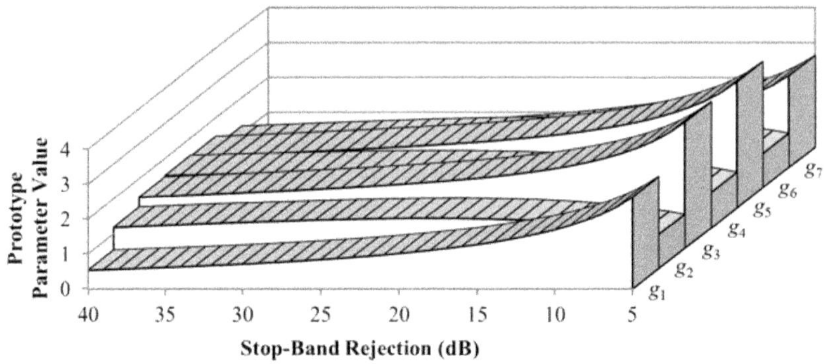

Figure 6.5 Chebyshev prototype parameter values as a function of stop-band rejection for $N = 7$.

given directly by the reflection coefficient of the even-mode equivalent circuit. To achieve a Chebyshev Type II, low-pass response, we would like the even-mode equivalent circuit to be a terminated, Chebyshev Type I high-pass filter. Equation (6.1b) shows that the first two elements of this high-pass equivalent circuit must be identical, and (6.1g) places a similar constraint on the last two. This is not generally the case for a conventional Chebyshev filter, except for a unique choice of the ripple factor, ε, which in turn determines the reflection ripple (or what would be the peak stop-band rejection of the reflectionless filter).

6.3.1 Chebyshev Prototype Parameters

Consider the conventional Chebyshev prototype parameters for a seventh-order filter, plotted as a function of the reflection ripple in Figure 6.5 (remember that the reflection ripple for a conventional Type I filter becomes the stop-band rejection for a reflectionless Type II filter). At the far left of the plot, where the ripple is nearly vanishing, the prototype parameters approach that of a Butterworth filter, with the outer parameter values (g_1 and g_7) smallest and increasing monotonically towards the middle. As the ripple is allowed to increase, the odd-numbered parameters become large while the even-numbered parameters, in general, become smaller. There is a special point, near about 13.4 dB, where the values of the first two parameters cross. At that point — and only at that point — the two parameters are equal. Further, since the list of parameter values for this type of filter is symmetric, the same can be said of the last two parameter values.

The ripple factor and corresponding prototype element values listed in Table 1.1 were chosen to have this precise feature, in anticipation of the results of this chapter. Now let us take a closer look at how that special ripple factor was obtained.

6.3.2 Ripple Factor

The formulae for the Chebyshev prototype element values are given in (1.39). Equating the first two prototype parameters,

$$g_1 = g_2 \tag{6.2a}$$

$$\frac{2a_1}{\gamma} = \frac{4a_1 a_2}{b_1 g_1} = \frac{2\gamma a_2}{b_1} \tag{6.2b}$$

$$\gamma^2 = \frac{a_1 b_1}{a_2} = \frac{\sin\left(\frac{\pi}{2N}\right)\left(\gamma^2 + \sin^2\left(\frac{\pi}{N}\right)\right)}{\sin\left(\frac{3\pi}{2N}\right)} \tag{6.2c}$$

$$\gamma = \sqrt{\frac{\sin^2\left(\frac{\pi}{N}\right)\sin\left(\frac{\pi}{2N}\right)}{\sin\left(\frac{3\pi}{2N}\right) - \sin\left(\frac{\pi}{2N}\right)}} \tag{6.2d}$$

$$\beta = 2N\sinh^{-1}(\gamma) = 2N\sinh^{-1}\left(\sqrt{\frac{\sin^2\left(\frac{\pi}{N}\right)\sin\left(\frac{\pi}{2N}\right)}{\sin\left(\frac{3\pi}{2N}\right) - \sin\left(\frac{\pi}{2N}\right)}}\right) \tag{6.2e}$$

$$\varepsilon = \sqrt{e^{4\tanh^{-1}\left(e^{-\beta}\right)} - 1} \tag{6.2f}$$

Therefore, (6.2f) gives just the ripple factor needed to derive a Chebyshev prototype in which the first two (and, by symmetry, the last two) element values are identical. These particular Chebyshev prototypes are tabulated in Table 1.1.

Perhaps surprisingly, the rest of the Chebyshev prototype element values fall neatly into place within our generalized reflectionless topology. No other compromise to the properties of the filter are required. Specifically, we may set the above element values to those of a standard prototype as follows,

$$h_1 = h_2 = h_3 = g_1 = g_2 \tag{6.3a}$$

$$h_4 = g_3 - g_2 \tag{6.3b}$$

$$h_5 = h_7 = g_4 \tag{6.3c}$$

$$h_6 = g_3 - g_4 \tag{6.3d}$$

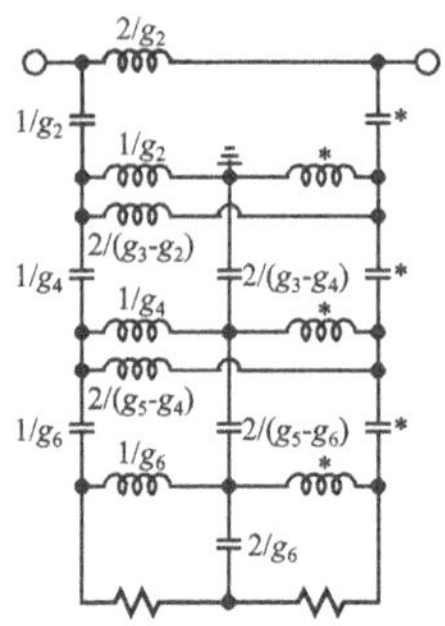

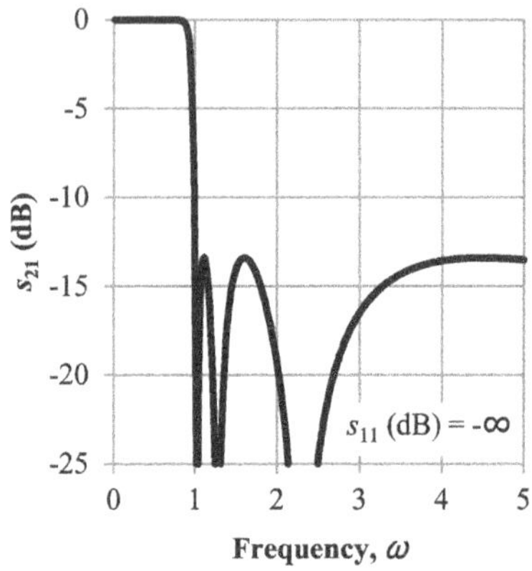

Figure 6.6 (a) Seventh-order, low-pass, Chebyshev Type II reflectionless filter topology. Elements labeled with an asterisk have the same value as their symmetric counterparts. (b) Frequency response.

$$h_8 = g_5 - g_4 \tag{6.3e}$$

$$h_9 = h_{11} = h_{12} = g_6 = g_7 \tag{6.3f}$$

$$h_{10} = g_5 - g_6 \tag{6.3g}$$

(In fact, any array of prototype element values, g_k, may be used, not just the Chebyshev parameters, provided that the first two and last two values are equal, and that the odd-numbered values are greater than or equal to the adjacent even-numbered values, lest some of the expressions above become negative.) The reader is left to verify that the assignments in (6.3) satisfy all the duality-constrained relationships in (6.1).

The final seventh-order, low-pass, Chebyshev Type II reflectionless filter is thus shown in Figure 6.6, along with its frequency response. Moreover, the topology may be generalized to any odd order that is desired, as shown in Figure 6.7. The same general limitations apply; namely, that the first two and last two elements are equal-valued, and that the odd-numbered prototype element values equal or exceed the adjacent even-numbered values to avoid negative components in the final filter. Fortunately, all Chebyshev Type II prototypes meet these requirements given the ripple factor specified in (6.2).

The importance of this result cannot be overstated. Beginning with a simple procedure in Chapter 3 for deriving symmetric reflectionless filters, nesting them in Chapter 4 to construct more sophisticated topologies, and then, in this chapter, stripping away some of the basic assumptions that went into the development of

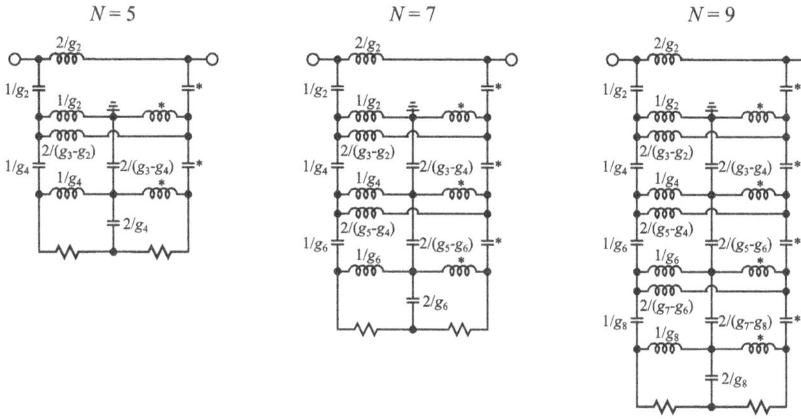

Figure 6.7 Low-pass, reflectionless Chebyshev Type II filters of several orders.

those topologies that made them reflectionless in order to derive more general requirements with better freedom to tailor the frequency response, we have arrived at last at a topology that is capable of achieving a classically optimized frequency response of arbitrary (odd) order which, nonetheless, has identically zero reflection coefficients at all frequencies from DC to infinity [1]. This is truly a remarkable achievement for reflectionless filters.

Also of note is the fact that the third-order form of this new generalized topology reverts back to the original reflectionless filter first derived in Chapter 3, only now the prototype element values, g_k, have been scaled and inverted in order to associate them with the Chebyshev Type I parameters found in many textbooks [2–5].

6.3.3 Stop-Band Rejection

In Section 6.3.2, the topology for the Chebyshev Type II reflectionless filters placed a constraint on the ripple factor, which in turn fixed the stop-band rejection to a certain level. That level is calculated from the ripple factor as follows,

$$IL_{\min}(dB) = 10\log_{10}\left(1+\varepsilon^{-2}\right) \tag{6.4}$$

The stop-band rejection above is tabulated in Table 6.2. These values are only very weakly dependent on the order of the filter, settling within a couple of decimal places of 13.4 dB for all orders above $N = 5$. It can be shown, in fact, that the limiting ripple factor and stop-band rejection for an arbitrarily high order is given by

Table 6.2

Limiting Stop-Band Attenuation of Reflectionless Chebyshev Type II Filters

	$N = 3$	$N = 5$	$N = 7$	$N = 9$	$N = 11$	$N = 13$	$N = 15$
$\varepsilon =$	0.1925	0.2164	0.2187	0.2192	0.2194	0.2194	0.2195
$IL_{min} =$	14.47 dB	13.49 dB	13.41 dB	13.39 dB	13.38 dB	13.38 dB	13.38 dB

Table 6.3

Prototype Parameters for Reflectionless Chebyshev Type II Filters, $\omega_0 = 1$ dB Corner

N	ε	g'_1	g'_2	g'_3	g'_4	g'_5	g'_6	g'_7	g'_8	g'_9
3	0.1925	1.788	1.788							
5	0.2164	1.568	1.568	2.537						
7	0.2187	1.496	1.496	2.477	1.627					
9	0.2192	1.464	1.464	2.435	1.611	2.501				
11	0.2194	1.448	1.448	2.411	1.597	2.490	1.615			
13	0.2194	1.438	1.438	2.396	1.588	2.480	1.611	2.493		
15	0.2195	1.432	1.432	2.387	1.582	2.472	1.607	2.490	1.612	
17	0.2195	1.428	1.428	2.380	1.578	2.466	1.604	2.486	1.611	2.491

Table 6.4

Prototype Parameters for Reflectionless Chebyshev Type II Filters, $\omega_0 = 3$ dB Corner

N	ε	g'_1	g'_2	g'_3	g'_4	g'_5	g'_6	g'_7	g'_8	g'_9
3	0.1925	1.521	1.521							
5	0.2164	1.470	1.470	2.379						
7	0.2187	1.446	1.446	2.394	1.573					
9	0.2192	1.434	1.434	2.384	1.577	2.450				
11	0.2194	1.428	1.428	2.377	1.575	2.455	1.593			
13	0.2194	1.424	1.424	2.372	1.572	2.455	1.595	2.468		
15	0.2195	1.421	1.421	2.369	1.570	2.453	1.595	2.471	1.600	
17	0.2195	1.420	1.420	2.366	1.569	2.451	1.594	2.471	1.601	2.476

$$\lim_{N\to\infty} \varepsilon = \sqrt{e^{4\tanh^{-1}\left(e^{-\pi\sqrt{2}}\right)} - 1} = 0.219486930877... \tag{6.5a}$$

$$\lim_{N\to\infty} IL_{\min} = 10\log_{10}\left(1+\varepsilon^{-2}\right) \approx 13.37616264 \text{ dB} \tag{6.5b}$$

6.3.4 Normalized Element Scaling

Recall from Section 3.2 that the standard Chebyshev prototype parameters have to be modified in order to specify the cutoff frequency of a Type II filter in terms of

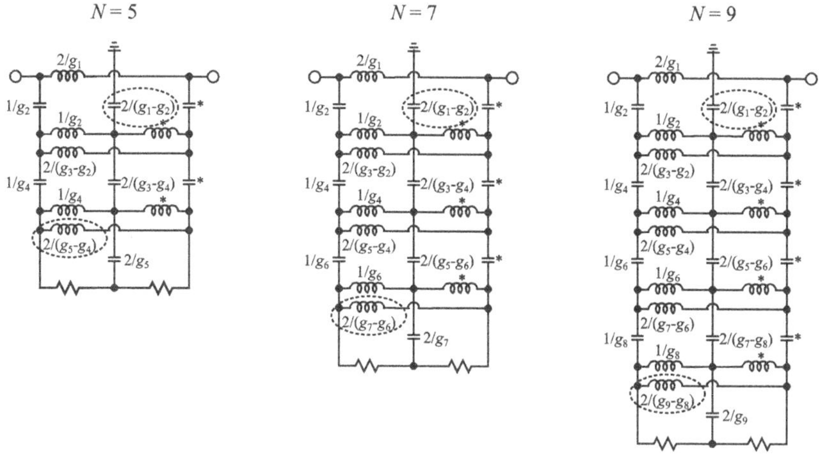

Figure 6.8 Low-pass, reflectionless Chebyshev Type II filters with adjustable ripple factor. The auxiliary elements added to remove the ripple factor constraint are encircled with dotted lines.

its 1 dB or 3 dB cutoff corner rather than the beginning of its stop-band, the former being more common in practice. This amounted to a constant scaling factor given by (3.15), and has no effect on the above constraints regarding the relative magnitude of the even- and odd-numbered parameters. For convenience, we present in Tables 6.3 and 6.4 a modified set of Chebyshev prototype parameters, scaled such that the normalized frequency corresponds to the 1 dB and 3 dB cutoff corners, respectively, with stop-band ripple satisfying (6.2). Since the list of parameters in each row is known to be symmetric about the middle, only the first half are shown.

6.4 EXPANDED TOPOLOGY FOR LARGER RIPPLE FACTOR

The reflectionless Chebyshev Type II topology of Section 6.3 was found by generalizing upon the element values of compound reflectionless filters of arbitrary rank. One might reasonably ask whether the constraint on the ripple factor can be removed by generalizing upon the topology as well — that is, by adding auxiliary elements to provide greater freedom from duality in assigning those values.

6.4.1 Generalized Auxiliary Elements

One way to do this is to add elements that only appear in one or the other equivalent half-circuit. Close examination of Figure 6.7 reveals that the horizontal inductors skipping over the vertical center line appear only in the odd-mode equivalent circuit, combining in parallel with those that do connect to the center

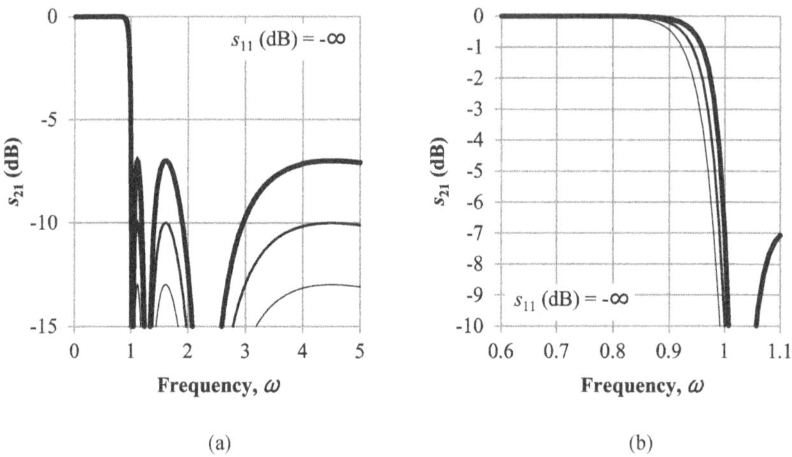

Figure 6.9 Plot of the frequency response for seventh-order reflectionless Chebyshev Type II filters with varying ripple factors. (a) Overall response. (b) Detail of cutoff corner.

line. On the other hand, the vertical capacitors on the center line appear only in the even-mode equivalent circuit, combining in series with those on the input and output sides. The equality of the first-two and last-two prototype element values demanded by duality may be broken by adding one more inductor across the center line parallel with the termination resistors, and one more capacitor along the center line at the top near the ports. These additional elements are shown in the topologies of Figure 6.8, enclosed in dotted circles.

With these topologies, one may realize Chebyshev Type II reflectionless filters of any arbitrary (odd) order, of any type (low-pass, high-pass, band-pass, and band-stop) given the appropriate frequency transformation, and with any ripple factor *larger* than that shown in Table 6.2. For example, Figure 6.9 is a plot of the frequency response achieved for several ripple factors (corresponding to peak stop-band rejections of 7, 10, and 13 dB) for seventh-order using these topologies. The advantage of larger ripple factor is steeper cutoff slope near the pass-band, as shown in Figure 6.9(b), at the expense of less rejection in the stop-band.

Note that at the limiting ripple factor, the auxiliary elements added in this section become infinite in value (a short-circuit in case of the series capacitor, and an open-circuit in the case of the parallel inductor). They revert back to the original topologies of Figure 6.7. At lower ripple factors, the auxiliary elements would have to take on negative values.

6.4.2 Transformer-Coupled Generalized Topology

It was noted earlier that the third-order reflectionless filter of Figure 3.2 was the

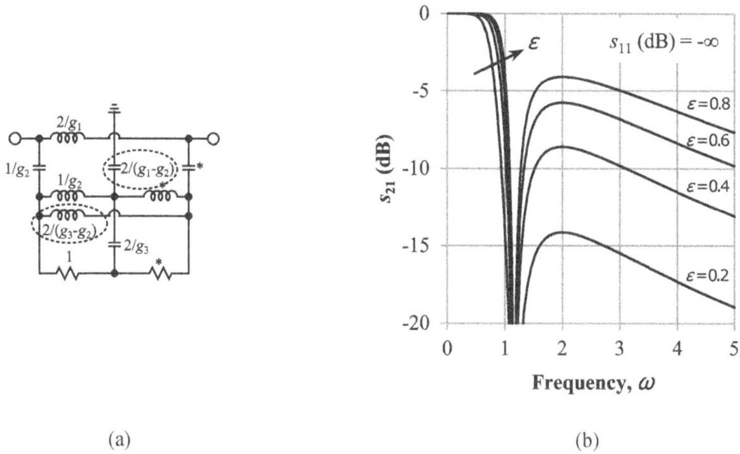

Figure 6.10 (a) Generalized third-order Chebyshev Type II filter reflectionless filter topology. (b) Frequency response for $\varepsilon = 0.2$ to 0.8.

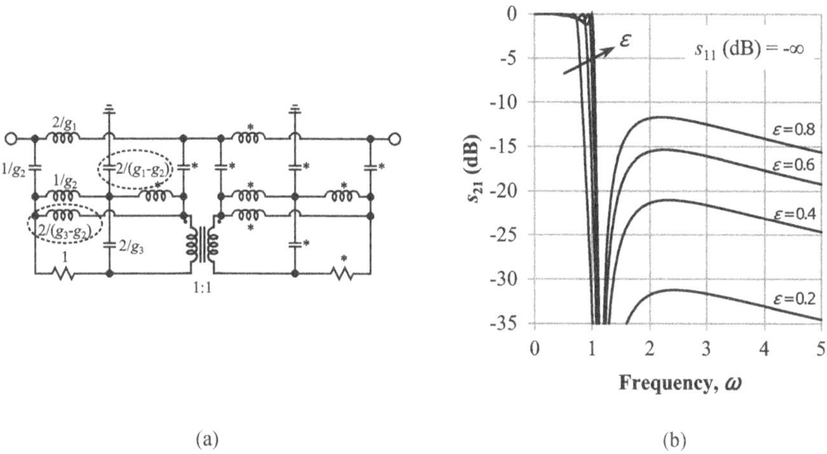

Figure 6.11 (a) Transformer-coupled reflectionless filter topology using third-order Chebyshev Type II cells. (b) Frequency response for $\varepsilon = 0.2$ to 0.8.

only one to realize an optimized response prior to the element-value generalization introduced in this chapter. The duality constraints at that time limited it to having all equal-valued components, rendering its ripple factor fixed at $\varepsilon = 0.1924$, as specified in Table 6.2. The generalized auxiliary elements introduced in this section release the ripple factor from this constraint. The more flexible third-order reflectionless Chebyshev Type II filter topology is thus shown (along with its frequency response) in Figure 6.10.

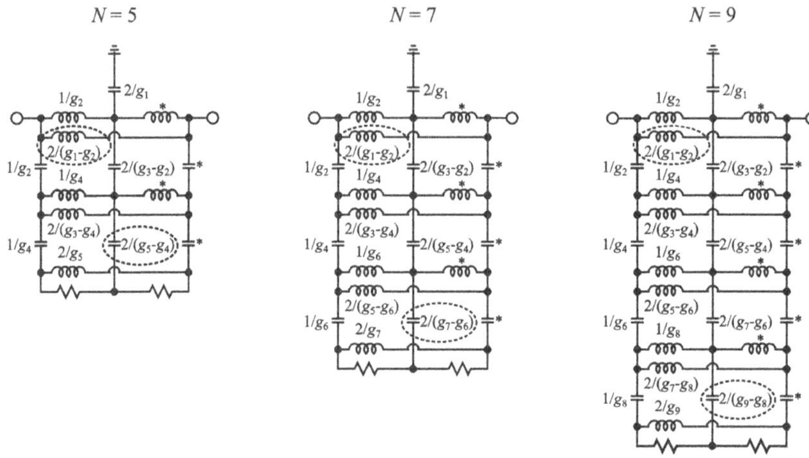

Figure 6.12 Dual topologies for low-pass, reflectionless Chebyshev Type II filters with adjustable ripple factor.

This is important because the third-order topology was a fundamental building block for other reflectionless filter topologies. One particular example is the transformer-coupled topology from Figure 4.4, wherein the internally facing stop-band terminations of cascaded third-order cells are cross-connected differentially. If the more general third-order cells are used, this allows greater freedom to tailor the pseudo-elliptic frequency response for transition steepness in lieu of stop-band rejection, as shown in Figure 6.11.

6.5 DUAL TOPOLOGY

As we have seen, virtually every reflectionless filter has a dual with equivalent performance (one exception is the $N = 1$ filter in Figure 3.2, which happens to be its own dual). For the general Chebyshev Type II topologies in Figure 6.8, the duals are shown in Figure 6.12. These may be derived by substitution of dual elements, as was illustrated in Chapter 5, or by stacking the dual third-order filters directly. As before, the elements encircled in dotted lines represent auxiliary elements that were added to mitigate the ripple factor constraint. Should minimum ripple (maximum stop-band attenuation) be desired (as given in Table 6.2), these elements become infinite and may be omitted; that is, the auxiliary inductor is left as an open circuit, and the auxiliary capacitor is replaced with a short.

Similar to the third-order networks, the dual Chebyshev Type II filters may also be derived by exchanging the positions of the ports and the internal terminations. As drawn, then, the topologies in Figure 6.12 are simply the topologies in Figure 6.8 upside-down, where the original ground node is reassigned as the common-node between the new termination resistors, and the former common-node between terminations is the new signal ground.

The performance (e.g., Figure 6.6(b) or Figure 6.9) is of course unchanged.

6.6 HIGH-PASS AND BAND-PASS FILTERS

Now armed with a generic Chebyshev Type II topology, transformation to other configurations (such as high-pass and band-pass) is a straightforward application of Table 1.2, and overcomes the limitations for these types described in Section 6.1.1. Topologies for seventh-order reflectionless Chebyshev Type II high-pass and band-pass filters are shown in Figure 6.13(a) and Figure 6.13(b). Their theoretical frequency responses for maximum stop-band rejection are shown in Figure 6.13(c) and Figure 6.13(d), respectively.

Let us work through a practical example of the band-pass design. Suppose that we want a pass-band of 1.5–2.5 GHz, as defined by the 1 dB cutoff corners. We will therefore use the prototype parameters found in Table 6.3, and the formulae from the schematic of Figure 6.13(b), with a center frequency of

$$f_c = \sqrt{f_1 f_2} = \sqrt{(1.5\text{GHz})(2.5\text{GHz})} = 1.936 \text{ GHz} \tag{6.6}$$

and a bandwidth parameter of

$$\Delta = \frac{f_2 - f_1}{f_c} = \frac{(2.5\text{GHz}) - (1.5\text{GHz})}{(1.936\text{GHz})} = 0.5164 \tag{6.7}$$

We may calculate the required elements as follows. For the inductors,

$$L_1 = \frac{2}{g_1 \Delta} \frac{Z_0}{\omega_c} = \frac{2}{(1.496)(0.5164)} \frac{50\Omega}{2\pi(1.936\text{ GHz})} = 10.64 \text{ nH} \tag{6.8a}$$

$$L_2 = g_2 \Delta \frac{Z_0}{\omega_c} = (1.496)(0.5164) \frac{50\Omega}{2\pi(1.936\text{ GHz})} = 3.174 \text{ nH} \tag{6.8b}$$

$$L_3 = \frac{(g_1 - g_2)\Delta}{2} \frac{Z_0}{\omega_c} = \frac{(1.496 - 1.496)(0.5164)}{2} \frac{50\Omega}{2\pi(1.936\text{ GHz})} = 0 \tag{6.8c}$$

$$L_4 = \frac{1}{g_2 \Delta} \frac{Z_0}{\omega_c} = \frac{1}{(1.496)(0.5164)} \frac{50\Omega}{2\pi(1.936\text{ GHz})} = 5.321 \text{ nH} \tag{6.8d}$$

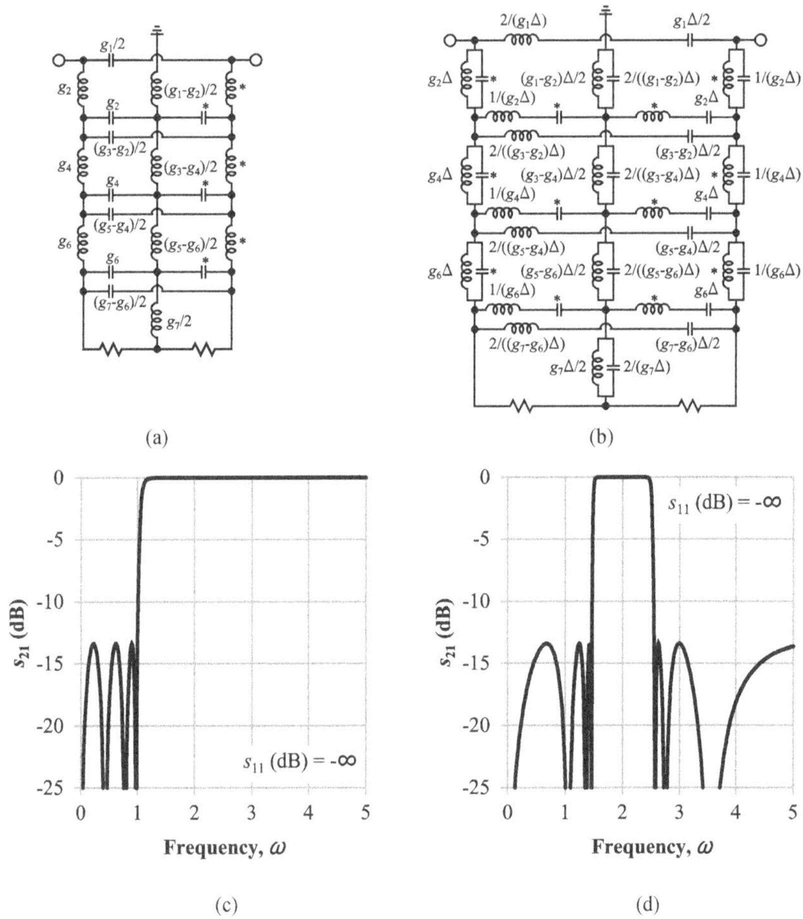

Figure 6.13 (a) High-pass and (b) band-pass seventh-order reflectionless Chebyshev Type II filter topologies. (c) High-pass frequency response. (d) Band-pass frequency response.

$$L_5 = \frac{2}{(g_3 - g_2)\Delta}\frac{Z_0}{\omega_c} = \frac{2}{(2.477 - 1.496)(0.5164)}\frac{50\Omega}{2\pi(1.936\,\text{GHz})} = 16.22\text{ nH} \quad (6.8e)$$

$$L_6 = g_4\Delta\frac{Z_0}{\omega_c} = (1.627)(0.5164)\frac{50\Omega}{2\pi(1.936\,\text{GHz})} = 3.453\text{ nH} \quad (6.8f)$$

$$L_7 = \frac{(g_3 - g_4)\Delta}{2}\frac{Z_0}{\omega_c} = \frac{(2.477 - 1.627)(0.5164)}{2}\frac{50\Omega}{2\pi(1.936\,\text{GHz})} = 0.901\text{ nH} \quad (6.8g)$$

$$L_8 = \frac{1}{g_4 \Delta} \frac{Z_0}{\omega_c} = \frac{1}{(1.627)(0.5164)} \frac{50\Omega}{2\pi(1.936\,\text{GHz})} = 4.89\ \text{nH} \tag{6.8h}$$

$$L_9 = \frac{2}{(g_5 - g_4)\Delta} \frac{Z_0}{\omega_c} = \frac{2}{(2.477 - 1.627)(0.5164)} \frac{50\Omega}{2\pi(1.936\,\text{GHz})} = 18.74\ \text{nH} \tag{6.8i}$$

$$L_{10} = g_6 \Delta \frac{Z_0}{\omega_c} = (1.496)(0.5164) \frac{50\Omega}{2\pi(1.936\,\text{GHz})} = 3.174\ \text{nH} \tag{6.8j}$$

$$L_{11} = \frac{(g_5 - g_6)\Delta}{2} \frac{Z_0}{\omega_c} = \frac{(2.477 - 1.496)(0.5164)}{2} \frac{50\Omega}{2\pi(1.936\,\text{GHz})} = 1.041\ \text{nH} \tag{6.8k}$$

$$L_{12} = \frac{1}{g_6 \Delta} \frac{Z_0}{\omega_c} = \frac{1}{(1.496)(0.5164)} \frac{50\Omega}{2\pi(1.936\,\text{GHz})} = 5.321\ \text{nH} \tag{6.8l}$$

$$L_{13} = \frac{2}{(g_7 - g_6)\Delta} \frac{Z_0}{\omega_c} = \frac{2}{(1.496 - 1.496)(0.5164)} \frac{50\Omega}{2\pi(1.936\,\text{GHz})} = \infty \tag{6.8m}$$

$$L_{14} = \frac{g_7 \Delta}{2} \frac{Z_0}{\omega_c} = \frac{(1.496)(0.5164)}{2} \frac{50\Omega}{2\pi(1.936\,\text{GHz})} = 1.587\ \text{nH} \tag{6.8n}$$

Note that the capacitors in every case are given by

$$C_k = \frac{1}{\omega_c^2 L_k} \quad 1 \le k \le 14 \tag{6.9}$$

therefore,

$$C_1 = \frac{1}{\omega_c^2 L_1} = \frac{1}{(2\pi \cdot 1.936\,\text{GHz})^2 (10.64\ \text{nH})} = 0.6347\ \text{pF} \tag{6.10a}$$

$$C_2 = \frac{1}{\omega_c^2 L_2} = \frac{1}{(2\pi \cdot 1.936\,\text{GHz})^2 (3.174\ \text{nH})} = 2.128\ \text{pF} \tag{6.10b}$$

$$C_3 = \frac{1}{\omega_c^2 L_3} = \frac{1}{(2\pi \cdot 1.936\,\text{GHz})^2 (0)} = \infty \tag{6.10c}$$

$$C_4 = \frac{1}{\omega_c^2 L_4} = \frac{1}{(2\pi \cdot 1.936\,\text{GHz})^2 (5.321\,\text{nH})} = 1.269\ \text{pF} \tag{6.10d}$$

$$C_5 = \frac{1}{\omega_c^2 L_5} = \frac{1}{(2\pi \cdot 1.936\,\text{GHz})^2 (16.22\,\text{nH})} = 0.4164\ \text{pF} \tag{6.10e}$$

$$C_6 = \frac{1}{\omega_c^2 L_6} = \frac{1}{(2\pi \cdot 1.936\,\text{GHz})^2 (3.453\,\text{nH})} = 1.956\ \text{pF} \tag{6.10f}$$

$$C_7 = \frac{1}{\omega_c^2 L_7} = \frac{1}{(2\pi \cdot 1.936\,\text{GHz})^2 (0.901\,\text{nH})} = 7.497\ \text{pF} \tag{6.10g}$$

$$C_8 = \frac{1}{\omega_c^2 L_8} = \frac{1}{(2\pi \cdot 1.936\,\text{GHz})^2 (4.89\,\text{nH})} = 1.381\ \text{pF} \tag{6.10h}$$

$$C_9 = \frac{1}{\omega_c^2 L_9} = \frac{1}{(2\pi \cdot 1.936\,\text{GHz})^2 (18.74\,\text{nH})} = 0.3604\ \text{pF} \tag{6.10i}$$

$$C_{10} = \frac{1}{\omega_c^2 L_{10}} = \frac{1}{(2\pi \cdot 1.936\,\text{GHz})^2 (3.174\,\text{nH})} = 2.128\ \text{pF} \tag{6.10j}$$

$$C_{11} = \frac{1}{\omega_c^2 L_{11}} = \frac{1}{(2\pi \cdot 1.936\,\text{GHz})^2 (1.041\,\text{nH})} = 6.489\ \text{pF} \tag{6.10k}$$

$$C_{12} = \frac{1}{\omega_c^2 L_{12}} = \frac{1}{(2\pi \cdot 1.936\,\text{GHz})^2 (5.321\,\text{nH})} = 1.269\ \text{pF} \tag{6.10l}$$

$$C_{13} = \frac{1}{\omega_c^2 L_{13}} = \frac{1}{(2\pi \cdot 1.936\,\text{GHz})^2 (\infty)} = 0 \tag{6.10m}$$

$$C_{14} = \frac{1}{\omega_c^2 L_{14}} = \frac{1}{(2\pi \cdot 1.936\,\text{GHz})^2 (1.587\,\text{nH})} = 4.257\ \text{pF} \tag{6.10n}$$

Note that, as expected because we used the limiting ripple factor, L_3, C_3, L_{13}, and C_{13} are all either 0 or ∞ and are situated within the circuit in such a way that they may simply be omitted. A proposed circuit board layout for discrete-element implementation is shown in Figure 6.14(a), and the simulated performance for $Q_L = 100$ and $Q_C = 400$ is shown in Figure 6.14(b).

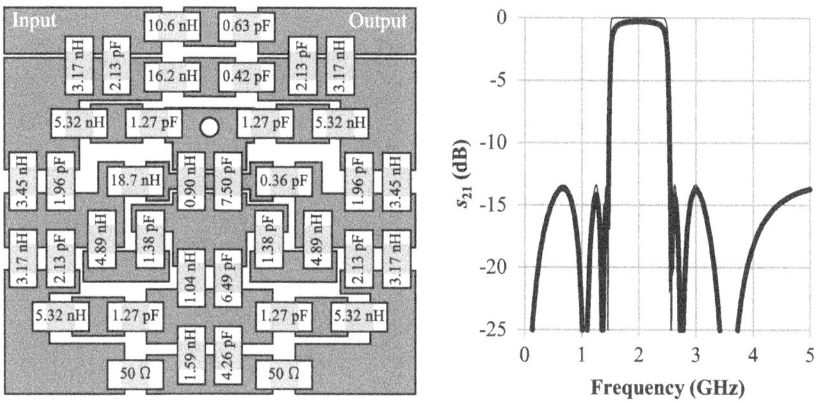

(a) (b)

Figure 6.14 (a) Circuit board layout for discrete-element band-pass Chebyshev Type II reflectionless filter. (b) Simulated performance (thick line) with inductor $Q_L = 100$ and capacitor $Q_C = 400$. Ideal-element model is shown with the thin line.

6.7 CHEBYSHEV TYPE I REFLECTIONLESS FILTERS

Although the ability to realize Chebyshev Type II filters with a reflectionless topology is a significant improvement with respect to Chapters 3–5, Chebyshev Type I filters are more common in industry. This may be because Chebyshev Type II filters are more difficult to realize using conventional (reflective) topologies. After all, ripples should be more readily accepted in the stop-band than in the pass-band of an electronic system. Perhaps the Chebyshev Type II response will become more popular as a result of the ease with which it can be realized in the newly discovered reflectionless topologies.

On the other hand, we have seen that the peak stop-band rejection is more limited with these topologies, at least without cascading several of them. That too may limit their usefulness in some applications. For those purposes, a Chebyshev Type I reflectionless response would be most desirable.

Fortunately, there is a relatively straightforward way to do this using the topologies we have already seen. Recall from the Chapter 1 that the poles of a Chebyshev Type I response are the inverse of those of a Chebyshev Type II. The zeros of the Type I are also located at infinity instead of at finite points along the pure frequency axis.

Now let us look back at the full four-port scattering parameter matrix of the basic third-order reflectionless filter given in (4.3). The pass-band response, $H(s)$, as we now know, is that of a Chebyshev Type II filter. The subnetwork coupling

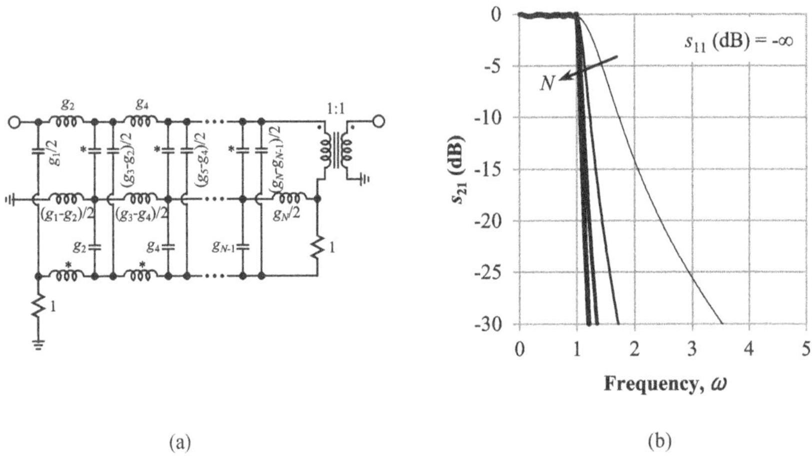

Figure 6.15 (a) Chebyshev Type I reflectionless filter topology and (b) frequency response for $N = 3$, 5, 7, and 9. For each curve, the limiting ripple factor as given in Table 6.2 is used.

function, on the other hand, has the same denominator, but a monomial numerator. Thus, it has the same poles as the pass-band response, but zeros all located at the origin. What would be the pole-zero configuration of the stop-band coupling function for a high-pass transformation of this topology? In fact, because the low-pass to high-pass transformation entails an inversion of the frequency variable, this would mean that the poles are replaced by their inverses, and the zeros (formerly located at the origin) are cast out to infinity — the exact pole-zero configuration of a low-pass Chebyshev Type I filter!

In fact, the full generalized topologies derived in this chapter yield actual Chebyshev Type I responses for their stop-band coupling function with variable ripple factor in the same way as the basic third-order topology. This exploits a fundamental relationship between the Type I and Type II filters, wherein one comprises the residual signal energy that is not passed by the other — that is, the portion of the signal that is reflected from a conventional filter, or absorbed by a reflectionless filter — after an appropriate inverse frequency transformation (such as low-pass to high-pass or band-pass to band-stop).

To exploit this property, we simply need to write down a high-pass version of the generalized reflectionless filter, such as that given in Figure 6.13(a), and exchange the output port with the input-side termination. Since the termination is differential in this topology, the port must be coupled out via a transformer, as shown in Figure 6.15(a) (those accustomed to using transformers know that many configurations are possible besides that shown here; what matters is that the common-mode impedance looking into the transformer terminals from the differential port is very high). The frequency response for several orders is shown in Figure 6.15(b). While the need for a transformer is sometimes inconvenient,

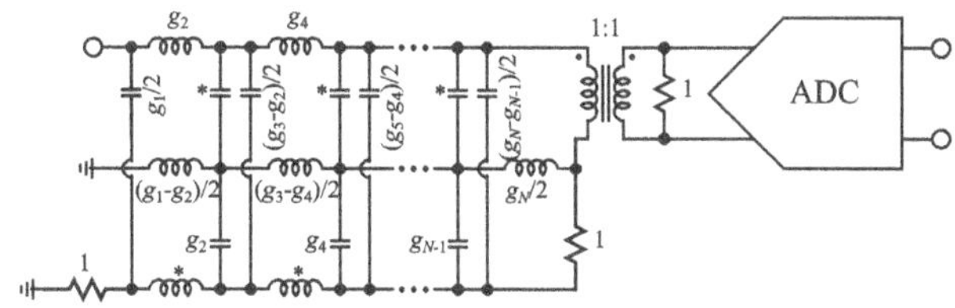

Figure 6.16 Schematic of Type I anti-aliasing reflectionless filter driving an Analog-to-Digital Converter differentially.

topologies such as these may be particularly useful as anti-aliasing filters since the input to a high-speed sampler is very often differential, necessitating a transformer anyway [1]. This application of Type I reflectionless filters is shown in Figure 6.16.

At this time, it is worth contrasting this result with the classical constant-resistance filters introduced in Section 2.5. Recall that the limitations of the then-known first- and second-order topologies created the need for excess insertion loss so that the individual factors would be realizable. The consequent amplitude response of Chebyshev Type I filters of seventh order is shown in Figure 6.17(a) for a range of ripple factors. As described in Chapter 2, the added loss is in excess of 25 dB, and increases with larger ripple factors. The exact same filter responses implemented with the reflectionless filter topology of this section are shown in Figure 6.17(b), where for any ripple factor the excess insertion loss is identically zero.

6.8 ZOLOTAREV REFLECTIONLESS FILTERS

Of course, as Section 6.2.2 showed, the range of filter functions that can be realized with the generalized topologies of this chapter is broader than just the Chebyshev Type I and Type II functions. Any set of prototype parameters, $g_1 \ldots g_N$, may be utilized, subject only to the constraint that the odd-numbered parameters are larger than the even-numbered parameters, so that any topological element involving their subtraction is positive. No matter what their values, the arrangement of those elements (given for the low-pass versions in Figures 6.8, 6.12, and 6.15(a)) is sufficient to ensure that they are reflectionless.

One of the more interesting classes of filters that is compatible with the above requirements is the so-called Zolotarev or Achieser-Zolotarev filter [6, 7]. Similar to Chebyshev Type I, these have ripples in the pass-band and are monotonic in the stop-band. Whereas the Chebyshev filters have all equal ripples, however, the Zolotarev filters have one relatively large ripple followed by equal and smaller ripples. The larger first ripple may be tolerable if the pass-band does not need to extend down to DC (in the low-pass case), while offering steeper rejection in the stop-band.

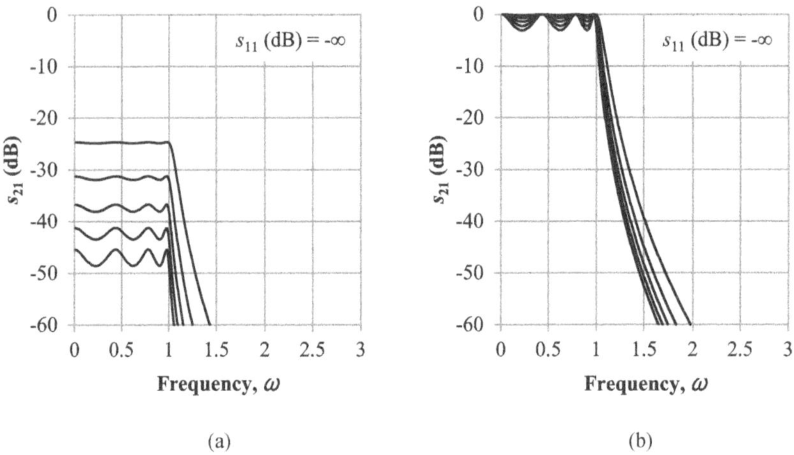

Figure 6.17 Seventh-order Chebyshev Type I filters with ripple factors ranging from $\varepsilon = 0.22$ to $\varepsilon = 1.02$, using (a) classical constant-resistance networks, and (b) the reflectionless filter topology illustrated in this section.

Zolotarev filter functions build upon Chebyshev functions as a basis and share the same design parameters, such as the order, N, and ripple factor, ε. However, they have an additional parameter, λ, which corresponds to the fractional bandwidth occupied by the larger-than-nominal first ripple. As in the Chebyshev case, for reflectionless filters there is a minimum λ and ε for which the odd prototype parameters are larger than the even parameters, yielding positive-only differences as required by the topology. Prototype parameters for Zolotarev filters meeting these conditions are tabulated in Appendix B.

A comparison of the Zolotarev and Chebyshev Type I filter frequency response for $N = 7$ and $\varepsilon = 0.3493$ is shown in Figure 6.18. For the Zolotarev filter, $\lambda = 0.5$. Note that the topology is the same for both, only the element values differ. The larger first ripple of the Zolotarev filter is evident, as is the slightly increased slope. The advantage of these filters, however, has more to do with the parameter values than the slope of the transition. In particular, the Zolotarev parameter values offer the reflectionless topology access to smaller ripple factors over a limited pass-band than would be available with a Chebyshev response. Recall from Table 6.2 that the limiting ripple factor for a seventh-order reflectionless Chebyshev filter is $\varepsilon = 0.2187$. The tables in Appendix B show that ripple factors as low as $\varepsilon = 0.048$ are obtainable with reflectionless Zolotarev filters of the same order.

For reflectionless filters, it is easy to generalize the Achieser-Zolotarev functions so that both Type I and Type II filters are realizable, analogous to the Chebyshev Type I and Type II frequency responses, where the ripples appear in the pass-band or stop-band, respectively. Examples of the Zolotarev Type II

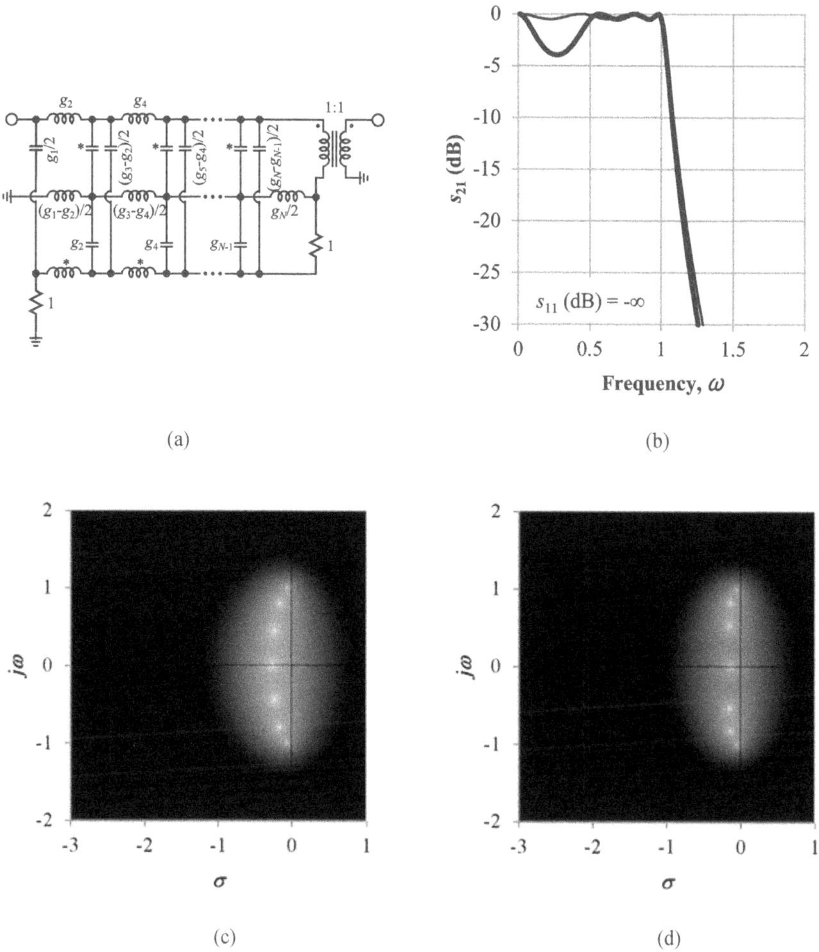

Figure 6.18 Comparison of Chebyshev and Zolotarev Type I filters. (a) Topology, which is common to both. (b) Frequency response for $N = 7$, $\varepsilon = 0.3493$, and, for the Zolotarev filter (thick line), $\lambda = 0.5$. (c) Pole-zero configuration for Chebyshev filter. (d) Pole-zero configuration for Zolotarev filter.

frequency response were already shown, though not called out as such, in Figure 6.4.

6.9 DELTA-WYE TRANSFORMATION

Whichever response one chooses, situations arise with these topologies wherein the element values become impractical. The most problematic elements are

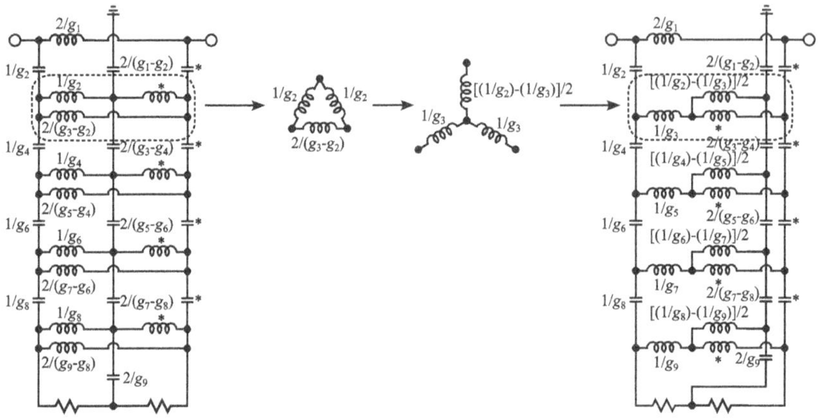

Figure 6.19 Application of delta-wye transformations to a ninth-order generalized reflectionless filter topology.

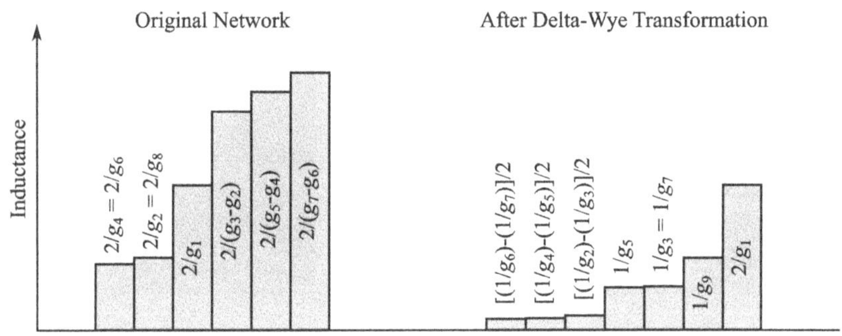

Figure 6.20 Relative inductor values for the topology in Figure 6.19 before and after delta-wye transformations.

usually those that derive from differences in the Chebyshev (or Zolotarev) prototype values. Consider the inductors of the form $2/(g_{k+1}-g_k)$ in Figure 6.19. These can become very large compared to the rest of the elements; large-valued inductors (in practice) suffer from greater losses and parasitics, limiting their applicable frequency range. They may also be larger physically, which is especially inconvenient in planar, monolithic layouts.

One solution is to recognize that these elements frequently occur in a three-element ring, otherwise known as a *delta* configuration. One may therefore apply a delta-wye transformation [8] to each such ring to arrive at an equivalent, alternate topology with different element values, as illustrated in Figure 6.19. The modified inductors now have the form $[(1/g_k)-(1/g_{k+1})]/2$, which is usually much smaller. To better illustrate the point, the prototype inductor values for the ninth-

(a) (b)

Figure 6.21 Fifth-order Chebyshev Type II filters implemented with (a) conventional topology and (b) reflectionless topology. The element values shown are (unitless) normalized quantities.

order low-pass topology in Figure 6.19 are shown in Figure 6.20, before and after the delta-wye transformations. The values are sorted from smallest to largest in both cases for ease of comparison. Maximum ripple factor was selected, and those elements which become 0 or ∞, resulting in shorts or opens, have been omitted.

6.10 COMPONENT SENSITIVITY

Having now constructed reflectionless filters that achieve the same frequency response of classically optimized, conventional filters, it is instructive to compare their sensitivity to component values. As a first basis of comparison, let us use a fifth-order Chebyshev Type II filter, shown using a conventional ladder topology in Figure 6.21(a) and a reflectionless topology in Figure 6.21(b). Both models were simulated with the nominal, normalized values shown, but iterated 30 times with random variations in the element values having a standard deviation of 5%. The frequency response variations are shown in Figure 6.22(a) and Figure 6.22(b). The dark solid lines represent the nominal performance, while the gray lines represent performance over many element values.

Interestingly, though the reflectionless filter requires more components, it is less sensitive to small variations in its elements than the conventional topology having an identical transfer function. The spread in performance around the nominal, both in frequency and in amplitude, is highlighted by the arrows. Note that the spread in 3 dB cutoff frequency for the reflectionless filter is roughly half as wide as it is for the conventional filter. Qualitatively, the same is true of the stop-band rejection around the secondary peak, having a 2 dB spread for the conventional filter and only 1 dB spread for the reflectionless filter. The first peak nearest to the transition is especially sensitive in the conventional filter, peaking almost 3 dB higher than it does for the reflectionless filter.

Of course, the transmission response tells only part of the story. The reflection coefficient for both filters is shown in Figure 6.22(c) and Figure

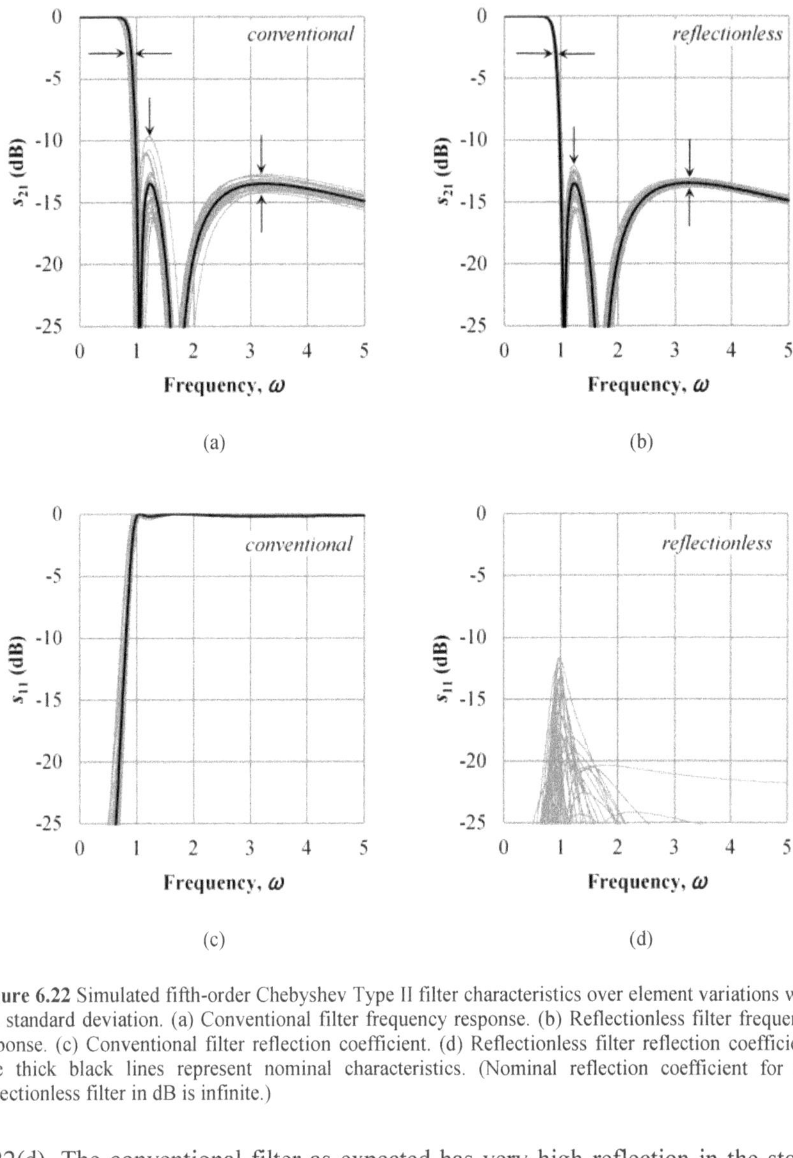

Figure 6.22 Simulated fifth-order Chebyshev Type II filter characteristics over element variations with 5% standard deviation. (a) Conventional filter frequency response. (b) Reflectionless filter frequency response. (c) Conventional filter reflection coefficient. (d) Reflectionless filter reflection coefficient. The thick black lines represent nominal characteristics. (Nominal reflection coefficient for the reflectionless filter in dB is infinite.)

6.22(d). The conventional filter as expected has very high reflection in the stop-band, and small variations around the nominal curve. The reflectionless filter, on the other hand, while having s_{11} (dB) = ∞ in the nominal case, is revealed to be most sensitive around the transition frequency, with peaks in this case approaching 11 dB. This localized sensitivity around the transition frequency is a typical feature of most reflectionless filters.

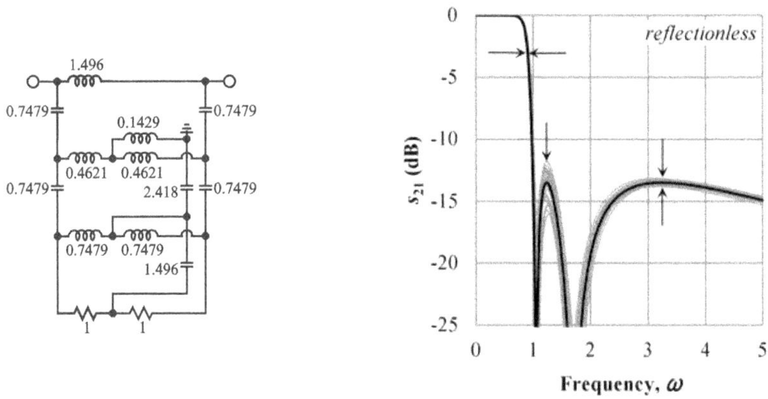

(a) (b)

Figure 6.23 (a) Delta-wye transformed fifth-order reflectionless filter topology. (b) Simulated frequency response over element variations with 5% standard deviation.

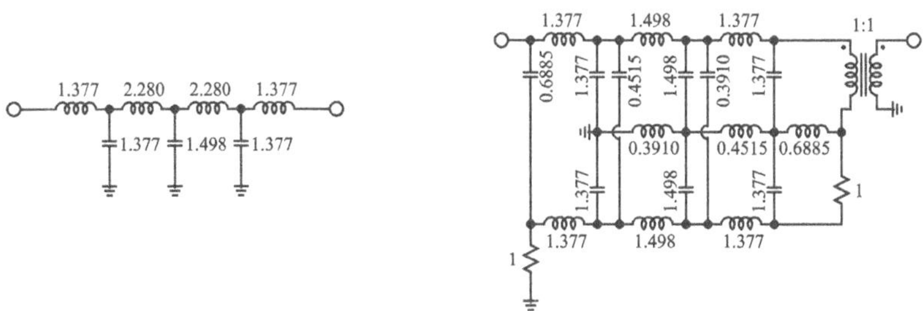

(a) (b)

Figure 6.24 Seventh-order Chebyshev Type I filters implemented with (a) conventional topology and (b) reflectionless topology. The element values shown are (unitless) normalized quantities.

The same sensitivity analysis is shown in Figure 6.23 for the delta-wye transformed version of the reflectionless filter; again, it has the exact same transfer function. The results are qualitatively unchanged, with the reflectionless filter exhibiting only half the spread in performance compared to that of the conventional filter design.

Finally, let us look at the relative sensitivity to component values of conventional and reflectionless Chebyshev Type I filters, such as those drawn in Figure 6.24. Once again, each design was iterated 30 times with random variations about the nominal values shown with 5% standard deviation. The results are shown in Figure 6.25. Quite similar to the Chebyshev Type II case, the spread in cutoff frequencies for the conventional filter is roughly twice that of the reflectionless filter for identical transfer functions.

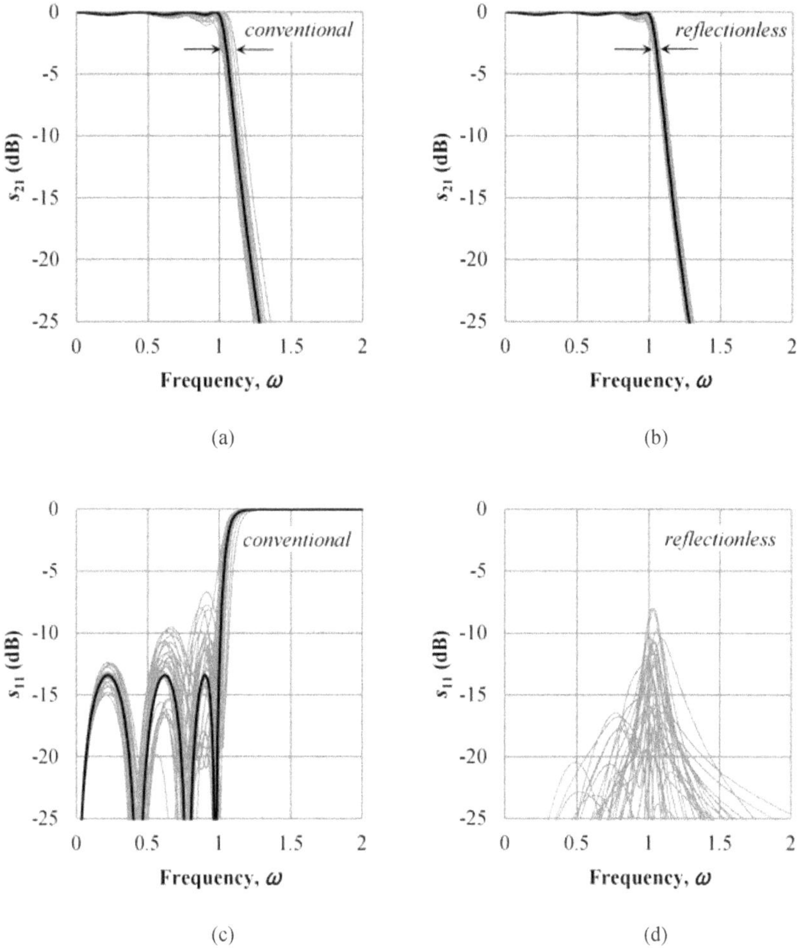

Figure 6.25 Simulated seventh-order Chebyshev Type I filter characteristics over element variations with 5% standard deviation. (a) Conventional filter frequency response. (b) Reflectionless filter frequency response. (c) Conventional filter reflection coefficient. (d) Reflectionless filter reflection coefficient. The thick black lines represent nominal characteristics. (Nominal reflection coefficient for reflectionless filter in dB is infinite.)

Another way of examining the sensitivity to element values is to consider the standard deviation of the transmission response across frequency, as shown in Figure 6.26(a) for the seventh-order conventional and reflectionless Chebyshev Type I filters. In this case, each model was iterated 1000 times to accumulate better statistical significance. It is evident from the figure that the conventional filter is more sensitive to component values at all frequencies. The pass-band in

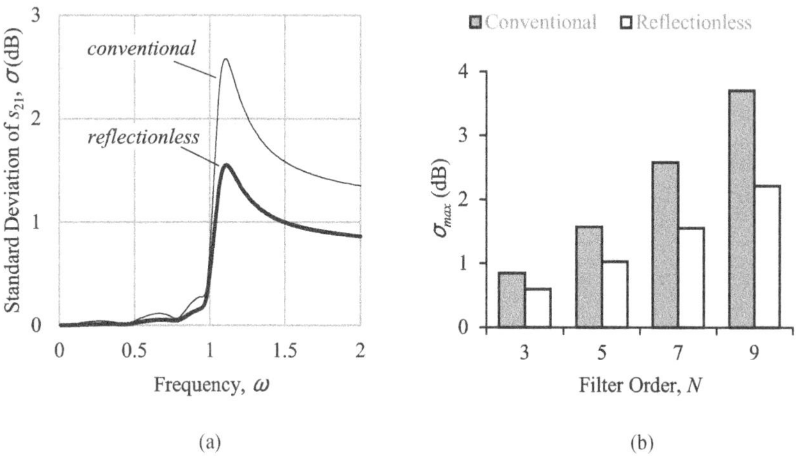

Figure 6.26 (a) Standard deviation of transmission coefficient for seventh-order Chebyshev Type I filters. (b) Worst-case standard deviation as a function of filter order.

particular is prone to developing larger ripples than would nominally be the case, whereas the reflectionless filter is more flat.

Note that in both cases, conventional and reflectionless, the worst-case standard deviation occurs just above the cutoff frequency. In Figure 6.26(b), this peak variation is plotted as a function of filter order for the Chebyshev Type I frequency response. The sensitivity advantage of the reflectionless filter is seen to grow at higher orders.

References

[1] Morgan, M., *Optimal Response Reflectionless Filters*, U.S. Patent Application No. 15/298,459, October 20, 2016, PCT Application No. PCT/US16/57829, October 20, 2016.

[2] Pozar, D., *Microwave Engineering*, 4th ed., New York: Wiley, p. 406, November 2011.

[3] Matthaei, G., Young, L., and Jones, E., *Microwave Filters, Impedance-Matching Networks, and Coupling Structures*, Norwood, MA: Artech House, pp. 98–102, 1980.

[4] Lee, T., *Planar Microwave Engineering: A Practical Guide to Theory, Measurements and Circuits*, New York: Cambridge University Press, pp. 782–783, 2004.

[5] Collin, R., *Foundations for Microwave Engineering*, 2nd ed., , New York: IEEE Press, p. 598, 2001.

[6] Levy, R., "Characteristics and element values of equally terminated Achieser-Zolotarev quasi-low-pass filters," *IEEE Trans. Circuit Theory*, vol. 18, no. 5, September 1971.

[7] Cameron, R., Kudsia, C., and Mansour, R., *Microwave Filters for Communications Systems*, New Jersey: John Wiley & Sons, pp. 436–438, 2007.

[8] Nilsson, J., and Riedel, S., *Electric Circuits*, 5th ed., Reading, MA: Addison-Wesley, pp. 78–79, 1996.

Chapter 7

Transmission-Line Reflectionless Filters

Up to now, our exploration of reflectionless filter topologies has been limited to lumped-element realizations. The validity of the lumped-element model is based on the assumption that the circuit dimensions may be kept small compared to the wavelength at any frequency under consideration, pass-band or stop-band. As the frequency increases, this becomes progressively more difficult, and the quality of the components realized becomes ever poorer. As one moves into the millimeter-wave band, a new framework becomes convenient — namely, that of the transmission lines where the circuit elements are now electrically small only in two of their dimensions. The third dimension is the realm of wave propagation. Despite this change in perspective, and perhaps even more so because of it, the need for reflectionless filtering remains strong. In this chapter, we explore ways of applying the concepts of the previous chapters within the transmission-line framework.

7.1 DERIVATION

7.1.1 Richard's Transformation

The fundamental characterization and mathematical treatment of transmission lines was reviewed in Chapter 1. Although transmission-line circuits may be developed directly from that basis, a convenient approach for importing the results of lumped-element circuit theory is the use of a relationship known as Richard's transformation [1–3]. It describes the equivalence in impedance between lumped reactances and transmission-line stubs on different frequency domains. Recall from (1.66) that the impedance seen looking into a lossless, short-circuited stub of characteristic impedance Z_1 is

$$Z_{in,sc} = jZ_1 \tan(\beta l) = jZ_1 \tan\left(\frac{\omega l}{v_p}\right) \tag{7.1}$$

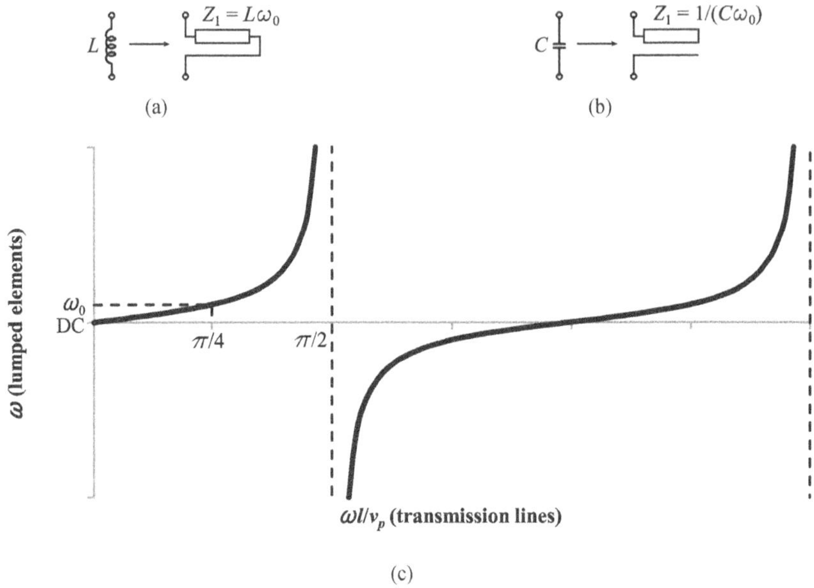

Figure 7.1 Richard's transformation applied to a lumped (a) inductor and (b) capacitor. (c) Mapping of frequencies from transmission-line to lumped-element networks.

where β is the lossless propagation constant, and v_p is the phase velocity of the propagating wave. Much like the high-pass, band-pass, and band-stop frequency transformations of Chapter 1, Richard's transformation involves a substitution of frequency variables,

$$\tan\left(\frac{\omega l}{v_p}\right) \rightarrow \frac{\omega}{\omega_0} \tag{7.2}$$

Substituting (7.2) in the stub's impedance, we have $Z_{in,sc} = j\omega Z_1/\omega_0$, which is recognizable as the impedance of a simple inductor having value $L = Z_1/\omega_0$. Similarly, the input impedance of an open-circuited stub under the same transformation is equivalent to that of a simple capacitor having value $C = Y_1/\omega_0 = 1/(\omega_0 Z_1)$. These equivalences are illustrated in Figure 7.1.

Transmission-line networks derived by application of Richard's transformation to lumped-element prototypes fall into a class of circuits known as commensurate-line networks [1]. These are characterized by the property of having all transmission line stubs and segments of equal length, or integer multiples of that length. Note that the transformation in (7.2) preserves DC ($\omega = 0$), while the frequency for which the argument of the tangent function becomes

Figure 7.2 Kuroda's identity for transmission-line stubs.

$\pi/4$ (i.e., the transmission line is one-eighth of a wavelength) corresponds to $\omega = \omega_0$ in the lumped-element prototype network.

On the other hand, the frequency for which the argument of the tangent function becomes $\pi/2$ (i.e., the transmission line is a quarter wavelength long) is mapped to infinity. Higher frequencies on the transmission line map back to negative frequencies on the lumped elements, then return to zero again. The impedance and scattering parameters of a commensurate-line network are then seen to be periodic in frequency. As such, the four basic types of filter responses (low-pass, high-pass, band-pass, and band-stop) reduce to only two in the transmission-line framework. Limiting our focus to the first two transition frequencies, we see that all commensurate-line filter networks are inherently band-pass or band-stop, where the center frequency of the band is that for which the transmission line is a quarter wavelength long. In Figure 7.1, as elsewhere in this book, all transmission lines are assumed to be a quarter wavelength long at the center frequency of interest, unless otherwise labeled.

7.1.2 Kuroda's Identity

Direct application of Richard's transformation to most lumped-element circuits will produce networks that comprise both series-connected and parallel-connected stubs, but no cascaded segments. This is problematic for two reasons. First, series-connected stubs are difficult to realize in many of the most commonly used transmission line geometries, such as microstrip or stripline. Those techniques admit only parallel-connected stubs easily. Second, since the terminals of all those stubs coincide with the connection points of a prototype lumped-element network, all the stubs in the design will have to emanate from an electrically small area, making it difficult to keep adjacent stubs in these open-air geometries from coupling to one another. Some method of altering the network structure so that the series stubs are eliminated while the remaining or replacement stubs are separated from one another must be found.

The most common way to do this is by substitution of transmission-line identities — two distinct transmission-line topologies that have identical impedance characteristics at all frequencies. Many such identities are known (see Appendix C), but one of the most useful is known as Kuroda's identity [2–7], which is illustrated in transmission-line form in Figure 7.2. It shows that a series-connected, short-circuited stub may be replaced by a parallel-connected, open-circuited stub at the other end of an equal length transmission line, subject to the

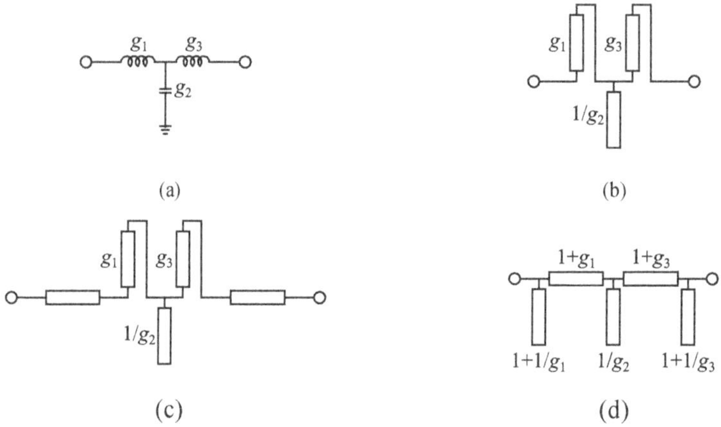

Figure 7.3 Derivation of conventional (reflective) band-stop filter. (a) Low-pass filter prototype. (b) Application of Richard's transformation. (c) Addition of reference plane extensions. Unlabeled transmission lines have characteristic impedance of unit normalized value. (d) Application of Kuroda's identity to eliminate series-connected stubs.

following equalities,

$$z_1 = z_a\left(1+\frac{z_a}{z_b}\right) \tag{7.3a}$$

$$z_2 = z_a + z_b \tag{7.3b}$$

or, in the reverse direction,

$$z_a = \frac{z_1 z_2}{z_1 + z_2} \tag{7.4a}$$

$$z_b = \frac{z_2^2}{z_1 + z_2} \tag{7.4b}$$

Before attempting to derive a reflectionless filter, let us first see how this applies to the conventional case. Take as an example the third-order, low-pass, Cauer-topology filter in Figure 7.3(a). The direct application of Richard's transformation yields the set of stubs shown in Figure 7.3(b). Since the original structure was low-pass, the frequency-periodic extension of it while preserving the DC behavior is a band-stop filter. In the next step, Figure 7.3(c), transmission lines matched to the characteristic impedance of the system are added to both

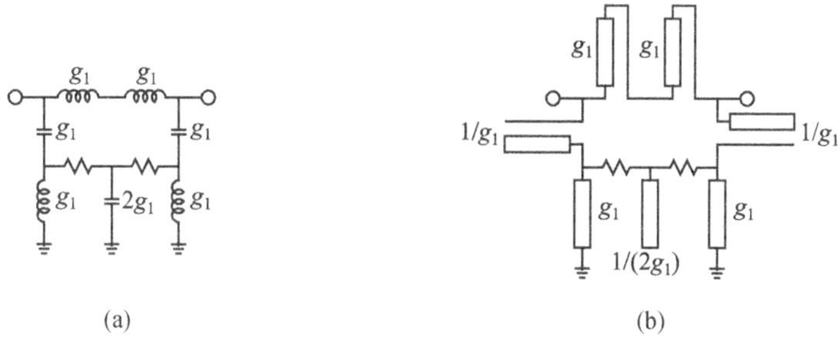

Figure 7.4 First attempt to derive a reflectionless transmission-line filter. (a) Lumped-element prototype. (b) Transmission-line equivalent after application of Richard's transformation. The series stubs between the two ports are very difficult to eliminate.

ports. Known as a reference plane extension, this has the effect of increasing the phase delay through the circuit, but without changing its amplitude response or the flatness of the group delay (the magnitude of the group delay, however, is changed). Finally, in Figure 7.3(d), Kuroda's identity is applied to the outermost series-connected stubs, exchanging them for parallel-connected stubs while simultaneously spacing them out so that stray coupling between them is minimized. Clearly, higher-order filters with many more stubs will require additional matched transmission lines at the ports and repeated application of Kuroda's identity.

There are, of course, other identities which are useful for realizing conventional filters of other kinds. See Appendix C for an extensive listing.

7.1.3 Application to Reflectionless Filters

A direct application of Richard's transformation to the final reflectionless filter prototypes described in the previous chapters is problematic [8]. To see why, consider even the relatively simple third-order reflectionless filter shown in Figure 7.4. After Richard's transformation, we are left with a mixture of series and parallel stubs that are tightly connected at multiple junctions in an electrically small cluster. Some may be eliminated by clever use of transmission-line identities, but none have yet been found that can totally eliminate them all. The band-pass version (starting from a high-pass prototype) is no better. Either way, the series stubs lying directly between the two ports are especially difficult to get rid of, as they are trapped between two junctions where there is no possibility of adding a reference plane extension without negatively impacting the duality of the equivalent even-/odd-mode circuits, losing the reflectionless property.

To address this issue, the extra line length must be added earlier, before the prototype circuit is finalized. Consider the sequence of steps outlined in Figure 7.5. This closely parallels the original derivation of the lumped-element reflectionless filters in Section 3.1.2.

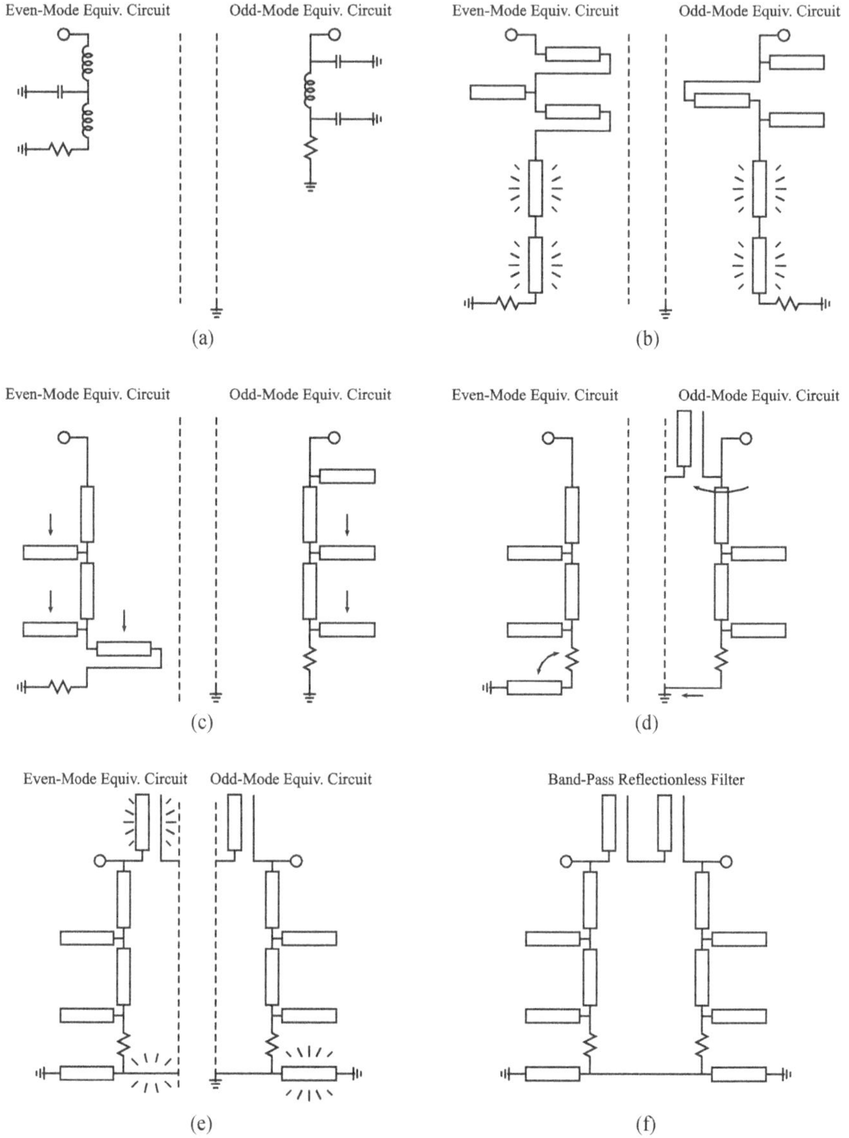

Figure 7.5 Derivation of a band-pass reflectionless transmission-line filter. (*After*: Morgan, M., and Boyd, T., IEEE [9].)

We start with the typical dual, lumped-element even- and odd-mode equivalent circuits. In the past, we have started with terminated high-pass filters here, so that the reflectionless filter derived from them would be low-pass. In this

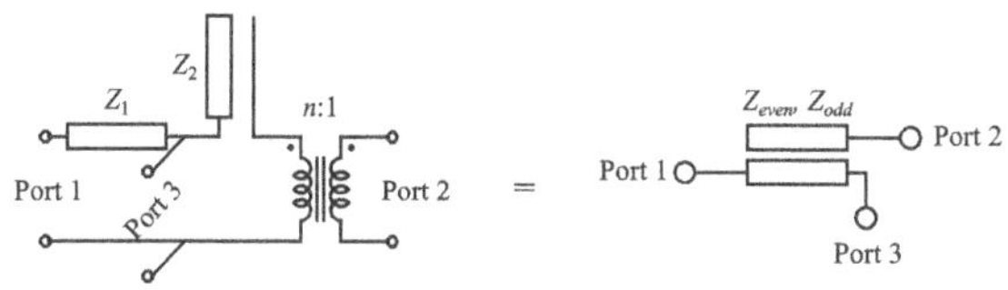

Figure 7.6 A transmission-line identity that will prove critical to finalization of the reflectionless filter derivation.

case, we start with terminated low-pass equivalent circuits. In the lumped-element regime, this would yield a high-pass reflectionless filter, but since we are going to transform the circuit to a commensurate-line network, the final periodic response could more accurately be described as band-pass.

Richard's transformation applied at this stage replaces the reactive components with series- and parallel-connected stubs. At this stage, we also add a series of dual-impedance transmission lines leading the termination resistors. This yields the equivalent circuits shown in Figure 7.5(b).

Next, in Figure 7.5(c), Kuroda's identity is applied on both sides to separate the stubs and convert most of the series stubs into parallel stubs. One series stub is left in place at the bottom of the even-mode equivalent circuit, in series with the termination resistor. In Figure 7.5(d), the order of these last two elements is exchanged, effectively grounding the series stub, making it a shunt stub instead. The first parallel stub on the odd-mode side is disconnected from absolute ground and connected instead to the virtual ground of the symmetry plane, turning this one into a series-connected stub. The termination resistor on the odd-mode side is likewise connected to the virtual ground node.

To complete the restoration of symmetry in Figure 7.5(e), a connector line is drawn from the termination resistor on the even-mode side to the open-circuit symmetry plane, and inactive stubs are added to the top of the even-mode side and bottom of the odd-mode sides, respectively.

Now we have a fully symmetric circuit (shown in Figure 7.5(f)), which is a viable reflectionless filter in its own right, but the series stubs directly between the two ports remain. Their presence is an inconvenience for fabrication, but elimination of these stubs requires a rather little-known three-port transmission-line identity.

7.1.4 A Three-Port Transmission-Line Identity

The identity we need is illustrated in Figure 7.6. It bears some resemblance to a more widespread two-port identity reported in many textbooks for coupled-line sections [6], but in this case, we have a third port emanating from the junction between the cascade transmission line and the series stub. In the commonly reported identity, port 3 is not present, or equivalently, open-circuited. To use in our filter, however, we must ensure that the full three-port network parameters are identical in both circuits, so that they apply under all load conditions.

The relationships between the element values in both circuits are prescribed by the identity as follows,

$$Z_{odd} = Z_1\left(1 - \frac{1}{n}\right) \tag{7.5a}$$

$$Z_{even} = Z_1\left(1 + \frac{1}{n}\right) \tag{7.5b}$$

$$n = \sqrt{\frac{Z_2}{Z_1} + 1} \tag{7.5c}$$

To see that these two networks are equivalent, it is sufficient to show that the full three-port impedance parameters (including all trans-impedances) are the same in both cases. The impedance (Z-) parameters are characterized by the voltage that is developed at one port due to a current injected at that or another port, when all ports not driven by a current source are open-circuited.

In the case of the coupled-line section on the right-hand side of Figure 7.6, the Z-parameters are simply those of a general four-port coupled-line section with the fourth port omitted (left open). These, in turn, may be derived from the superposition of the even- and odd-mode equivalent transmission lines. That is,

$$Z_{11} = Z_{22} = Z_{33} = \tfrac{1}{2}\left(Z_{even} + Z_{odd}\right)\coth(\gamma l) \tag{7.6a}$$

$$Z_{12} = Z_{21} = \tfrac{1}{2}\left(Z_{even} - Z_{odd}\right)\operatorname{csch}(\gamma l) \tag{7.6b}$$

$$Z_{13} = Z_{31} = \tfrac{1}{2}\left(Z_{even} + Z_{odd}\right)\operatorname{csch}(\gamma l) \tag{7.6c}$$

$$Z_{23} = Z_{32} = \tfrac{1}{2}\left(Z_{even} - Z_{odd}\right)\coth(\gamma l) \tag{7.6d}$$

For the purposes of this identity, we have assumed there is no differential dispersion between the even- and odd-modes (i.e., $\gamma_{even} = \gamma_{odd} = \gamma$). Substituting (7.5) into these equations, we have

$$Z_{11} = Z_{22} = Z_{33} = Z_1 \coth(\gamma l) \tag{7.7a}$$

$$Z_{12} = Z_{21} = \tfrac{1}{n} Z_1 \operatorname{csch}(\gamma l) \tag{7.7b}$$

$$Z_{13} = Z_{31} = Z_1 \operatorname{csch}(\gamma l) \tag{7.7c}$$

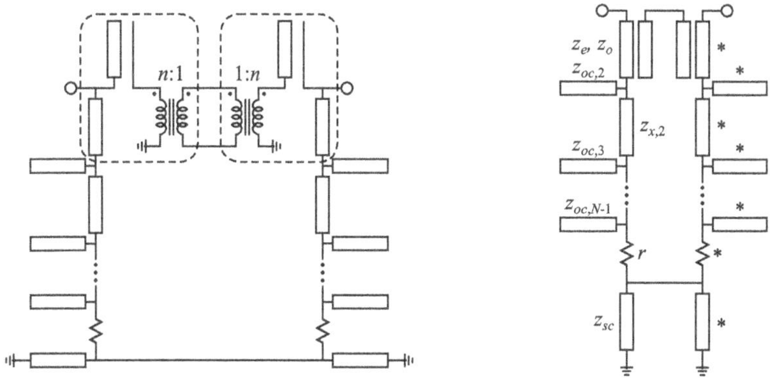

Figure 7.7 Final steps in derivation of band-pass transmission-line reflectionless filter. (a) Insertion of back-to-back transformers. (b) Application of three-port identity from Figure 7.6. As elsewhere in this book, elements labeled with an asterisk have values equal to their symmetric counterparts.

$$Z_{23} = Z_{32} = \tfrac{1}{n} Z_1 \coth(\gamma l) \tag{7.7d}$$

To prove equivalence, we must now derive the same expressions for the line and stub network on the left-hand side of Figure 7.6. If port 2 is left open, the remaining two-port network between ports 1 and 3 is simply that of a transmission line having characteristic impedance Z_1. Therefore, from (1.70) we have

$$Z'_{11} = Z'_{33} = Z_1 \coth(\gamma l) \tag{7.8a}$$

$$Z'_{13} = Z'_{31} = Z_1 \operatorname{csch}(\gamma l) \tag{7.8b}$$

matching (7.7c) and part of (7.7a). If ports 1 and 3 are left open, port 2 is looking into the series connection of two open-circuited stubs, modified by the transformer turns ratio,

$$Z'_{22} = \frac{1}{n^2}\left(Z_1 \coth(\gamma l) + Z_2 \coth(\gamma l)\right) = \frac{Z_1 + Z_2}{\frac{Z_2}{Z_1} + 1} \coth(\gamma l) = Z_1 \coth(\gamma l) \tag{7.9}$$

which verifies the remainder of (7.7a). For the remaining parameters, consider the voltage that is developed at ports 1 and 3 when a current, I_2, is injected into port 2. The transformer reduces the current by a factor n, and since the stub is in series, the same current flows into the cascade transmission line having characteristic impedance Z_1. We are once again left with a two-port transmission line equivalent, only with the current source at the driving end reduced by a factor of n,

$$Z'_{12} = Z'_{21} = \tfrac{1}{n} Z_1 \operatorname{csch}(\gamma l) \tag{7.10a}$$

$$Z'_{23} = Z'_{32} = \tfrac{1}{n} Z_1 \coth(\gamma l) \tag{7.10b}$$

This confirms the expressions in (7.7b) and (7.7d), thus completing the proof of the identity [9].

7.1.5 Final Topology and Component Values

In order to apply this identity to the reflectionless filter network in Figure 7.5(f), we must, notionally at least, insert two transformers back to back, as shown in Figure 7.7(a). They essentially cancel each other, so this insertion is valid. The identity is then applied twice, once for each side, yielding the completed network in Figure 7.7(b). Note also that we have made arbitrary the number of stubs in this figure, as it relates to the order of the original prototype filter, but does not otherwise change the derivation in any way.

The transmission-line elements in Figure 7.7(b) have been labeled with their normalized impedance values. To determine the duality constraints, we must first consider the even- and odd-mode equivalent circuits. These are shown in Figure 7.8. A few previously described transmission-line identities are needed to put them into recognizable dual forms. In particular, the two-port coupled-line sections at the inputs of both equivalent circuits may be replaced with transmission-line and stub equivalents using some of the identities given in Appendix C. Kuroda's identity is also used in the even-mode equivalent circuit to transform each of the open-circuited shunt stubs into short-circuited series stubs (e.g., the two elements enclosed in dashed circles). This was done to make the duality of the even- and odd-mode sides more obvious. Note also that in the final step on the even-mode side, the transmission lines are labeled with their normalized admittances instead of their impedances. The normalized admittances on the even-mode side may therefore be directly equated to the normalized impedances on the odd-mode side to ensure duality.

In terms of the original network transmission-line parameters, the characteristic immittances of the lines and stubs in Figure 7.8 are

$$y_{x,k} = \begin{cases} z_{oc,1}^{-1} + 2(z_e + z_o)^{-1} & k = 1 \\ z_{oc,k+1}^{-1} + z_{x,k}^{-1} & 1 < k < N-1 \end{cases} \tag{7.11a}$$

$$z_{oc,1} = 2 z_e z_o \frac{z_e + z_o}{(z_e - z_o)^2} \tag{7.11b}$$

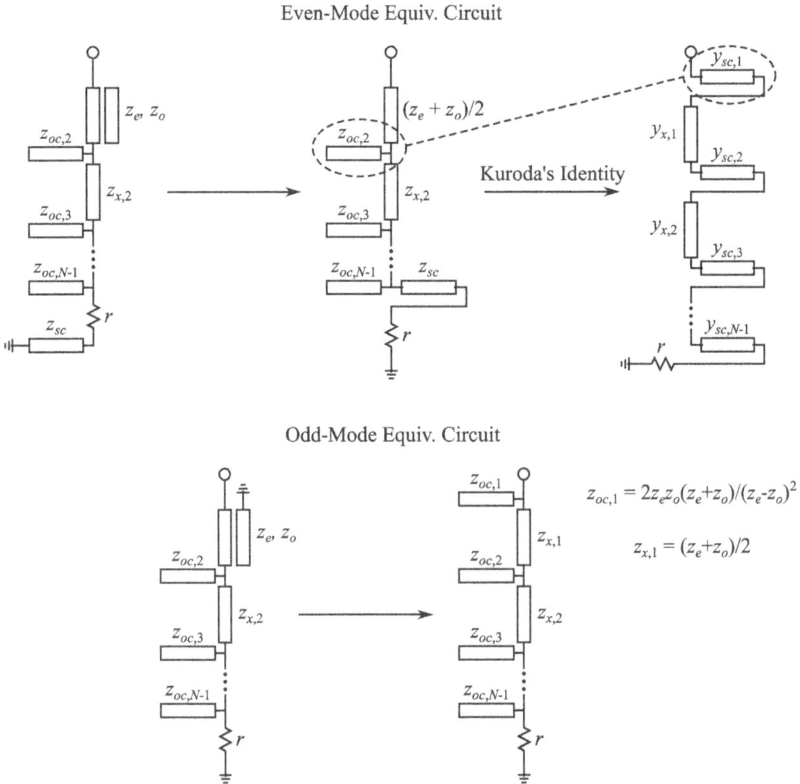

Figure 7.8 Even- and odd-mode equivalent circuits for band-pass transmission-line reflectionless filter.

$$z_{x,1} = \tfrac{1}{2}\left(z_e + z_o\right) \tag{7.11c}$$

$$y_{sc,k} = \begin{cases} \dfrac{z_{oc,1} + \frac{1}{2}\left(z_e + z_o\right)}{\frac{1}{4}\left(z_e + z_o\right)^2} & k = 1 \\ \dfrac{z_{oc,k+1} + z_{x,k}}{z_{x,k}^2} & 1 < k < N-1 \\ \dfrac{1}{z_{sc}} & k = N-1 \end{cases} \tag{7.11d}$$

Duality requires that the normalized characteristic admittance of the series-connected, short-circuited stubs in the even-mode equivalent circuit must be equal to the normalized characteristic impedance of the parallel-connected, open-circuited stubs in the odd-mode equivalent circuit, and further that the normalized characteristic immittances of the cascade lines in both equivalent circuits are the inverses of one another. Thus,

$$y_{sc,k} = z_{oc,k} \text{ for } 1 \le k < N \tag{7.12a}$$

$$y_{x,k} = z_{x,k} \text{ for } 1 \le k < N\text{-}1 \tag{7.12b}$$

Substituting the values from (7.11) into (7.12), we find that

$$\frac{z_{oc,k+1} + z_{x,k}}{z_{x,k}^2} = z_{oc,k} \text{ for } 1 < k < N\text{-}1 \tag{7.13a}$$

$$z_{oc,k+1}^{-1} + z_{x,k}^{-1} = z_{x,k} \text{ for } 1 < k < N\text{-}1 \tag{7.13b}$$

Let us rename $z_{x,k} = x$, and apply this to (7.13),

$$z_{oc,k+1} = \frac{1}{x - x^{-1}} \text{ for } 1 < k < N\text{-}1 \tag{7.14a}$$

$$z_{oc,k} = \frac{\left(x - x^{-1}\right)^{-1} + x}{x^2} = \frac{1}{x - x^{-1}} = z_{oc,k+1} \text{ for } 1 < k < N\text{-}1 \tag{7.14b}$$

We see, therefore, that all of the open-circuited stubs in the original network have the same value. Equation (7.13b) then permits two solutions for each $z_{x,k}$, but only one positive, thus the cascade transmission lines must also all have the same value, namely x. The short-circuited stub in the original network is trivially shown by (7.11d) and (7.12a) to be the inverse of the open-circuited stub values, or $z_{sc} = x - x^{-1}$. By inspection, we have that $r = r^{-1} = 1$. This leaves only the determination of the even- and odd-mode characteristic impedance of the coupled lines. Substituting into (7.12) for the $k = 1$ case,

$$\frac{\left(x - x^{-1}\right)^{-1} + \frac{1}{2}\left(z_e + z_o\right)}{\frac{1}{4}\left(z_e + z_o\right)^2} = 2 z_e z_o \frac{z_e + z_o}{\left(z_e - z_o\right)^2} \tag{7.15a}$$

$$x - x^{-1} = \tfrac{1}{2}\left(z_e + z_o\right) - 2\left(z_e + z_o\right)^{-1} \tag{7.15b}$$

$$\frac{\left(\frac{1}{2}\left(z_e + z_o\right) - 2\left(z_e + z_o\right)^{-1}\right)^{-1} + \frac{1}{2}\left(z_e + z_o\right)}{\frac{1}{4}\left(z_e + z_o\right)^2} = 2 z_e z_o \frac{z_e + z_o}{\left(z_e - z_o\right)^2} \tag{7.15c}$$

$$(z_e - z_o)^2 = z_e z_o\left((z_e + z_o)^2 - 4\right) \tag{7.15d}$$

$$z_e z_o = 1 \tag{7.15e}$$

Let $\rho = z_e/z_o$,

$$x - x^{-1} = \tfrac{1}{2\sqrt{\rho}}(\rho+1) - 2\sqrt{\rho}(\rho+1)^{-1} \tag{7.16a}$$

$$x = \tfrac{1}{2\sqrt{\rho}}(\rho+1) \tag{7.16b}$$

$$\rho - 2x\sqrt{\rho} + 1 = 0 \tag{7.16c}$$

$$\rho = \left(x + \sqrt{x^2 - 1}\right)^2 \tag{7.16d}$$

To summarize, then, we have for the normalized impedance values of the original network,

$$z_e = \sqrt{\rho} \tag{7.17a}$$

$$z_o = 1/\sqrt{\rho} \tag{7.17b}$$

$$z_{x,k} = x \text{ for all } k \tag{7.17c}$$

$$z_{oc,k} = \left(x - x^{-1}\right)^{-1} \text{ for all } k \tag{7.17d}$$

$$z_{sc,k} = x - x^{-1} \text{ for all } k \tag{7.17e}$$

$$r = 1 \tag{7.17f}$$

where

$$\rho = \left(x + \sqrt{x^2 - 1}\right)^2 \tag{7.18}$$

and where $x>1$ is a free parameter [10].

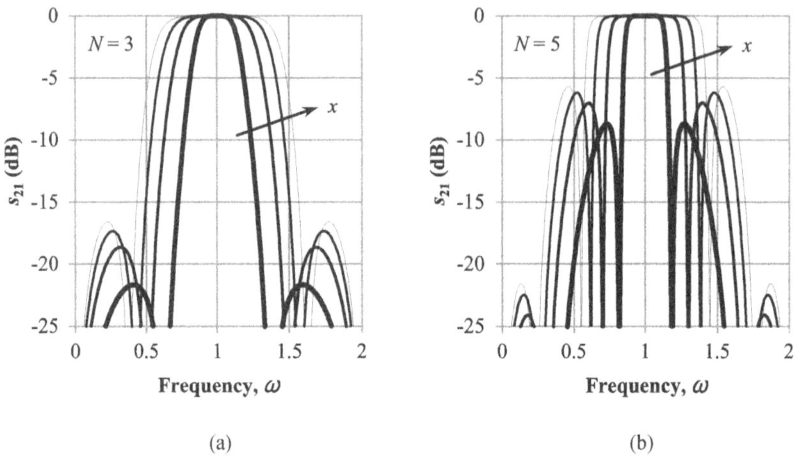

Figure 7.9 Frequency response of transmission line reflectionless filters for (a) $N = 3$ and (b) $N = 5$. The free parameter x = 1.2, 1.4, 1.6, and 1.8, increasing in the direction of the arrow. As with the lumped-element reflectionless filters, the theoretical return loss in dB is infinite.

7.1.6 Frequency Response

The frequency response of the basic transmission-line reflectionless filter for order $N = 3$ and $N = 5$ is shown in Figure 7.9. Note that as the parameter x increases, the bandwidth gets wider and the peak stop-band rejection becomes less. Also, as was the case with the first lumped-element reflectionless filters developed in Chapter 3, the higher-order designs have higher stop-band peaks, or less stop-band rejection.

7.1.7 Lumped-Element Correspondence

There is a subtlety here that is easy to miss. Working backward with the element values derived in (7.17), we find that the cascade transmission lines added as a reference plane extension to the terminations in Figure 7.5(b) are not matched to the system impedance for any permissible value of the free parameter, x. Inevitably, this alters the frequency response of the even- and odd-mode equivalent circuits first written down in Figure 7.5(a), beyond just the change in frequency variables associated with Richard's transformation. In fact, the added quarter-wave transformer in this step increases the order of the filter from that which we started with.

Contrast this with the original lumped-element reflectionless filter derivation, where all changes made to the even- and odd-mode equivalent circuits were carefully chosen to avoid affecting the equivalent circuit response. Were we to

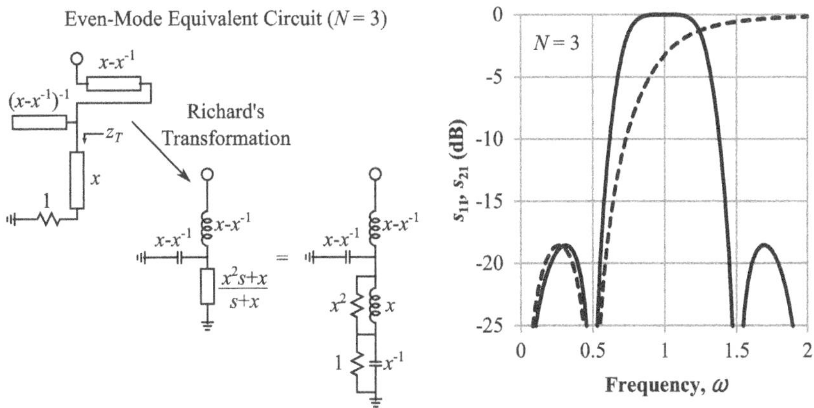

Figure 7.10 Even-mode equivalent circuit for third-order band-pass transmission-line reflectionless filter, and its corresponding high-pass lumped-element prototype. The plot on the right shows the transmission response of the complete transmission-line filter (solid line) and the even-mode reflection coefficient of the lumped-element prototype (dashed).

maintain that same rigor here, then we would have to conclude that the added cascade lines in front of the termination resistors in this step must have unity normalized characteristic impedance. Doing so, it turns out, precludes the possibility of a symmetric and dual assignment of element values later on.

But if our goal is simply to come up with a reflectionless filter structure, it is not strictly necessary to avoid changing the equivalent-circuit response at this stage. We need only to avoid spoiling the duality conditions. To that end, we allowed that the transmission lines added in this step do not have unity normalized impedance, but rather have impedances that are dual to each other on the even- and odd-mode sides.

It may seem as if we should have started with a lumped-element prototype that accurately represents the response of the final filter. In principle, this could have been done, but a lumped-element prototype that corresponds (through Richard's transformation) to the transmission-line filter we ultimately derived is quite counter-intuitive, and not something we would have been likely to try had we not found this solution by other means.

Consider the form of the even-mode equivalent circuit for a third-order filter, with element values included, shown in Figure 7.10. The termination resistor has unit normalized value, but the cascade transmission line added to provide electrical length for later application of Kuroda's identity has a normalized characteristic impedance of x. There is no easy way to apply the stub/lumped-element substitutions from Figure 7.1 to revert this back to a lumped-element equivalent. Instead, we must synthesize an equivalent network algebraically. The impedance looking into the cascade line, z_T, may be found by application of (1.65) in normalized form,

$$z_T = x\frac{1 + jx\tan(\beta l)}{x + j\tan(\beta l)} \tag{7.19}$$

Using the algebraic form of Richard's transformation in (7.2), and converting to the Laplace frequency domain, we have

$$z'_T = x\frac{1 + xs}{x + s} = \frac{x^2 s + x}{s + x} = \left(\frac{1}{x^2} + \frac{1}{xs}\right)^{-1} + \left(\frac{s}{x} + 1\right)^{-1} \tag{7.20}$$

which is a positive-real function. The first term in parenthesis may be recognized as the admittance of a resistor in parallel with an inductor, while the second term in parenthesis corresponds to a resistor in parallel with a capacitor. Hence, the corresponding lumped-element prototype circuit is shown in the final schematic of Figure 7.10. A comparison of the reflection coefficient for the even-mode lumped-element equivalent (dashed line) and the frequency response of the transmission-line filter (solid line) is shown in the plot. These curves are equivalent except for the change of frequency domains resulting from Richard's transformation.

Clearly, the lumped-element prototype that actually matches our final filter is too obscure to have been happened upon by chance. This reinforces the point that was made in Section 3.7 about the difficulty of discovering new topologies by direct application of the even-/odd-mode synthesis approach taught in Chapter 3. Less direct ways of approaching the problem are required. In many cases (such as this one), it is fruitful to consider the element values as arbitrary in the beginning, maintaining only the topological duality in each step. Like the generalized filters of Chapter 6, the conditions for numerical duality can be examined later, only after completion of the symmetric two-port network.

7.2 ALTERNATIVE FORMS

The impedances required to implement the reflectionless filters derived in Section 7.1 may be very difficult to realize in practice. A good rule of thumb for most printed-circuit technologies is that any impedance within a factor of two of the system impedance (0.5 to 2.0 in normalized values) is achievable. Larger than that often requires traces that are too thin to reliably fabricate, and have a lot of series resistance in any case, whereas smaller impedances require traces that may be so wide as to become overmoded — with the possible exception of the odd-mode impedance of coupled lines, which are made small by the very close electric-wall boundary condition.

Unfortunately, there is no value of x for which all of the impedances required are within that 0.5–2.0 range, though $x \approx 1.3$ comes close. The coupling factor of the coupled lines, ρ, is also quite large over much of the parameter space. Consequently, alternative topologies are required for the transmission-line

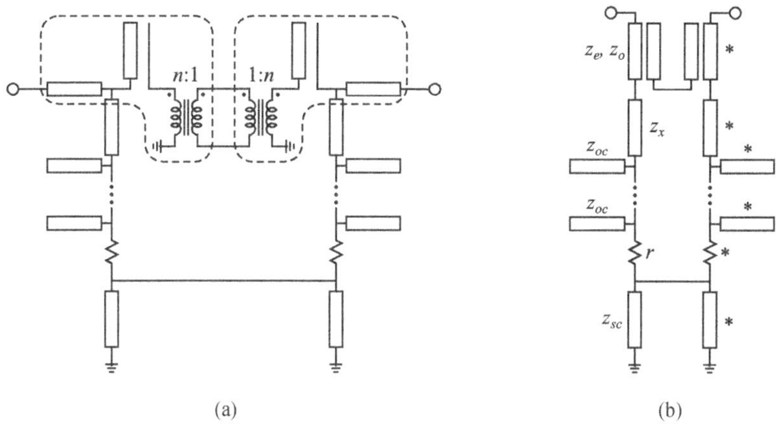

Figure 7.11 (a) Application of three-port identity with leading transmission lines, and (b) resulting alternative band-pass reflectionless filter topology.

reflectionless filter to be practical in most cases.

7.2.1 Reversed Input Coupling

One alternative concerns the orientation of the coupled-line section at the input of the filter. Recall that in the original derivation a three-port identity described in Section 7.1.4 was applied to a pair of series stubs branching off directly from the port nodes. It could just as easily have been applied to these stubs in combination with leading transmission lines separating them from the ports, as shown in Figure 7.11(a). The resulting alternative topology in Figure 7.11(b) is almost the same, except that the coupled lines are now oriented in the opposite direction, and the duality constraints as they apply to the coupled lines are different. Consider the even- and odd-mode equivalent circuits shown in Figure 7.12. The normalized values of the leading transmission line and stub on the odd-mode side are

$$z_1 = 2z_e z_o \frac{z_e + z_o}{\left(z_e - z_o\right)^2} \tag{7.21a}$$

$$z_2 = \tfrac{1}{2}\left(z_e + z_o\right) \tag{7.21b}$$

Duality requires these to be equal to the admittances on the even-mode side. Therefore,

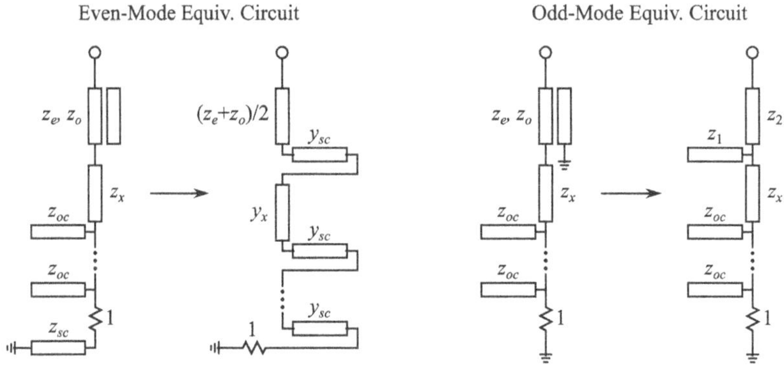

Figure 7.12 Even- and odd-mode equivalent circuits for the topology shown in Figure 7.11(b).

$$\tfrac{1}{2}(z_e + z_o) = \frac{1}{\tfrac{1}{2}(z_e + z_o)} = 1 \tag{7.22a}$$

$$2z_e z_o \frac{z_e + z_o}{(z_e - z_o)^2} = z_{oc} = \frac{1}{x - x^{-1}} \tag{7.22b}$$

$$\frac{4\rho}{(\rho-1)^2} = \frac{1}{x - x^{-1}} \tag{7.22c}$$

$$\rho^2 - 2\rho(1 + 2x - 2x^{-1}) + 1 = 0 \tag{7.22d}$$

$$\rho = 1 + 2(x - x^{-1}) \pm \sqrt{(1 + 2(x - x^{-1}))^2 - 1} \tag{7.22e}$$

$$\rho = 1 + 2(x - x^{-1}) + 2\sqrt{(x - x^{-1})(1 + x - x^{-1})} \tag{7.22f}$$

The last step follows because ρ is required to be positive. To satisfy (7.22a), then, the even- and odd-mode characteristic impedances of the coupled lines are given by

$$z_e = \frac{2\rho}{\rho + 1} \tag{7.23a}$$

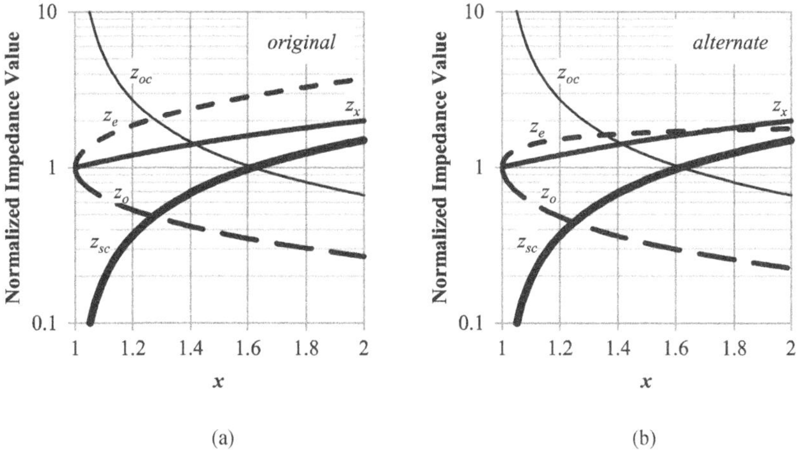

Figure 7.13 Normalized characteristic impedances for (a) original and (b) alternative transmission-line reflectionless filter topologies.

$$z_o = \frac{2}{\rho + 1} \tag{7.23b}$$

These formulae and those from (7.17) are plotted in Figure 7.13. The performance curves are the same in both cases, shown earlier in Figure 7.9, but the even-mode characteristic impedance of the coupled lines for the alternative configuration is not as high, permitting all normalized impedances (except z_o) to lie within a 0.5–2.0 range for some values of the free parameter, x.

7.2.2 Cascade Resonators

Continuing in a manner parallel to the development of the lumped-element reflectionless filters, we find that the derivation presented in Section 7.1 is not unique. The direct use of Richard's transformation on our lumped even-mode equivalent circuit prototype gave us quarter-wave stubs as the resonating elements in the filter, but there are other types. For example, a simple quarter-wave section of cascade transmission line may act as a resonator as well.

Consider the topologies shown in Figure 7.14. The input coupled-line section and the short-circuited stubs connected to the termination resistors are the same as before, but the intervening open-circuited stubs have been replaced by quarter-wave cascade line resonators having a normalized characteristic impedance of x^{-1}. Interestingly, the formulae from (7.17), (7.22f), and (7.23) ensuring the duality of the equivalent half-circuits need no modification. The even-mode equivalent circuit (not shown) comprises a set of quarter-wave transformers with a stub at the

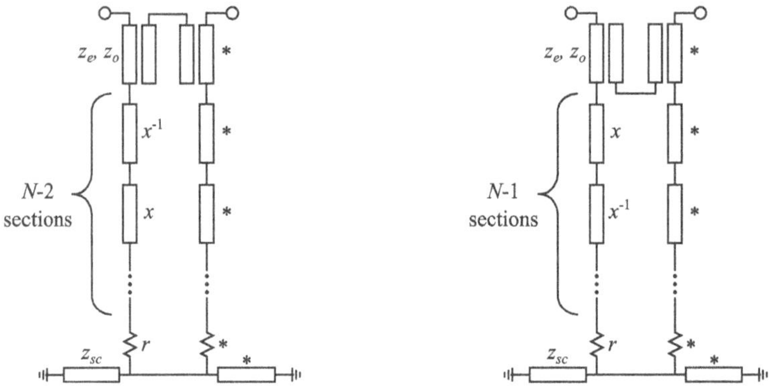

Figure 7.14 Two equivalent topologies for a band-pass transmission-line reflectionless filter having cascade quarter-wave resonators.

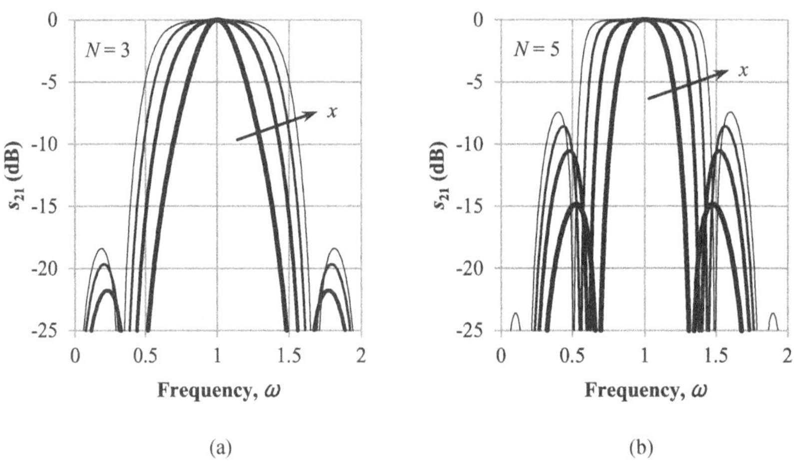

Figure 7.15 Frequency response of reflectionless filters with quarter-wave cascade resonators. (a) $N = 3$. (b) $N = 5$. The parameter $x = 1.2$, 1.4, 1.6, and 1.8, increasing in the direction of the arrow.

end near the termination resistor, whereas the odd-mode equivalent circuit has the stub at its input, but none after the quarter-wave transformers near the resistor. Given the same values previously derived, these may be shown to be duals of one another via repeated application of Kuroda's identity. The resulting filter transfer function, however, is slightly different. Examples for $N = 3$ and $N = 5$ are shown in Figure 7.15.

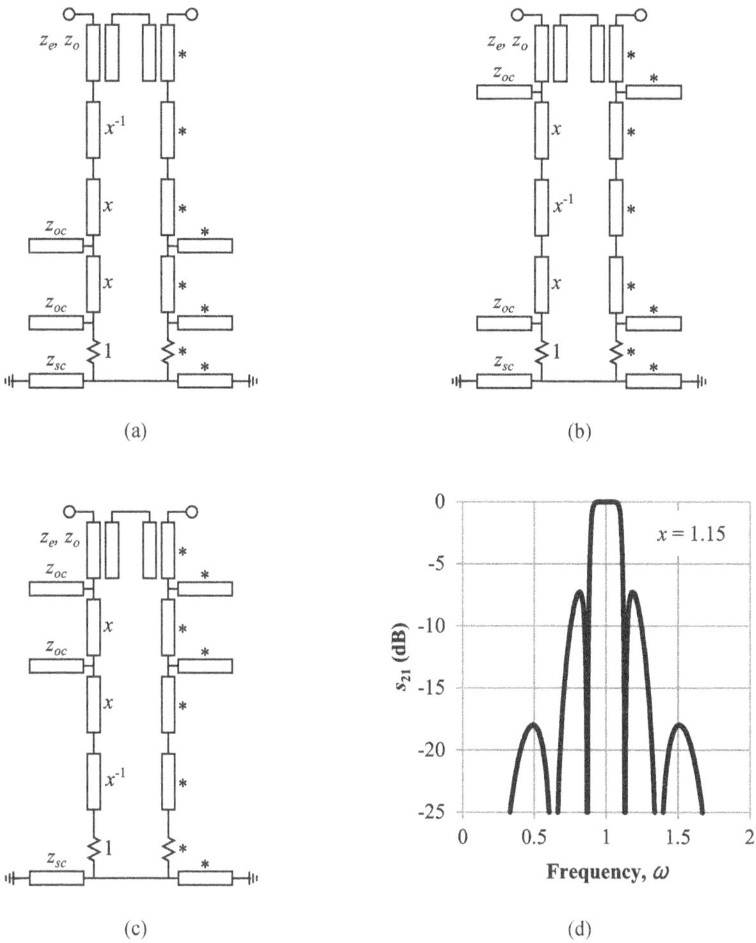

Figure 7.16 Three versions, (a)-(c), of a seventh-order band-pass reflectionless filter, all of which have identical frequency response, (d).

7.2.3 Mixed Resonators

In fact, open-circuited stubs and quarter-wave sections may be used interchangeably and in any combination. Each stub has a normalized characteristic impedance given by (7.17d), and each quarter-wave section has impedance x^{-1}. Between each resonator (stub or cascade line) is a line of impedance x. As a general rule, the more stubs one uses instead of quarter-wave cascade lines, the flatter the pass-band will be, while more cascade resonators lead to lower sidelobes. The sequence of stub- and cascade-resonators does not matter, only the relative number. As an example, the three topologies shown Figure 7.16 have identical frequency response.

7.3 DESIGN EXAMPLE

As an example, consider the alternative form with reversed input coupling shown in Figure 7.11(b). Let us use a bandwidth parameter $x = 1.4$ and select the order $N = 3$. (Recall that x is the characteristic impedance of the cascade transmission lines in the reflectionless filter topology, and is a convenient quantity for parameterization of the remaining elements.) From (7.22f) and (7.23) we have for the coupled line parameters, scaled to the characteristic impedance of the system, $Z_0 = 50\Omega$,

$$\rho = 1 + 2\left(x - x^{-1}\right) + 2\sqrt{\left(x - x^{-1}\right)\left(1 + x - x^{-1}\right)} = 4.522 \tag{7.24a}$$

$$Z_e = Z_0 \frac{2\rho}{\rho + 1} = \left(50\Omega\right)\left(\frac{2 \cdot 4.522}{4.522 + 1}\right) = 82\Omega \tag{7.24b}$$

$$Z_o = Z_0 \frac{2}{\rho + 1} = \left(50\Omega\right)\left(\frac{2}{4.522 + 1}\right) = 18\Omega \tag{7.24c}$$

The impedance of all the remaining transmission lines may then be calculated from (7.17),

$$Z_x = z_x Z_0 = x Z_0 = \left(1.4\right)\left(50\Omega\right) = 70\Omega \tag{7.25a}$$

$$Z_{oc} = z_{oc} Z_0 = \frac{Z_0}{x - x^{-1}} = \frac{\left(50\Omega\right)}{\left(1.4 - 1/1.4\right)} = 73\Omega \tag{7.25b}$$

$$Z_{sc} = z_{sc} Z_0 = \left(x - x^{-1}\right) Z_0 = \left(1.4 - 1/1.4\right)\left(50\Omega\right) = 34\Omega \tag{7.25c}$$

$$R = r Z_0 = \left(1\right)\left(50\Omega\right) = 50\Omega \tag{7.25d}$$

As a commensurate-line network, all transmission lines should be the same length; in this case, a quarter-wavelength at the center of the pass-band.

We thus arrive at the schematic shown in Figure 7.17(a). A possible layout in microstrip is shown in Figure 7.17(b). The line widths and spacings are determined by the above impedances, while the lengths are primarily a function of

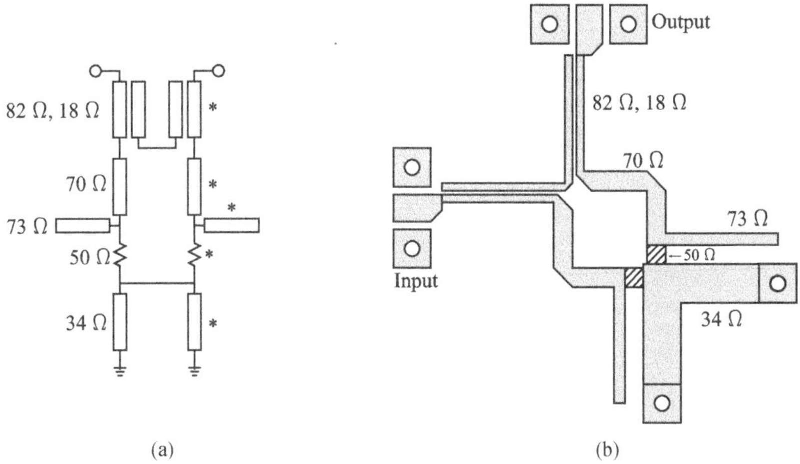

Figure 7.17 (a) Schematic of a band-pass transmission-line reflectionless filter. (b) Possible monolithic layout.

the substrate material and center frequency.

References

[1] Richard, P., "Resistor-transmission line circuits," *Proc. IRE*, vol. 36, no. 2, pp. 210–220, Feb 1948.

[2] Lee, T., *Planar Microwave Engineering: A Practical Guide to Theory, Measurements and Circuits*, New York: Cambridge University Press, pp. 795–800, 2004.

[3] Pozar, D., *Microwave Engineering*, 4th ed., New York: Wiley, pp. 415–419, November 2011.

[4] Matthaei, G., Young, L., and Jones, E., *Microwave Filters, Impedance-Matching Networks, and Coupling Structures*, Norwood, MA: Artech House, pp. 219–228, 1980.

[5] Levy, R., and Cohn, S., "A history of microwave filter research, design, and development," *IEEE Trans. Microw. Theory Techn.*, vol. 32, no. 9, pp. 1055–1067, Sep. 1984.

[6] Wenzel, R., "Exact design of TEM microwave networks using quarter-wave lines," *IEEE Trans. Microw. Theory Techn.*, vol. 12, no. 1, pp. 94–111, Jan. 1964.

[7] Ozaki, H., and Ishii, J., "Synthesis of a class of strip-line filters," *IRE Trans. Circuit Theory*, vol. 5, pp. 104–109, June 1958.

[8] Jackson, C., "Transmission line replacements for a lumped element reflectionless filter," in *Radio Wireless Symp.*, San Diego, CA, Jan. 2014.

[9] Morgan, M., and Boyd, T., "Reflectionless filter structures," *IEEE Trans. Microw. Theory Tech.*, vol. 63, no. 4, pp. 1263–1271, April 2015.

[10] Morgan, M., "Transmission line reflectionless filters," U.S. Patent Application No. 14/927,881, Oct. 30, 2015.

Chapter 8

Reflectionless Multiplexers

Much was made at the beginning of this book of the fact that multiplexers are difficult to design properly using conventional topologies. A closed-form solution is rarely adequate, forcing one to rely instead on numerical optimization in order to provide the component values individually for each specific case. Had multiplexers been easier to develop in general form, they would have provided a ready-made solution to creating reflectionless filters. In this chapter, we find that the reverse is true; reflectionless filters, it turns out, suggest a natural and convenient way to create multiplexers [1].

8.1 DIPLEXERS

One of the coincidental features of almost all the reflectionless filter topologies described in this book is that there is a single termination resistor equal to the characteristic impedance of the system that absorbs all of the stop-band energy from one input port, and only from that port, and another symmetric resistor that absorbs all of and only the stop-band energy from the opposite port. This is not a fundamental feature, nor was it prescribed by our derivation; it was not true, for example, of the filters in Chapter 4 that had attenuators in the subnetworks, and others not yet shown. See, for example, Figure 8.1. Although a true reflectionless filter, in that its input match is perfect at all frequencies, and its transmission coefficient is frequency-selective, the stop-band energy it absorbs is distributed among all three resistors, each of which in turn absorbs some energy from both ports[1].

But aside from some contrived examples such as these, the general rule of *one isolated stop-band resistor per input port* typically holds. Most reflectionless filters, then, are a kind of dual-directional diplexer anyway, where the termination

---

[1] To satisfy the curious reader, this particular example is simply our basic, third-order reflectionless filter from Chapter 3, Figure 3.2, with a delta-wye transformation applied to the two resistors and their adjoining capacitor. See Appendix A for a listing of mixed-element delta-wye transformations.

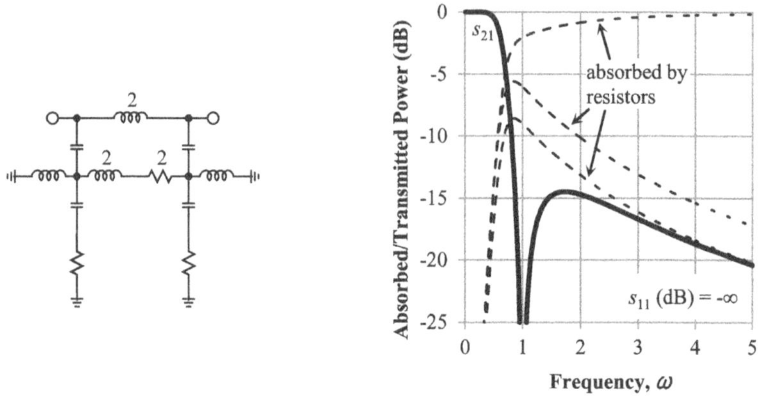

(a) (b)

Figure 8.1 (a) A reflectionless filter that does *not* behave as a diplexer. (b) Frequency response (solid line), and absorption characteristics (dashed lines) for each of the three resistors for inputs from one side.

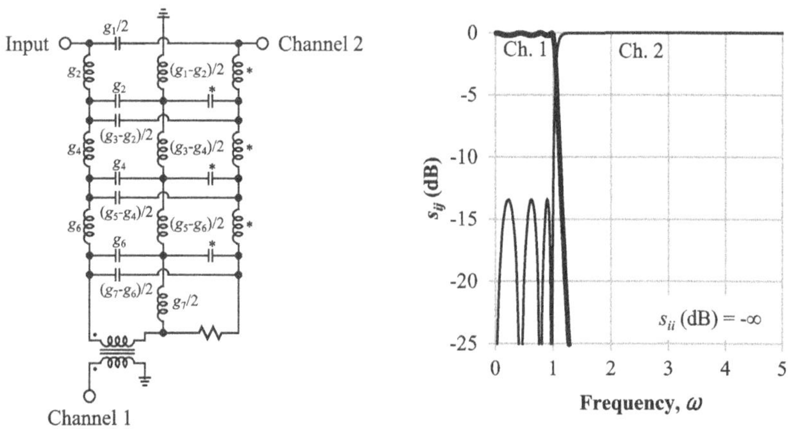

(a) (b)

Figure 8.2 (a) A reflectionless diplexer based on a high-pass filter, and (b) its frequency response.

resistors may be considered as additional ports. We have already seen some evidence of this in Chapter 4 in the context of subnetwork embedding, and again in Chapter 6 when we coupled out one of the terminations as a port in order to realize a Chebyshev Type I filter. In the latter example, had three ports been connected to the structure (as in Figure 8.2), we would have had a natural diplexer with a Chebyshev Type I filter response in one signal path and a Chebyshev Type

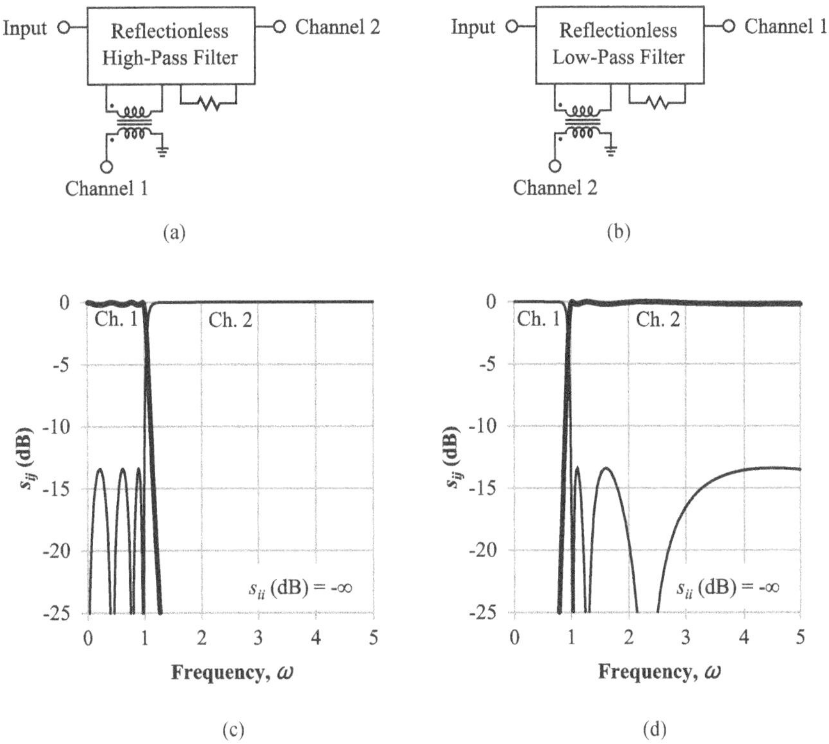

Figure 8.3 Diplexers using (a) reflectionless high-pass and (b) reflectionless low-pass filters. (c) Frequency response for the diplexer in (a). (d) Frequency response for the diplexer in (b). In both cases, the thick line corresponds to the channel that is coupled-out through a transformer, having a Chebyshev Type I response.

II filter in the other path. Moreover, the input return loss from any one of the three ports would be theoretically infinite at all frequencies — a true, reflectionless diplexer.

Whether the core reflectionless filter in this example is a high-pass or low-pass depends on which channel one wishes to have the Chebyshev Type I response, as opposed to Type II. This may be illustrated most simply by the block diagrams in Figure 8.3, where the internal topologies are assumed to be that of a reflectionless Chebyshev filter with $N = 7$. The filter on the left (which is the same as the filter shown more explicitly in Figure 8.2) yields a Type I response in the lower-frequency channel and a Type II response in the upper, whereas the reverse is true of the filter on the right.

The transformers used in these examples, like the other components, are assumed to be ideal at this stage. A practical consideration is that no real transformer can operate down to DC. If channel 1 is truly meant to be low-pass,

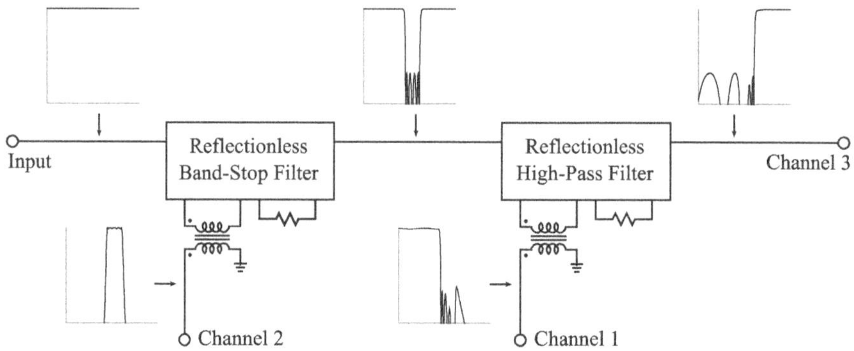

Figure 8.4 Schematic of a triplexer utilizing two reflectionless filters and two transformers. The signal spectra at various points are shown.

then this would suggest that the diplexer has to be of the second type shown in Figure 8.3(b),(d). In other cases, if channel 1 is not actually required to cover baseband, then the first type shown in Figure 8.3(a),(c) may be acceptable. In fact, if strict DC operation is not required, then the diplexer utilizing a high-pass reflectionless filter might actually be better, because the transformer is then only required to work well in channel 1, and good quality transformers are harder to make at higher frequencies.

8.2 TRIPLEXERS

Assuming more than two output channels are required, it makes no sense to try combining multiple reflectionless filters by joining them in parallel at a common junction, as is often the case with conventional multiplexers [2]. Although the input impedance of the reflectionless filter is constant, and therefore better-behaved across frequency than conventional filters, that impedance is not what one wants when paralleling filters in this way. If the system characteristic impedance is 50Ω, and three reflectionless filters are combined in parallel, then the input port would see a combined impedance of (50Ω)/3 = 16.7Ω, causing a frequency-independent but still very large reflection. That could be repaired by scaling each individual filter to match 150Ω, but then each output channel would have to have that characteristic impedance. More importantly, though, the two subfilters that are not active in any particular frequency band would absorb two-thirds of the power that should have been passed through by the remaining filter.

Instead, we create additional channels by cascading reflectionless filters, as shown in Figure 8.4. Figure 8.4 shows that the first filter is configured as a band-stop, which extracts the central channel and directs it to the differential output. This leaves a band hole in the spectrum passed on to the next filter, which in this

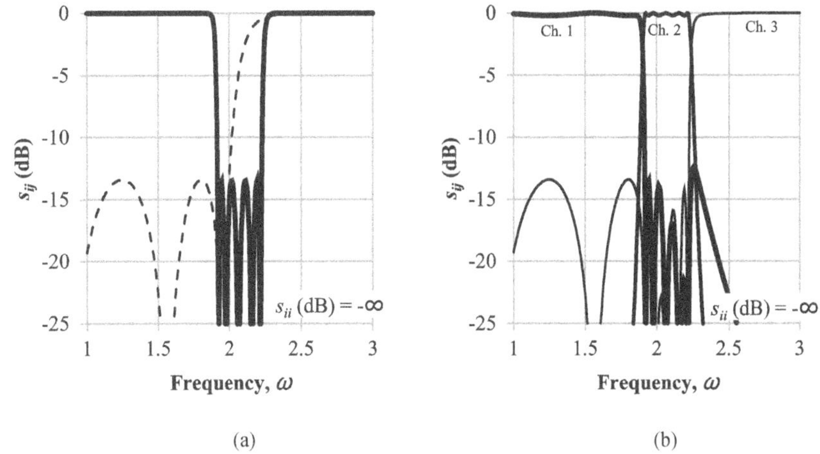

Figure 8.5 (a) High-pass filter (dashed line) applied to the band-suppressed spectrum (solid line) after the first stage of the triplexer. (b) Three final triplexer outputs.

case is a high-pass filter. (See Figure 8.5(a).) The upper half of the spectrum, which corresponds to channel 3, is therefore passed on to the filter's primary output, while the lower half is directed toward its differential port. Assuming generalized Chebyshev filters are used (which is not a requirement), then channel 2 will exhibit a pure Chebyshev Type I band-pass response, channel 3 will be a high-pass Type II (squared), and channel 1 will be the low-pass product of a Type I and Type II. These three outputs are illustrated in Figure 8.5(b).

Virtually all of the selectivity in this design is in the band-stop filter. By the time the spectrum is passed on to the secondary filter, the two remaining channels are separated by a gap, as shown in Figure 8.5(a). This relaxes the selectivity requirement on the second filter, so that less expensive or smaller components may be used. Whether a low-pass or high-pass design is utilized in this stage is essentially a free choice, subject to the same practical concerns discussed in Section 8.1.

8.2.1 Design Example

Let us work through this triplexer as an example. For the band-stop section, we will use a fifth-order Chebyshev Type II reflectionless filter, and the following high-pass section will be seventh order, also Chebyshev Type II. The overall topology is shown in Figure 8.6. The design of such filters has been described previously. In fact, the topology of both sections has been taken from Figure 6.7, along with the appropriate frequency transformation from Table 1.2. Symmetries in the Chebyshev prototype parameters have been exploited to reduce the number of element values that must be calculated. For example, since $g_2 = g_4$ when $N = 5$,

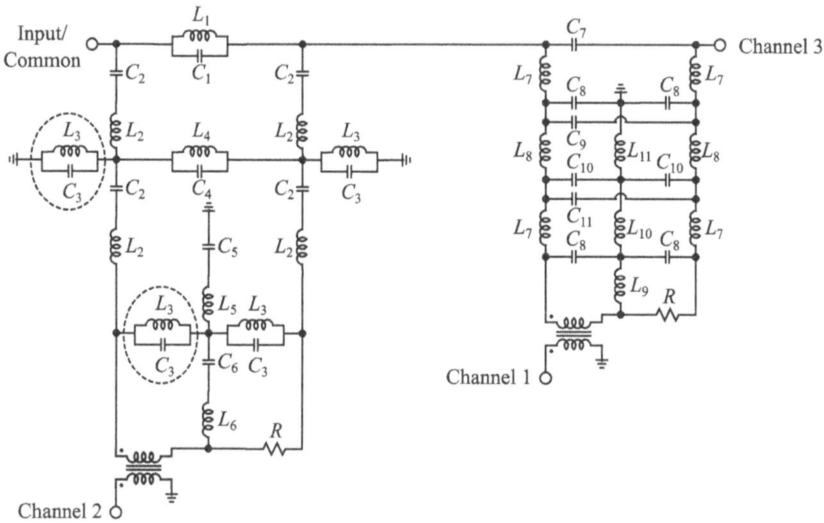

Figure 8.6 Topology for a reflectionless triplexer. The circled resonators are identical due to symmetry of the Chebyshev prototype parameters.

the resonator pairs circled in the band-stop section must be the same. This is but one example of many pairs of elements in these two topologies that exhibit this symmetry.

For the band-stop section, let us use a center frequency of 2.065 GHz, and a fractional bandwidth $\Delta = 0.15$. For the ripple factor, we use $\varepsilon = 0.2164$ in accordance with Table 1.1 and take the prototype parameters from the second row of that table. The element values are therefore given by

$$L_1 = \frac{2Z_0\Delta}{g_2\omega_0} = \frac{2(50\Omega)(0.15)}{(1.337)2\pi(2.065\text{GHz})} = 0.86 \text{ nH} \tag{8.1a}$$

$$C_1 = \frac{g_2Y_0}{2\omega_0\Delta} = \frac{(1.337)(50\Omega)^{-1}}{2\cdot 2\pi(2.065\text{GHz})(0.15)} = 6.87 \text{ pF} \tag{8.1b}$$

$$L_2 = \frac{g_2Z_0}{\omega_0\Delta} = \frac{(1.337)(50\Omega)}{2\pi(2.065\text{GHz})(0.15)} = 34.4 \text{ nH} \tag{8.1c}$$

$$C_2 = \frac{Y_0\Delta}{g_2\omega_0} = \frac{(50\Omega)^{-1}(0.15)}{(1.337)2\pi(2.065\text{GHz})} = 0.17 \text{ pF} \tag{8.1d}$$

$$L_3 = \frac{Z_0\Delta}{g_2\omega_0} = \frac{(50\Omega)(0.15)}{(1.337)2\pi(2.065\text{GHz})} = 0.43\text{ nH} \tag{8.1e}$$

$$C_3 = \frac{g_2 Y_0}{\omega_0\Delta} = \frac{(1.337)(50\Omega)^{-1}}{2\pi(2.065\text{GHz})(0.15)} = 13.7\text{ pF} \tag{8.1f}$$

$$L_4 = \frac{2Z_0\Delta}{(g_3 - g_2)\omega_0} = \frac{2(50\Omega)(0.15)}{(2.164 - 1.337)2\pi(2.065\text{GHz})} = 1.40\text{ nH} \tag{8.1g}$$

$$C_4 = \frac{(g_3 - g_2)Y_0}{2\omega_0\Delta} = \frac{(2.164 - 1.337)(50\Omega)^{-1}}{2\cdot 2\pi(2.065\text{GHz})(0.15)} = 4.25\text{ pF} \tag{8.1h}$$

$$L_5 = \frac{(g_3 - g_4)Z_0}{2\omega_0\Delta} = \frac{(2.164 - 1.337)(50\Omega)}{2\cdot 2\pi(2.065\text{GHz})(0.15)} = 10.6\text{ nH} \tag{8.1i}$$

$$C_5 = \frac{2Y_0\Delta}{(g_3 - g_4)\omega_0} = \frac{2(50\Omega)^{-1}(0.15)}{(2.164 - 1.337)2\pi(2.065\text{GHz})} = 0.56\text{ pF} \tag{8.1j}$$

$$L_6 = \frac{g_4 Z_0}{2\omega_0\Delta} = \frac{(1.337)(50\Omega)}{2\cdot 2\pi(2.065\text{GHz})(0.15)} = 17.2\text{ nH} \tag{8.1k}$$

$$C_6 = \frac{2Y_0\Delta}{g_4\omega_0} = \frac{2(50\Omega)^{-1}(0.15)}{(1.337)2\pi(2.065\text{GHz})} = 0.35\text{ pF} \tag{8.1l}$$

For the high-pass section, we'll use a transition frequency of 2 GHz, and a ripple factor $\varepsilon = 0.2187$ so that we may use the Chebyshev prototype parameters given in the third row of Table 1.1. Thus, we have

$$C_7 = \frac{g_2 Y_0}{2\omega_0} = \frac{(1.377)(50\Omega)^{-1}}{2\cdot 2\pi(2\text{GHz})} = 1.10\text{ pF} \tag{8.2a}$$

$$L_7 = \frac{g_2 Z_0}{\omega_0} = \frac{(1.377)(50\Omega)}{2\pi(2\text{GHz})} = 5.48\text{ nH} \tag{8.2b}$$

$$C_8 = \frac{g_2 Y_0}{\omega_0} = \frac{(1.377)(50\Omega)^{-1}}{2\pi(2\text{GHz})} = 2.19\text{ pF} \tag{8.2c}$$

$$L_8 = \frac{g_4 Z_0}{\omega_0} = \frac{(1.498)(50\Omega)}{2\pi(2\text{GHz})} = 5.96\text{ nH} \tag{8.2d}$$

$$C_9 = \frac{(g_3 - g_2)Y_0}{2\omega_0} = \frac{(2.280 - 1.377)(50\Omega)^{-1}}{2\cdot 2\pi(2\text{GHz})} = 0.72\text{ pF} \tag{8.2e}$$

$$L_9 = \frac{g_6 Z_0}{2\omega_0} = \frac{(1.377)(50\Omega)}{2\cdot 2\pi(2\text{GHz})} = 2.74\text{ nH} \tag{8.2f}$$

$$C_{10} = \frac{g_4 Y_0}{\omega_0} = \frac{(1.498)(50\Omega)^{-1}}{2\pi(2\text{GHz})} = 2.38\text{ pF} \tag{8.2g}$$

$$L_{10} = \frac{(g_5 - g_6)Z_0}{2\omega_0} = \frac{(2.280 - 1.377)(50\Omega)}{2\cdot 2\pi(2\text{GHz})} = 1.80\text{ nH} \tag{8.2h}$$

$$C_{11} = \frac{(g_5 - g_4)Y_0}{2\omega_0} = \frac{(2.280 - 1.498)(50\Omega)^{-1}}{2\cdot 2\pi(2\text{GHz})} = 0.62\text{ pF} \tag{8.2i}$$

$$L_{11} = \frac{(g_3 - g_4)Z_0}{2\omega_0} = \frac{(2.280 - 1.498)(50\Omega)}{2\cdot 2\pi(2\text{GHz})} = 1.56\text{ nH} \tag{8.2j}$$

The termination resistors are trivially given by $R = Z_0 = 50\Omega$. The transformers are both 1:1, and primarily need to have good performance in the pass-bands of the multiplexer channels 1 and 2 for which they are used. Fortunately, LTCC balun transformers are available in this frequency range with very small footprints, e.g., [3].

We may therefore lay out the multiplexer as shown in Figure 8.7. In practice, it would require very high-quality components in the band-stop section to keep the loss down, due to the narrow bandwidth of channel 2. This section is most critical, as it provides the bulk of the selectivity in this configuration.

8.3 MULTIPLEXERS

It should be clear by now that there are many ways in which reflectionless filters

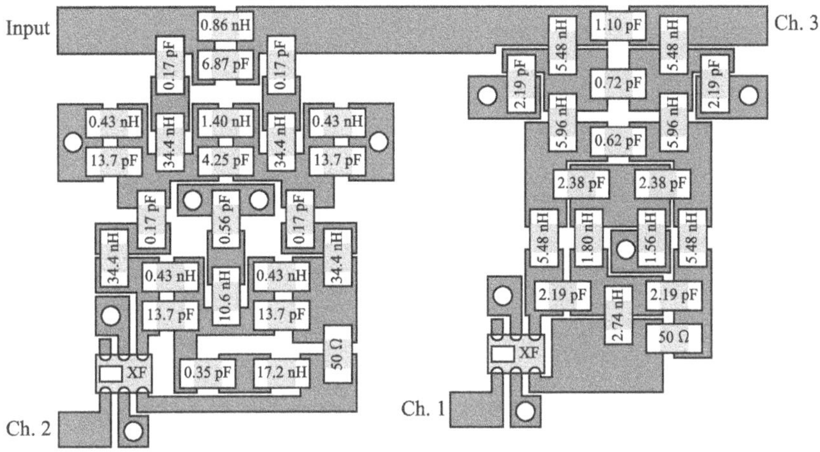

Figure 8.7 Sample layout for the reflectionless triplexer shown in Figure 8.6.

may be chained together to produce multiplexers of all kinds, but it is useful to come up with a standard schema for an arbitrary number of ports. To simplify, let us consider only two possibilities: an array of all low-pass filters, and an array of all high-pass filters.

8.3.1 Low-Pass Arrays

If all low-pass filters are cascaded at their single-ended ports, as has been the case in the examples so far, then they must be arranged in order by cutoff frequency from highest to lowest, lest some channels become blocked by filters earlier in the chain. This fits with our natural preference that the highest-frequency channels are brought out first, as this will minimize the total loss at those frequencies where the loss is typically worst. It also means that the lowest frequency filter need not have perfect performance over the full frequency range of the multiplexer, since only the lowest two frequency channels will reach it. Thus, for an M-channel multiplexer, we start with the highest frequency, channel M, and work our way down, as suggested in Figure 8.8. The baseband channel (channel 1) is emitted last and, not requiring a transformer, may include DC.

Although not explicit in Figure 8.8, the first filters in the array at the highest frequency may need to be of higher order, N, than the latter filters in order to maintain similar cutoff slopes in absolute frequency-space as opposed to relative frequency-space. This is illustrated in Figure 8.9. As the previous schematic shows, a total of four low-pass filters are required to complete a five-channel multiplexer. In Figure 8.9(a), all four multiplexers were of order $N = 7$. These filters in principle have identical pass-band shapes, but are scaled according to frequency. Therefore, the lowest frequency bands have apparently steeper slopes

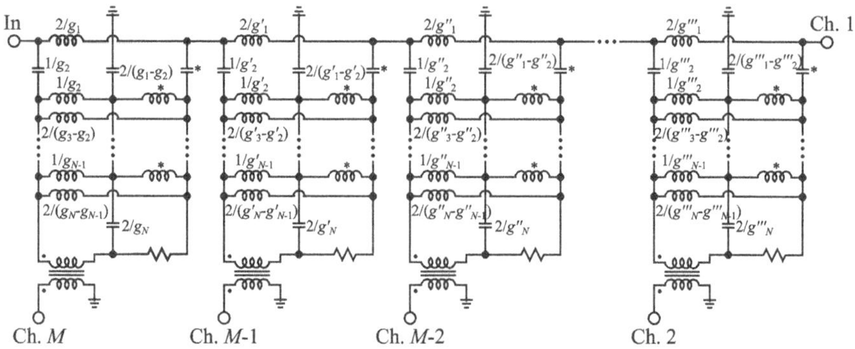

Figure 8.8 An M-channel multiplexer built up using M-1 reflectionless low-pass filters of order N, arranged from highest to lowest in frequency.

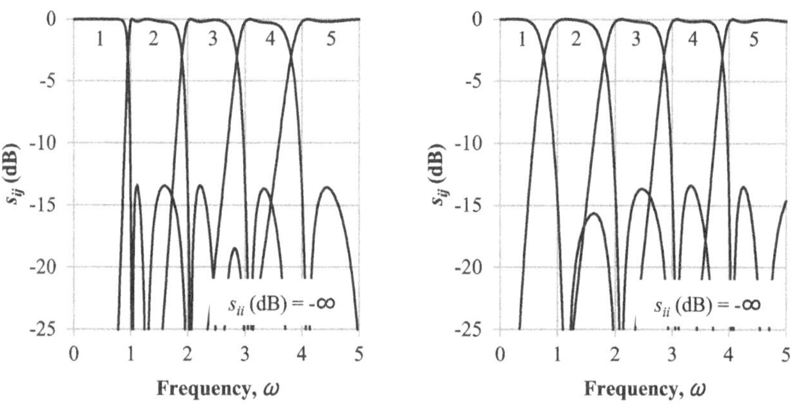

(a) (b)

Figure 8.9 Frequency response of a five-channel multiplexer using all low-pass filters, with (a) $N = 7$ in all cases, or (b) $N = 9$, 7, 5, and 3, respectively.

than the higher frequency bands.

In order to keep approximately equal pass-band shapes *translated* in frequency instead of scaled, the orders have to vary, as in Figure 8.9(b). In this case, the filters from highest to lowest (first to last) are of order $N = 9$, 7, 5, and 3, respectively.

8.3.2 High-Pass Arrays

If high-pass filters are used in an array, and they are cascaded at their single-ended ports, then the lowest frequency filter would have to come first, so that the higher

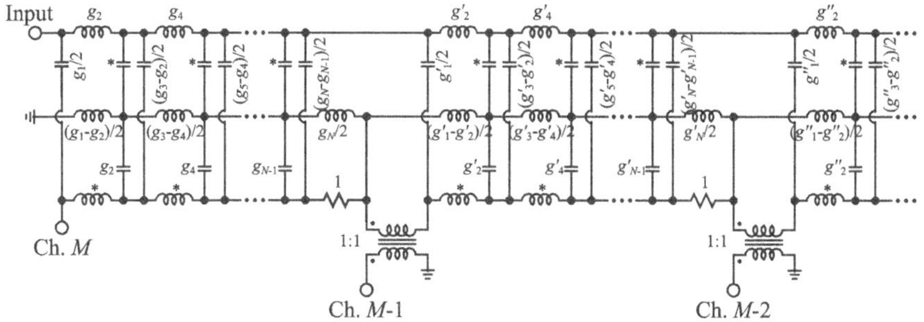

Figure 8.10 An M-channel multiplexer built up using M-1 reflectionless high-pass filters of order N, arranged from highest to lowest in frequency.

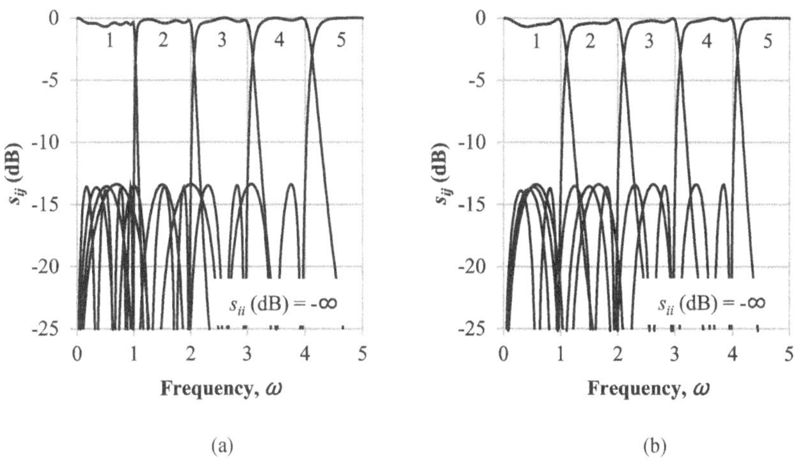

Figure 8.11 Frequency response of a five-channel multiplexer using all high-pass filters, with (a) $N = 9$ in all cases, or (b) $N = 11, 9, 7$, and 5, respectively.

frequency channels are still available to the filters downstream. This would mean that the highest frequency band would pass through all M-1 filters, requiring each one to cover the entire frequency range and maximizing the loss slope across the range of channels.

A better approach is to array the high-pass filters by cascading at their differential ports, as shown in Figure 8.10. This allows the highest-frequency channel to be coupled out first. Importantly, it does not mean that the last (lowest-frequency) channel has to pass through multiple transformers, as they are only needed when converting from differential to single-ended, which is not required in the interior of the multiplexer. It does, however, require the lowest-frequency channel to be coupled out via *one* transformer at its final output, which precludes

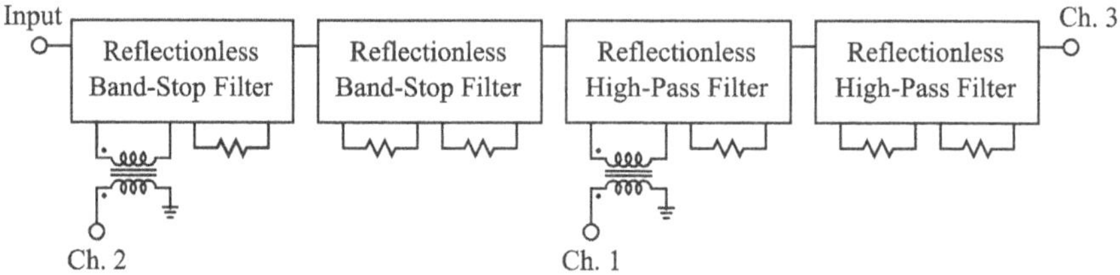

Figure 8.12 Cascaded reflectionless filters in a high-rejection triplexer configuration.

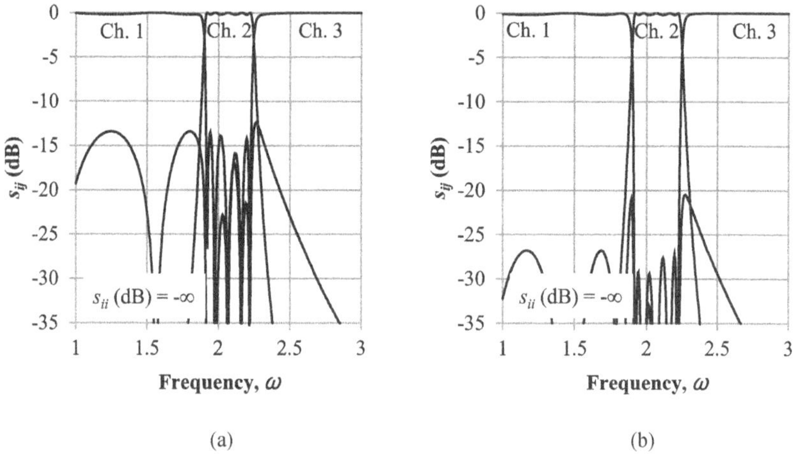

Figure 8.13 Comparison of triplexer performance (a) without, and (b) with cascaded filter sections. The band-stop and high-pass filters are fifth-order and seventh-order Chebyshev designs, respectively.

it in principle from operating at DC.

The same argument regarding fixed or varying filter orders discussed in the previous section applies here as well. A plot of the frequency response for both fixed and varying filter orders is shown in Figure 8.11.

8.4 HIGH-REJECTION MULTIPLEXERS

If higher rejection is needed than is available with a single reflectionless filter stage, then filter stages may be cascaded. Take the triplexer design discussed in Section 8.2. A modified block diagram is shown in Figure 8.12, where each stage has been inserted into the cascade twice. In such cases, the differential output channels should be taken from the first like-filter in the cascade, as that is where the signal strength in the selected band will be strongest. A comparison of triplexer responses with single-stage filters and two-stage filters using the Chebyshev topologies from Chapter 6 is shown in Figure 8.13. For this example, the band-stop filter is a fifth-order Chebyshev design, and the high-pass is a seventh-order Chebyshev.

8.5 CONTIGUOUS CHANNELS

Note that in every case here, the crossover points between multiplexer channels always occur at -3 dB. In that sense, the reflectionless multiplexers demonstrated in this chapter are among a class known as contiguous-channel multiplexers [4–6], where there is no loss (absorption or reflection) of signal at any frequency; all inputs at the common port, no matter what their frequency, are fully coupled to outputs. The remaining termination resistors in the reflectionless multiplexers only absorb energy in the reverse direction, from the channelized ports, maintaining their isolation while still being reflectionless.

References

[1] Morgan, M., *Optimal Response Reflectionless Filters*, U.S. Patent Application No. 15/298,459, October 20, 2016, PCT Application No. PCT/US16/57829, October 20, 2016.

[2] Matthaei, G., Young, L., and Jones, E., *Microwave Filters, Impedance-Matching Networks, and Coupling Structures*, Norwood, MA: Artech House, pp. 991–999, 1980.

[3] *Mini-Circuits datasheet*, NCS1-292+, available at http://www.minicircuits.com/pdfs/NCS1-292+.pdf, downloaded January 2016.

[4] Cristal, E. and Matthaei, G., "A technique for the design of multiplexers having contiguous channels," *IEEE Trans. Microwave Theory Techn.*, vol. 12, no. 1, pp. 88–93, January 1964.

[5] Rauscher, C., Kirchoefer, S., Pond, J., Guyette, A., and Jachowski, D., "A compact ridge-waveguide contiguous-channel frequency multiplexer," *IEEE Trans. Microwave Theory Techn.*, vol. 57, no. 3, March 2009.

[6] Rauscher, C., "Logarithmic-periodic contiguous-channel microwave multiplexers," *IEEE MTT-S Intl. Microw. Symp. Dig.*, vol. 2, pp. 675–678, June 1989.

Chapter 9

Applications of Reflectionless Filters

Having now established a rich theoretical framework for creating reflectionless filters of many kinds and with many different responses, it is time to take a look at how these can be applied to the benefit of our electronic systems. Fundamentally, this will deal with the safe dissipation of out-of-band signal power in areas where its presence could negatively impact performance. As novices we tend to think of out-of-band signals as something that can safely be ignored, but hard experience eventually teaches us that this is not the case. We will review some of those cases in this chapter and illustrate the benefits of using reflectionless filters.

9.1 WITH UP-/DOWN-CONVERTERS

One of the most common places to find a filter is next to an up- or down-converting mixer. Mixers, of course, are designed specifically to translate bandwidth from one spectral location to another, thus creating the need for a filter to suppress leakage into the wrong part of the spectrum, or to reject other spurious signals that mix in from other frequencies (e.g., image bands). Unfortunately, this inherent cross-coupling of frequencies is also why mixers' performance suffers from the often poorly controlled out-of-band terminations presented to them by conventional filter topologies.

9.1.1 Image Termination

In principle, just as the electrical behavior of a linear device depends on the embedding impedance presented to it at its ports, the behavior of any *non*linear device is dependent upon the termination impedance presented to it at its ports and at all harmonics and mixing products of the operational frequencies that are incident upon it [1–3]. In practice, depending on the strength of the nonlinearity, only the first few harmonics and/or mixing products need to be taken into account. This is the basis of the frequency-domain harmonic balance method of nonlinear

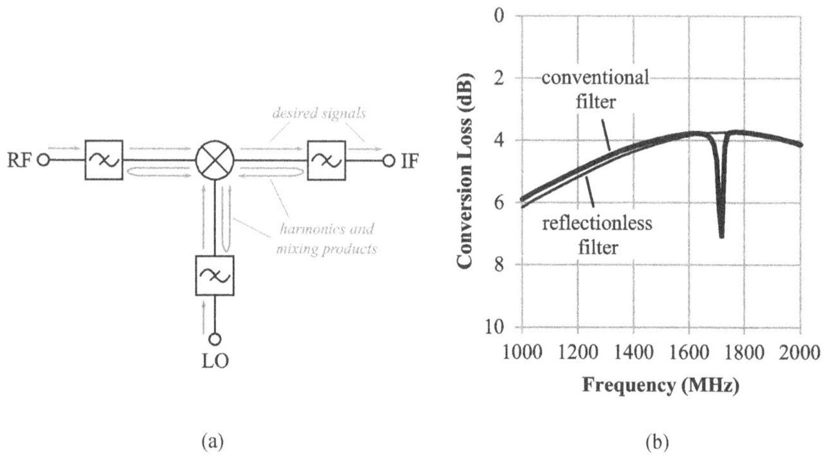

Figure 9.1 (a) Block diagram of a mixer with conventional filters on its ports. (b) Simulated conversion loss with standing waves on the IF port causing a null to appear.

analysis[1]. This technique works on the conservation of power entering and leaving the device at all its ports and at different frequencies, taking into account frequency-conversion effects as well as the signals reflected back into the device after being emitted from it in the first place. This is an iterative process. Because signals entering at any port are subject to conversion and re-emission at disparate frequencies, a potentially large number of conversions, reflections, and re-emissions must be accounted for before the solution converges upon a meaningful steady-state [2].

This illustrates why the behavior of mixers is critically dependent on the reactive loading presented to them out of band by the adjoining filters. The phase and the strength of the reflection plays a role in the convergence of the nonlinear steady state, thus making the ensemble sensitive to a number of factors, such as the uniformity of filter characteristics in multichannel systems, the electrical length of the connecting line between the components, the temperature stability of the basic elements, and even the isolation between ports, which in turn, is affected by the circuit housing.

The consequences of this sensitivity to fabrication and operating conditions can be subtle, affecting only the most stringent performance requirements, or they can be severe. (See the example in Figure 9.1.) With an unlucky choice of electrical length between the filters and the mixer, standing waves in the higher-harmonics or mixing products may interfere to create a null or *suckout* in conversion efficiency. Even if a deep null such as this does not fully develop, the

---

[1] There are other methods, such as the Volterra Series operating in the time domain [2], but these necessarily capture the same basic principles.

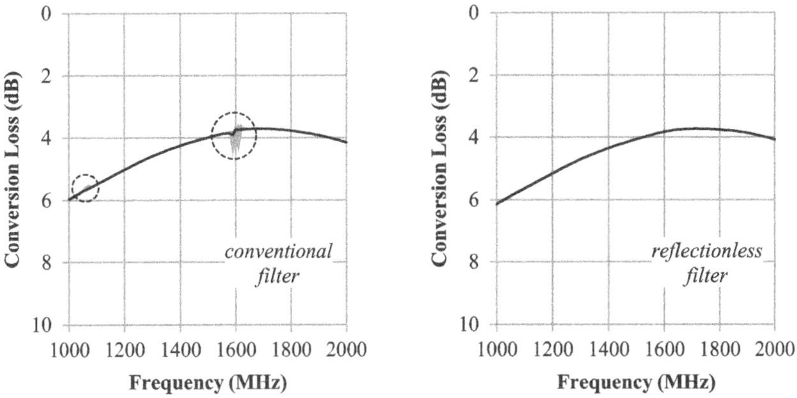

(a) (b)

Figure 9.2 Sensitivity analysis for the case when a deep null is *not* present. Electrical length between the mixer and IF filter was allowed to vary by 1%. (a) Conventional filter. (b) Reflectionless filter.

same effects can lead to extreme sensitivity of the frequency response to temperature or other operating conditions at distinct frequencies. Figure 9.2(a) shows the conversion loss for the same mixer where the electrical length between the mixer and filter is such that a deep null is not present, but the line length was allowed to vary by 1% from nominal. Extreme sensitivity to this line length is seen at around 1050 MHz and 1600 MHz. Using reflectionless filters, as in Figure 9.2(b), substantially mitigates these problems [4, 5].

9.1.2 Image Rejection Stability

The sensitivity of downconverter performance to out-of-band terminations is further illustrated by a practical example. Consider the downconverter shown in Figure 9.3(a)–(b). Configured as an I/Q-mixer, the in-phase and quadrature IF outputs were designed to be independently sampled, and then the upper and lower sidebands reconstructed digitally after a numerical calibration of the amplitude and phase balance.

In the first iteration, conventional anti-aliasing filters were used in conjunction with the I and Q sub-mixers. While the overall conversion loss of the downconverter was measured to be smooth as a function of frequency, the calibration data revealed complex frequency structure in the relative upper- and lower-sideband contributions to that overall curve. The calibration data is plotted in Figure 9.3(c) and Figure 9.3(d), with dotted lines for the initial calibration using conventional filters, and the solid lines for the revised module using reflectionless filters. The dotted lines show fast variations with frequency in both amplitude and phase, leading to instability in the calibration coefficients as a function of time and

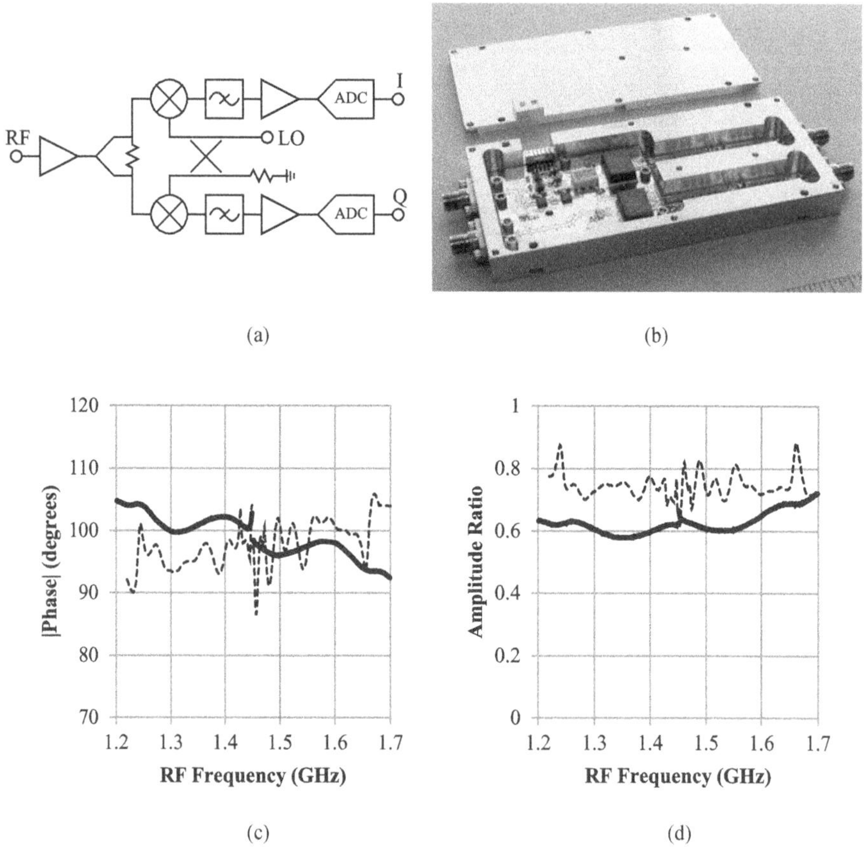

Figure 9.3 (a) Schematic of an I/Q downconverter. (b) Photograph of completed module. (c) Phase and (d) amplitude balance calibration curves with the initial module using conventional filters (dotted lines) and with the revised module using reflectionless filters (solid lines).

temperature.

After revising the module to use reflectionless filters, the downconverter achieved better than 50 dB sideband separation with exceptional calibration longevity. Tests showed that this level of sideband rejection was maintained over temperature swings of 12 degrees Celsius, and over a period of months, without recalibration [6].

9.1.3 Intermodulation Expansion

In addition to the direct impact on conversion efficiency, the recirculation of mixing products out of the mixer and back into it again effectively multiplies the opportunity for the generation of high-order spurious tones. This is sometimes

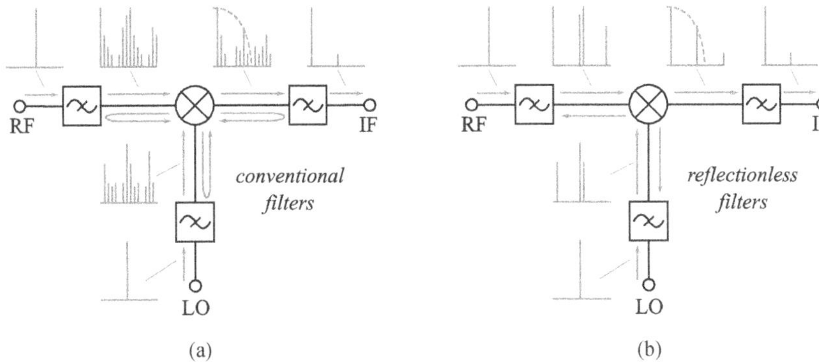

Figure 9.4 Illustration of intermodulation products with (a) conventional filters, leading to family members between the mixer and the filters, and with (b) reflectionless filters, leaving only the first-order tones and mixing products. Although the filtered output spectra may look very similar, this belies a much more intricate and unstable interaction between the mixer and conventional/reflective filters than with reflectionless filters.

referred to as *family members* [5] as indicated in Figure 9.4. The rich forest of spurs that develops in the pre-filtered output spectrum reveal a complex and spectrally diverse interaction between the mixer and its neighboring components, which is difficult to model or accurately predict. It is important to note that even if a conventional filter significantly cleans up the spectrum, hiding all but the strongest of spurs from appearing at the system output, this intricate interaction between components and frequencies is still present. The conventional filter has done nothing to alleviate the sensitivity of system performance to these spurious interactions; it has simply hidden the evidence of the cause.

Those spurs that do get through the filter (and that are still influenced by their interaction with the hidden high-order spurs, even at a low level) will directly impact the dynamic range with which sensitive measurements may be carried out. Intermodulation tests show that the third-order intermodulation product (IP3) of a mixer with a reflectionless filter attached is more consistent with that of the bare mixer than is the case when a conventional filter is used. The relatively smooth IP3 curve of the bare mixer develops a complex frequency structure when paired with a conventional filter, becoming better at some frequencies and worse at others, by as much as 5 to 10 dB [4, 5]. This forces the broadband system designer to utilize a much higher intrinsic-IP3 mixer than would otherwise be needed in order to meet system specifications. With the reflectionless filter, the IP3 of the mixer and of the mixer-filter combination are one and the same.

9.2 WITH MULTIPLIER CHAINS

Multiplier chains are another example in which filters are commonly used along-

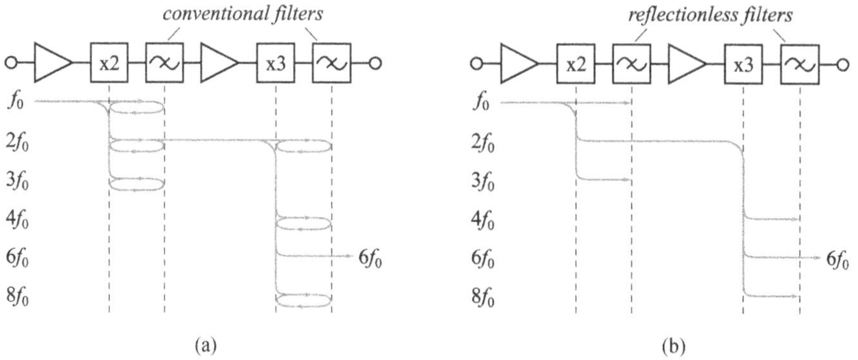

Figure 9.5 Illustration of an active multiplier chain with (a) conventional filters causing standing waves between the filters and the multipliers, and (b) with reflectionless filters having no standing waves.

side nonlinear, frequency-translating devices. It is often not practical to generate a high harmonic number within a single multiplier, for reasons of power and efficiency, as well as the potential for closely spaced spurious harmonics in the output. A better solution is to multiply in stages with doublers and/or triplers, re-amplifying before each multiplier in order to provide sufficient power, and filtering after each one so that each new multiplier sees only a single tone and therefore cannot generate closely spaced mixing products [7–9].

The problem with doing this using conventional filters is that the spurious harmonics do not disappear, but instead get reflected back into the multiplier. As with mixers, the multiplier behavior also is influenced by the reactive loading presented to it at the harmonic frequencies. Furthermore, the output return loss of multipliers is notoriously poor. This is a perfect setup for standing waves to develop between the multiplier and the filters, creating large ripples in the conversion efficiency and subjecting the system to the same kind of environmental sensitivity that was previously described in connection with mixers.

This situation is illustrated in Figure 9.5(a). After the first doubler, several harmonics are present, such as the fundamental, second, and third. The second harmonic is allowed to pass through the filter, but the fundamental and third harmonic are blocked. With a conventional filter, the power in these harmonics has nowhere to go but back toward the doubler, where it can either be reabsorbed or bounced back again toward the filter. Ultimately, the energy has to go somewhere. Some tiny bit may leak through the filter, or radiate away into the housing, neither of which is a particularly desirable outcome. A small fraction may get absorbed by the losses in the connecting line or the filter, but given that filters are often designed using the best-quality, low-loss components available to achieve the necessary selectivity, the standing wave would have to build up to a significant amplitude before reaching equilibrium. The lion's share of the power will almost certainly get re-absorbed by the doubler, where it will undoubtedly modulate the conversion efficiency with a strong frequency and level dependence.

Thus, the second harmonic finally leaves the doubler stage to be amplified with the ripples and other effects of environmental sensitivity imprinted upon it, only to go through another round of multiplication and multiple-reflection-modulation between the tripler and its associated filter. Perhaps it is not so surprising that flat conversion efficiency is so elusive in multiplier chains such as these.

Reflectionless filters at least offer a partial solution to these problems, as shown in Figure 9.5(b). While the in-band return loss of multipliers is still an issue, causing some interaction between the multiplier and the surrounding amplifiers, the spurious harmonics are at least well terminated with frequency- and temperature-stable absorptive loads that do not allow standing waves to develop, and certainly not to grow excessively in amplitude. The harmonic energy has been *removed* from the system by the filters instead of being trapped in it by them.

9.3 WITH BROADBAND AMPLIFIERS

It is not only the intentionally nonlinear devices that are improved by pairing with reflectionless filters. Linear gain blocks may also exhibit symptoms associated with poor out-of-band reactive terminations.

Consider that most amplifiers exhibit positive gain over significantly broader bandwidth than that which is their intended use, as is necessary to achieve the desired gain flatness. The roll-off in gain of general-purpose amplifiers is much slower than the typical filters with which they are paired. This too leads to the trapping of out-of-band signals as described above, including spurious harmonics, external interferers, or even thermal noise, which accumulates not only via the standing-wave mechanism, but by repeated reamplification if the output-to-input isolation of the amplifier or its housing is not very good. Under the right conditions, this *regenerative* amplification has been observed to create unusually strong noise bands at localized frequencies, if not outright oscillations, where the amplification loop completes a full rotation in phase. Even if oscillations are not present, or readily apparent, the presence of these strong noise bands can degrade dynamic range and introduce amplitude and phase drifts that undermine the overall system performance.

Again, reflectionless filters alleviate the problem by providing a mechanism for removing the unwanted signals before they grow to such a level that they can impact the other components. In high-performance systems, this has been found to be a critical aspect of meeting the most stringent amplitude and phase stability requirements [6].

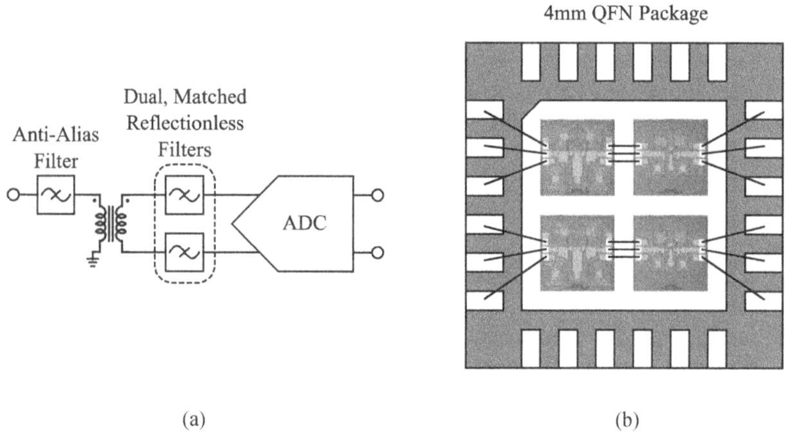

Figure 9.6 (a) Schematic of an ADC with a preceding dual-channel reflectionless filter to suppress out-of-band switching transients. (b) Bonding diagram for the reflectionless filter in a 4 mm QFN package using monolithic chips for channel matching.

9.4 WITH ANALOG-TO-DIGITAL CONVERTERS

In the front-end of most analog-to-digital converters (ADCs) is a sample-and-hold or track-and-hold amplifier that captures the input voltage at the instant of the clock edge and holds it steady for the subsequent conversion to a digital output. The switching action of this input stage inevitably presents an abrupt change in load conditions to the driving circuit and may additionally generate transient pulses that propagate backward into the analog system. These transients may extend well above the intended operating frequency range, such that a conventional anti-aliasing filter would reject them. Rather than eliminating these aliased signals, however, the filter unwittingly sends them directly back into the ADC where they can do the most damage. Switching transients such as these may become the most significant limiting factor in the spur-free dynamic range of high-precision samplers.

Reflectionless filters once again provide a solution to this problem. By absorbing out-of-band signals at both the input and output side, the filter removes these unwanted transients from the spectrum harmlessly, rather than bouncing them back into the ADC where they were first generated. An example is shown in Figure 9.6, where the ADC is preceded by a pair of reflectionless filters. By using two filters on the differential input of the ADC, both differential and common-mode transients are absorbed. The filters are implemented monolithically to ensure good matching in both channels, so that the input impedance presented to the ADC is well balanced. A sharper anti-aliasing filter using higher Q elements

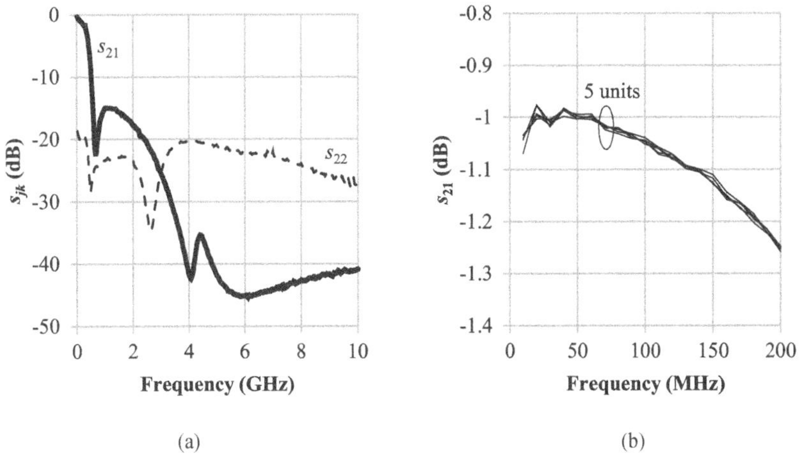

Figure 9.7 (a) Plot of measured transmission and reflection coefficient of dual-channel filter for ADC. (b) Detail of pass-band insertion loss for 5 units.

may be needed to provide maximum selectivity, which could be conventional or reflectionless, and may be single-ended as shown.

Measured data on the pair of filters chosen for this example is plotted in Figure 9.7(a) (this is essentially the same data as shown earlier in Figure 4.14). A high-frequency low-pass and a low-frequency low-pass filter are combined in each channel to provide good return loss with a fairly extended stop-band. In this particular example, the sample rate of the ADC is assumed to be only a few hundred MHz, however the filter provides absorption of transients up to 10 GHz (and beyond). The monolithic fabrication of the individual filter chips ensures good uniformity and balance between the two channels. This is best illustrated by the extreme repeatability for five different chips in Figure 9.7(b). The standard deviation of the insertion loss in the pass-band is less than 0.01 dB.

9.5 DISTRIBUTED FILTERING

As Sections 9.1–9.4 have hopefully made clear, the good broadband impedance match that reflectionless filters have makes them well-suited for pairing with almost any kind of device. This inherent cascadability [10] is a tremendous advantage when it comes to signal-path optimization, for it grants the designers complete freedom to locate the filters in their system where they can do the most good without fear of adverse interactions between components.

Consider a baseband filtering application such as that shown in Figure 9.8. Individual reflectionless filters may not have sufficient rejection on their own, leading one to cascade several together — something that, as was discussed in

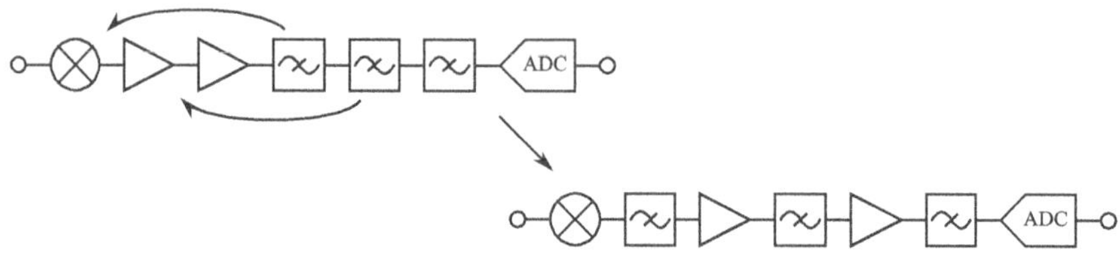

Figure 9.8 Example of distributed filtering in a baseband anti-aliasing application.

Sections 2.1 and 4.5, would not be practical with conventional filters. If we follow our usual instincts, these would all be combined in one place wherever they are perceived to do the most good. For example, one might choose to locate them after the final amplifier and before the ADC. This would tend to optimize the sensitivity by minimizing losses in front of the amplifiers. Alternatively, one could put them all ahead of the amplifiers to optimize dynamic range.

However, it is usually more optimal to avoid either of these extremes and spread the filters out in the signal path. This will help in a number of ways. For one, it keeps the strength of the out-of-band signals down at all points in the chain, rather than letting them build up to a maximum value before finally filtering them. This will inevitably improve dynamic range. It also provides a measure of isolation between the other components out of band, eliminating some harmful interactions that they might otherwise have had with one another.

9.6 WITH CONVENTIONAL FILTERS

In those situations where the performance of a reflectionless filter is not quite sufficient for the application, it is useful to think about cascading one with a conventional filter. The high-performance filter may provide the steep transition or deep rejection needed, while the reflectionless filter acts as a pad in the stop-band without adding any substantial loss to the pass-band. An example is shown in Figure 9.9. In this case, a low-pass Chebyshev Type I conventional filter of eleventh-order has been cascaded with a low-pass Chebyshev Type II reflectionless filter of seventh order.

In Figure 9.9(b) the cutoff frequency of the reflectionless filter has been tuned just slightly below that of the conventional filter, to ensure that good return loss is maintained at all frequencies. The transition-response is dominated by the seventh-order Chebyshev Type II filter. In Figure 9.9(c) the cutoff of the reflectionless filter is slightly higher than that of the conventional filter, allowing the Chebyshev Type I transition response to be realized, but allowing a reflection peak around that transition.

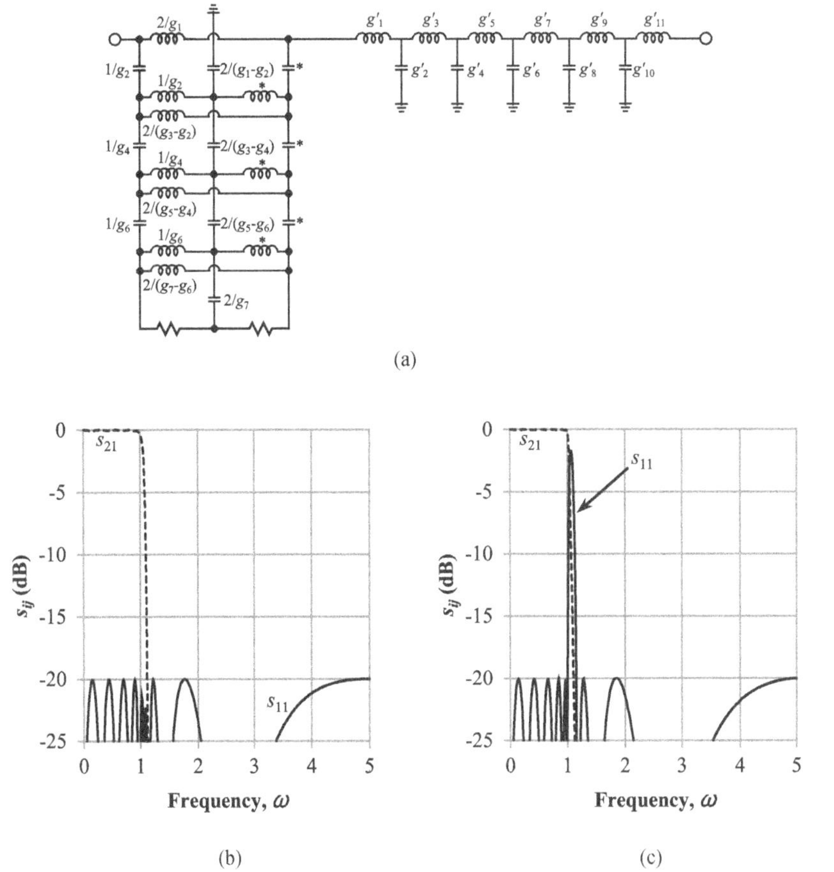

Figure 9.9 (a) Cascade of a low-pass Chebyshev Type II reflectionless filter of seventh order with a low-pass Chebyshev Type I conventional filter of eleventh order. (b) Frequency response and return loss when the reflectionless filter is tuned slightly lower than the conventional filter. (c) Frequency response and return loss when the reflectionless filter is tuned slightly higher than the conventional filter.

In both cases, the deep rejection in the stop-band is provided by the conventional filter while the reflectionless filter absorbs most of the stop-band energy. The ripple factor for the conventional filter has been set to $\varepsilon = 0.1$. This leads to a pass-band insertion loss ripple amplitude, IL_r, of

$$IL_r = 10\log\left(1+\varepsilon^2\right) = 10\log\left(1+0.1^2\right) \approx 0.043 \text{ dB} \tag{9.1}$$

and, by conservation of energy, a pass-band return loss amplitude, RL_r, of

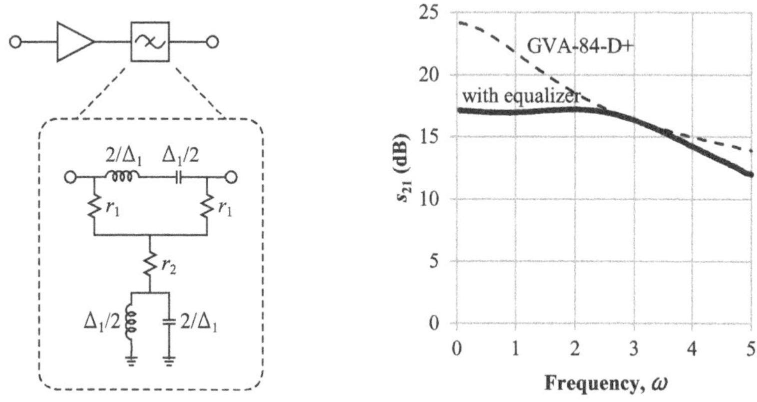

(a) (b)

Figure 9.10 (a) An amplifier paired with a lumped-element slope-equalizer. (b) Gain of the amplifier (dashed line) and the amplifier-equalizer pair (solid line).

$$RL_r = 10\log\left(\frac{1+\varepsilon^2}{\varepsilon^2}\right) = 10\log\left(\frac{1+0.1^2}{0.1^2}\right) \approx 20\text{ dB} \tag{9.2}$$

The return loss in the stop-band will be very high on the side of the conventional filter. On the side of the reflectionless filter, however, the return loss will depend primarily on the reflectionless filter's stop-band attenuation. To match the pass-band return loss of 20 dB, we only need 10 dB attenuation in the reflectionless filter, which according to the results of Section 1.4.4 corresponds to a ripple factor of $\varepsilon = 0.3333$. This is higher than the limiting ripple factor for reflectionless filters listed in Table 6.1 (for any order), so the ripple factor in this case has been deliberately relaxed. By relaxing the ripple factor, we get a steeper overall response in Figure 9.9(b), and a narrower reflection peak in Figure 9.9(c).

9.7 SLOPE EQUALIZERS

Section 4.5.5 showed us how to make slope equalizers out of very simple reflectionless filter networks. As an example of how these could be applied in a realistic application, consider a requirement for baseband amplification. Most single-stage amplifiers intrinsically have much higher gain at low frequency than they do at higher frequencies. For broadband applications, then, this could lead to slopes of 10 dB or more, which could seriously affect the dynamic range of the components downstream.

Take for example the GVA-84-D+ amplifier die available from Mini-Circuits [11]. A high gain amplifier, it nevertheless introduces a gain slope of about 8 dB from DC to 3 GHz, as indicated by the dashed line in Figure 9.10(b). If flatter gain was desired over this frequency range, the reflectionless network shown in Figure 4.16(b) provides a very simple solution. Tuned in this case for $A = 7$ dB, and $\Delta = 30/A \approx 4.3$, the amplifier-equalizer pair achieves a very flat gain of 16.8 ± 0.4 dB from DC to 3 GHz.

9.8 TIME-DOMAIN APPLICATIONS

Aside from the brief introduction of group delay in Chapter 1 and the measurements thereof in Chapter 3, little has been made of the time-domain response of reflectionless filters throughout this book. In part, this is because the time-domain response of a reflectionless filter is exactly the same as that achieved by a conventional filter implementing the same transfer function, so there is nothing new to say in this arena. On the other hand, the Chebyshev Type II response (which is the most readily achievable using reflectionless filters) has some unique time-domain properties that are not often reported in other works due to their comparative obscurity next to Chebyshev Type I filters. Additionally, the effects of signal distortion (which are the primary concern of time-domain applications) are themselves further aggravated by reflections between components. This problem is often overlooked in analytical treatments, but it is one that the reflectionless topologies in this book are uniquely capable of addressing.

9.8.1 Measures of Performance in the Time Domain

As a general rule, there is an inherent trade-off between frequency-domain and time-domain characteristics in filter design. That is, filters with sharper transitions in frequency will typically have larger group delays, especially near the band edge, and a tendency toward excessive ringing in the time-domain. Filters with very flat delays and little ringing, on the other hand, will have quite poor selectivity in the frequency-domain (e.g., the absorptive Gaussian filters described in Section 2.4.2). The best filter for a given situation will depend heavily on the specifications, and is almost always a compromise between frequency- and time-domain characteristics.

Reflectionless filters offer a new option to time-domain system designers, one that is capable of greater selectivity than the Gaussian or Bessel-Thompson filters, but with similar immunity to reflections as the absorptive implementations of these, and with better time-domain characteristics than comparably selective Chebyshev Type I filters or elliptic filters of conventional design.

It is a fact that Chebyshev Type I and Type II responses of the same order achieve the same selectivity given equal, flat insertion-loss and rejection limits.

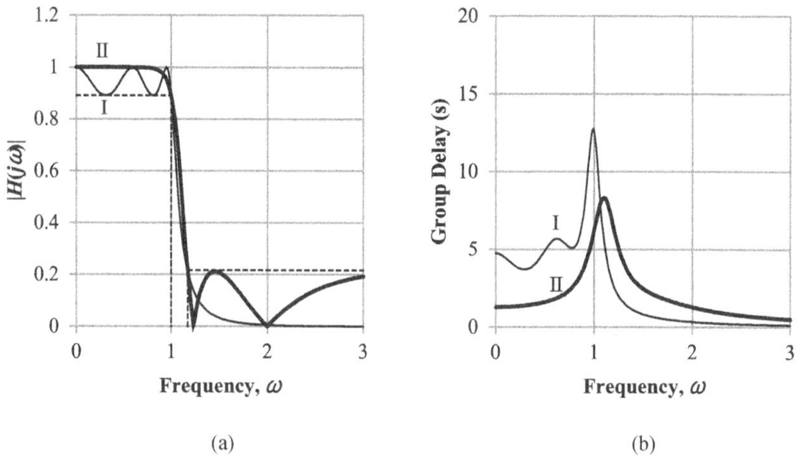

Figure 9.11 (a) Frequency response of Type I and Type II Chebyshev filters having equal, flat insertion-loss and rejection specifications, for $N = 5$. (b) Group delay for these two filters ($\omega_0 = 1$ rad/s.)

This is illustrated in Figure 9.11(a) for fifth-order filters of both types. While the selectivity is the same, Figure 9.11(b) reveals that there are some differences in the delay characteristics of the two filters. First, the Chebyshev Type I filter has ripples in its group delay that contribute to the overall distortion of time-domain signals, whereas the group delay of the Type II filter is smoother and monotonic in its pass-band. Second, the Type II filter has less delay overall, or latency, a feature that becomes even more prominent as the order (and hence selectivity) is increased.

Performance in the time domain is most often characterized by the impulse- and step-responses of the transfer function. The impulse response, $h(t)$, is the output of the system when the input is a unit impulse (mathematically, a Dirac delta function, $\delta(t)$) at time $t = 0$. It is given by the inverse Laplace transform of the transfer function,

$$h(t) = L^{-1}\{H\}(t) = \frac{1}{j2\pi}\int_{\sigma-j\infty}^{\sigma+j\infty} e^{st} H(s) ds \tag{9.3}$$

where σ is chosen such that path of integration lies in the region of convergence [12–14]. As a consequence of elementary system theory, the response of the filter to any arbitrary input waveform (in the time domain) is the convolution of that waveform with the impulse response.

On the other hand, the step response, $g(t)$, is the output of the system when its input is a unit step (or Heaviside step function, $u(t)$) at time $t = 0$. It is given by the

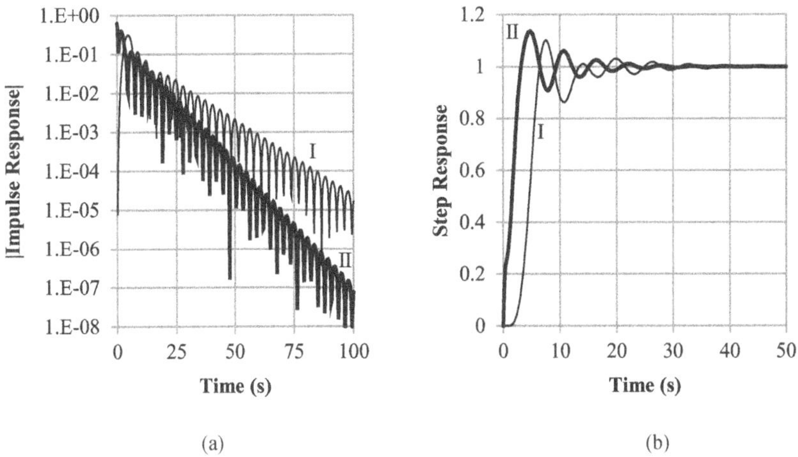

Figure 9.12 (a) Magnitude of the impulse response for fifth-order Chebyshev filters, Type I (thin line) and Type II (thick line). (b) Step responses.

integral of the impulse response, which by the properties of the Laplace transform is equivalent to taking the inverse transform of $s^{-1}H(s)$,

$$g(t) = L^{-1}\left\{\int_0^t h(\tau)d\tau\right\} = \frac{1}{j2\pi}\int_{\sigma-j\infty}^{\sigma+j\infty} e^{st}s^{-1}H(s)ds \tag{9.4}$$

The impulse- and step-responses of our two Chebyshev filters are shown in Figure 9.12. Somewhat unconventionally, the impulse response is shown here in magnitude only, so that it may be plotted on a logarithmic scale to better emphasize the different ringing characteristics of the two filters. This refers to the extended oscillatory response of the filters at long timescales, well beyond the point at which the input has stopped. It is quite clear from this plot that the ringing of the Type II filters decays more rapidly than that of the Type I.

The step responses in Figure 9.12(b) reveal that the first ripple, or overshoot, is about the same magnitude in both cases, or slightly larger for the Type II given these parameters (e.g., order and ripple factor). The step responses also highlight the faster response of the Type II filter as it first reaches the final output value in about half the time taken by the Type I filter. This is essentially the same feature that we gleaned from the group delay curves previously, and will be discussed in greater detail in the next section.

9.8.2 Latency

The transit-time for an in-band waveform to pass through the filter, or latency, is

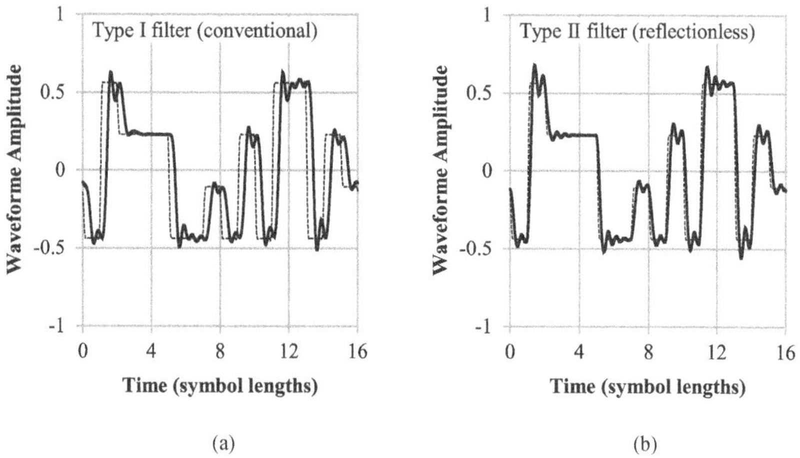

Figure 9.13 An arbitrary waveform passing through (a) Type I, and (b) Type II Chebyshev filters of fifth order. The input waveform is shown with the dashed lines, while the outputs are shown with solid lines.

of no consequence for many filter applications, but there are some real-time systems for which it matters a great deal (e.g., audio and feedback-control electronics). It was noted in Figure 1.9(b) that the overall group delay of a Chebyshev Type II filter actually decreases with increasing order, in stark contrast to the other canonical filters types described. This potentially makes the Chebyshev Type II the filter of choice where latency is a dominant concern. Combined with the antireflection properties of the Chebyshev Type II topologies described in this book, there is a compelling case for using such filters in real-time systems.

Consider the arbitrary waveform shown in Figure 9.13. It consists of a series of steps, alternating between four different amplitude levels at regular intervals. This is characteristic of a two-bit digital modulation. A wide variety of meaningful time-domain waveforms could have been used, but this is sufficient to illustrate the properties with which we are presently concerned. Note that in both the Type I and Type II filters, the output waveform tracks that of the input, with roughly the same degree of oscillation around the target point at each level. However, the output waveform from the Type I filter is shifted to the right, or delayed, relative to the input. The latency of the Type II filter is noticeably less, alluding to a system that can respond to its input much closer to real time.

9.8.3 Intersymbol Interference

In the preceding sections, it was illustrated that the overshoot, or initial ripples, in the impulse- and step-responses of the Type II filter were slightly larger than for

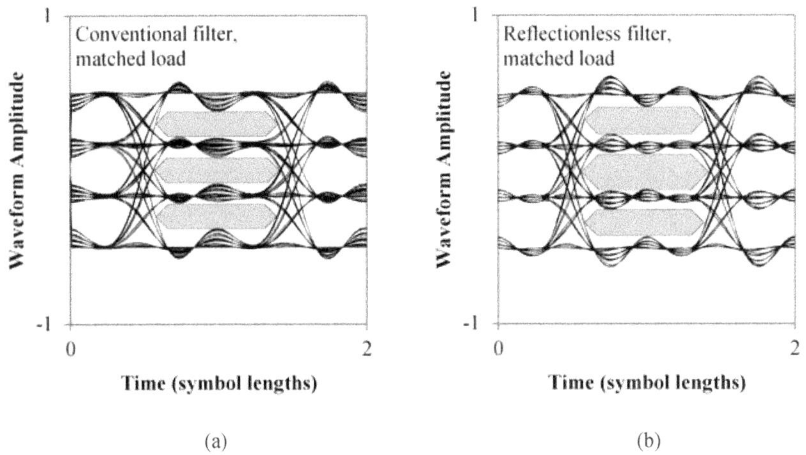

Figure 9.14 (a) Eye diagram for Chebyshev (a) Type I, conventional and (b) Type II, reflectionless filters using 2-bit modulation. In both cases, the port terminations of the filters are matched.

the Type I, but that the subsequent ripples decayed much faster. To see the impact this has on actual systems, we must consider how the past history of a waveform affects the current output at a given time.

A closer examination of the curves from Figure 9.13 will show that the ripples at a given level are dependent not only on the current or most recent input step, but on past input values as well. In digitally modulated systems, this property is known as intersymbol interference (ISI) [15]. However, it is important to point out that we highlight the ISI formalization here for illustration only. The same general principles will apply to a wide variety of systems and waveforms where information is contained in the time domain (e.g., multiple echoes from a radar or time-gated spectroscopic measurement).

The curves in Figure 9.13 were plotted as a function of time in units of symbol lengths; essentially, the dwell time of the signal at a given level. Another useful way to plot this information is to reduce the horizontal axis to only one or two symbol lengths, and then overlay multiple symbol periods on top of one another on the same graph. In this way, the clear distinction between digital levels in the aggregate case for a large number of bits can be visualized. This is known in digital communication circles as an eye diagram [16, 17].

Eye diagrams for the two types of filters with two-bit modulation are shown in Figure 9.14. The shaded grey areas represent the opening of the eyes at the critical time intervals during which a logic circuit downstream will have to determine which level the system is attempting to transmit. While the two plots are very similar, a close examination reveals that the eyes are slightly more open in Figure 9.14(b), representing the Type II reflectionless filter, especially the central eye between levels 2 and 3 of the waveform. This is because the previous

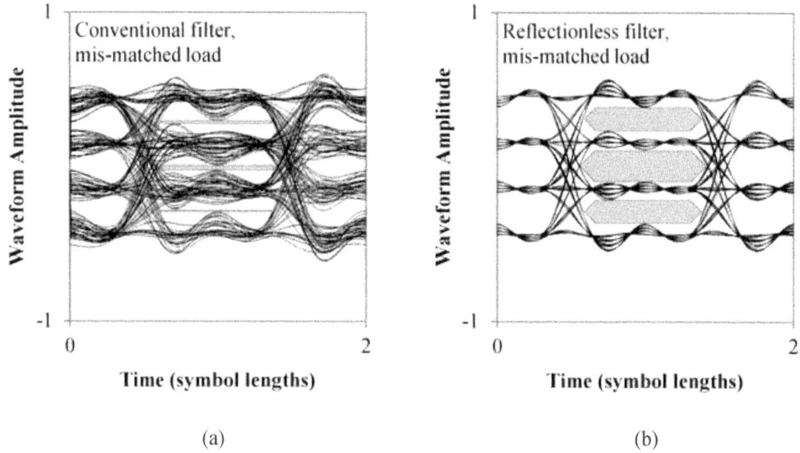

Figure 9.15 Eye diagrams for (a) conventional and (b) reflectionless filters with mismatched loads.

random symbols in the data stream, some of which are at levels 1 and 4 and thus of greater amplitude, have a lasting effect due to ringing.

However, that minor advantage is actually small compared to what happens when we consider that real-world filters are seldom terminated with a near-perfect impedance match. For example, if the load is not perfectly matched, then it will generate a reflection of the waveform back into the filter. If the filter itself is reflectionless, this backward-traveling duplicate of the waveform, or at least the high-frequency components of it, are harmlessly absorbed (the residual lower-frequency components would then be absorbed by the source on the input side of the filter). Conventional filters, on the other hand, have substantial reflections near the pass-band edge, returning the high-frequency components back towards the output after a delay associated with the electrical length between the filter and the load. In either high-speed or long-distance transmission links, this delay can be substantial, and it leads to additional intersymbol interference of a much higher order than the ringing of the filters themselves.

This effect is illustrated in Figure 9.15 wherein the same waveforms and the same filters as before are used, but where the output termination is mismatched, having a return loss of only 7 dB. That is a fairly severe mismatch, but it is a useful example for highlighting the immunity to such effects that a reflectionless filter affords the system designer. The output waveforms in Figure 9.15(a) corresponding to the conventional filter have been so badly corrupted that the eyes are almost completely closed, making it impossible for the receiver downstream to reliably distinguish the transmission levels. In contrast, the eyes in the waveforms passing through the reflectionless filter, Figure 9.15(b), are virtually unaffected, save a for small reduction in overall amplitude resulting from imperfect coupling from the source to the load.

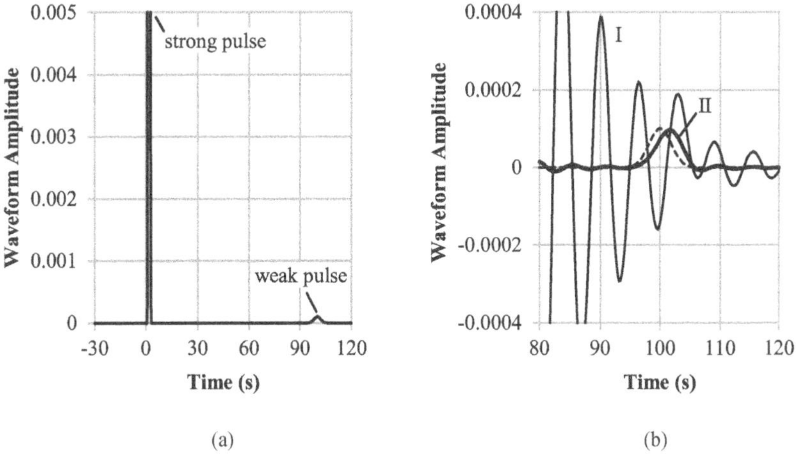

Figure 9.16 (a) Input waveform consisting of a strong and a weak Gaussian pulse. (b) Output waveforms for Chebyshev Type I and Type II filters (input waveform shown with a dashed line). Timescale shown is for 1 rad/s cutoff frequency.

It was stated earlier that the digital modulation waveform used in these examples was for illustration only, but that the same general principles apply to many different kinds of applications and waveforms. Take, for example, the input waveform shown in Figure 9.16(a), which comprises a pair of Gaussian pulses, the first very strong and the second very weak. These might, for example, be the echoes from one nearby and another more distant object in a radar system, or the primary and secondary returns from a time-gating measurement. Let us examine the Type I and Type II filter response to these waveforms, and magnify the portion around the weaker, secondary pulse as shown in Figure 9.16(b). The Type II filter has accurately reproduced the secondary pulse at its output with a very slight delay, while the Type I filter is still ringing to such a degree that the secondary pulse is completely obliterated.

Additionally, any mismatch effects from the external circuits at one of the filter ports would be ameliorated by the reflectionless filter, while it would be compounded by the band-edge reflection of the conventional filter, leading to multiple ghost echoes that would confuse the results of the measurement.

9.8.4 Low-Selectivity Time-Domain Filters

In many applications, the effects of ripple and phase distortion in the time-domain are so critical that selectivity is substantially compromised in order to avoid them. In these applications, Gaussian or Bessel-Thompson filters are often the filters of choice. The most competitive reflectionless filter in these cases will be the basic, third-order topology shown in the middle panel of Figure 3.2. It offers greater selectivity than can be achieved with the absorptive Gaussian filter of Figure 2.6 or with Bessel-Thompson filters of any order [18], but with still very competitive time-domain characteristics.

References

[1] Penfield, P., and Rafuse, R., *Varactor Applications*, Cambridge, MA: MIT Press, 1962.

[2] Maas, S., *Nonlinear Microwave Circuits*, New York: IEEE Press, pp. 81–207, 1997.

[3] Maas, S., *Microwave Mixers*, 2nd ed., Boston: Artech House, pp. 95–151, 1993.

[4] "Pairing mixers with reflectionless filters to improve system performance," *Mini-Circuits Application Note*, AN-75-007, available at http://www.minicircuits.com/app/AN75-007.pdf, October 2015.

[5] "Reflectionless filters improve linearity and dynamic range," *Microwave Journal*, vol. 58, no. 8, pp. 42–50, August 2015.

[6] Morgan, M., and Fisher, J., "Experiments with digital sideband-separating downconversion," *Publications of the Astronomical Society of the Pacific*, vol. 122, no. 889, pp. 326–335, March 2010.

[7] Bryerton, E., Morgan, M., Thacker, D., and Saini, K., "Maximizing signal-to-noise ratio in local oscillator chains for sideband-separating single-ended mixers," *18th Intl. Symp. on Space THz Technology*, Pasadena, CA, March 2007.

[8] Bryerton, E., Saini, K., Morgan, M., Stogoski, M., Boyd, T., and Thacker, D., "Development of electronically-tuned local oscillators for ALMA," *Joint 30th Intl. Conf. on Infrared and Millimeter-waves and 13th Intl. Conf. on THz Electronics*, Williamsburg, VA, September 2005.

[9] Morgan, M., Bryerton, E., Cesarano, P., Boyd, T., Thacker, D., Saini, K., and Weinreb, S., "A millimeter-wave diode-MMIC chipset for local oscillator generation in the ALMA telescope," *IEEE MTT-S Intl. Microwave Symp.*, Long Beach, CA, June 2005.

[10] "Advantages of cascading reflectionless filters," *Mini-Circuits Application Note*, AN-75-008, available at http://www.minicircuits.com/app/AN75-008.pdf, November 2015.

[11] http://www.minicircuits.com/pdfs/GVA-84-D+.pdf.

[12] Kuo, F., *Network Analysis and Synthesis*, 2nd ed., New York: Wiley, p. 136, 1962.

[13] Arfken, G. and Weber, H., *Mathematical Methods for Physicists*, 4th ed., San Diego, CA: Academic Press, pp. 877–916, 1995.

[14] Lathi, B., *Linear Systems and Signals*, Carmichael, CA: Berkeley-Cambridge Press, pp. 249–274, 1992.

[15] Wong, T., and Lok, T., "Intersymbol interference and equalization," *Theory of Digital Communications*, EEL6535 course notes, University of Florida, pp. 4.1–4.24, available at http://wireless.ece.ufl.edu/twong/Notes/Comm/ch4.pdf.

[16] "Understanding data eye diagram methodology for analyzing high speed digital signals," *ON Semiconductor Application Note*, AND9075/D, available at http://www.onsemi.com/pub_link/Collateral/AND9075-D.PDF.

[17] Behera, D., Varshney, S., Srivastava, S., and Tiwari, S., "Eye diagram basics: reading and applying eye diagrams," *EDN Network How-To Article*, December 2016, available at

http://www.edn.com/design/test-and-measurement/4389368/Eye-Diagram-Basics-Reading-and-applying-eye-diagrams.

[18] Morgan, M., and Boyd, T., "Theoretical and experimental study of a new class of reflectionless filter," *IEEE Trans. Microw. Theory Tech.*, vol. 59, no. 5, pp. 1214–1221, May 2011.

Chapter 10

Tools for Advanced Topology Creation

It is absolutely certain that the topologies developed thus far do not represent a comprehensive list. Indeed, I consider them merely the first small explorations of a vastly larger class of networks still waiting to be found. Admittedly, finding them is a difficult problem. Despite the many useful techniques described in Chapters 3–8, this ultimately remains a creative process, one driven as much by intuition as mathematical rigor.

The fact is that some desirable features and transfer functions remain out of our reach. For these, new tools and new methods of visualization will be necessary in guiding the process forward. This chapter introduces some of those tools, many of which have not yet borne fruit, but which I consider most promising and may ultimately hold the key to helping researchers discover the next generation of reflectionless filters.

10.1 SPLIT SYMMETRY-PLANE REPRESENTATION

Symmetry and duality are the foundational concepts upon which a wide variety of reflectionless filter structures are built [1, 2]. Symmetry is usually obvious at a glance; however, it is usually more difficult to discern from simple inspection whether topological duality of the even- and odd-mode equivalent circuits is met. It is useful, therefore, to draw a representation of a circuit in such a way that the even- and odd-mode equivalent circuits are more evident, while symmetry is merely prescribed (but not shown). I will refer to one such view as the split symmetry-plane representation.

10.1.1 Chebyshev Type II Example

Consider, for example, the seventh-order Chebyshev Type II reflectionless filter topology shown in Figure 10.1(a). As it is known to be symmetric (indeed, it was designed from the outset to be that way) we may simplify it somewhat by drawing only half the elements, as in Figure 10.1(b), and allowing the others to be implicit.

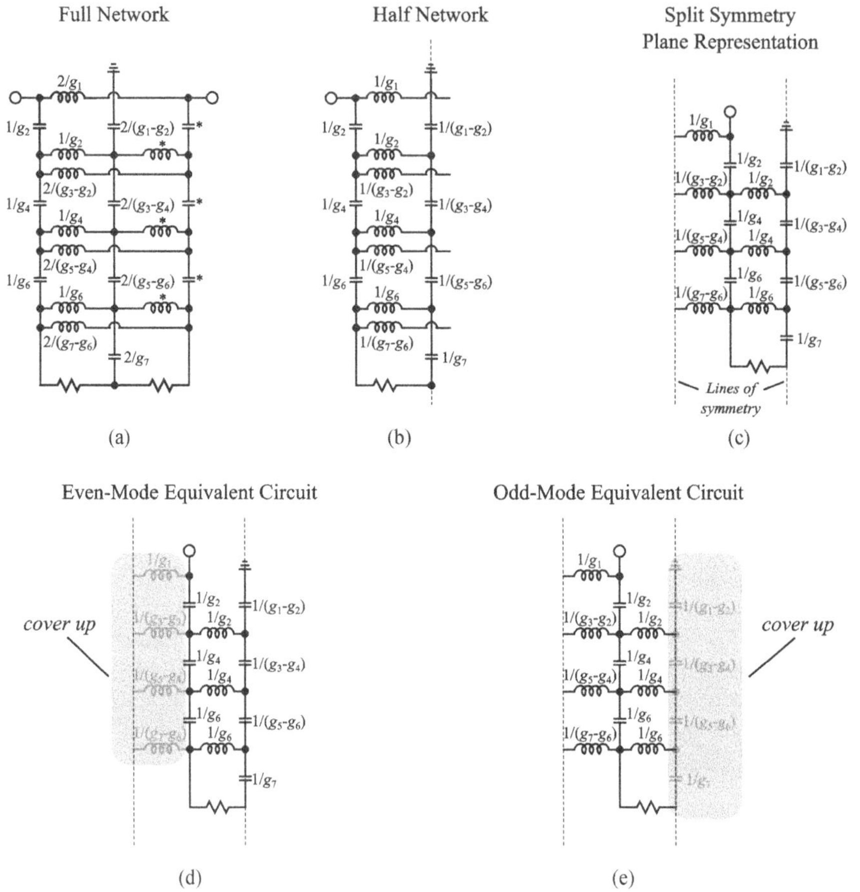

Figure 10.1 (a) Chebyshev Type II reflectionless filter of order $N = 7$. (b) Redrawn to show only half the elements. (c) Split symmetry plane representation. The dashed lines indicate nodes on the symmetry plane. (d) Even-mode, and (e) odd-mode equivalent circuits, as determined by the *cover-up* method.

In this view, the symmetry plane runs perpendicular to the page down a line on the right side of the network. We note that some elements lie along that line in the symmetry plane, while others merely cross over it; the latter intersect the symmetry plane along a different line at points above the page, but do not connect to the nodes beneath them. It is acceptable in a schematic view to unfold and flatten out the circuit by drawing these two lines of symmetry on different parts of the page. This is shown in Figure 10.1(c).

The advantage is that the even- and odd-mode equivalent circuits may now be visualized more quickly by merely covering up those parts of the drawing that are not relevant to them. For the even-mode equivalent circuit, Figure 10.1(d), the

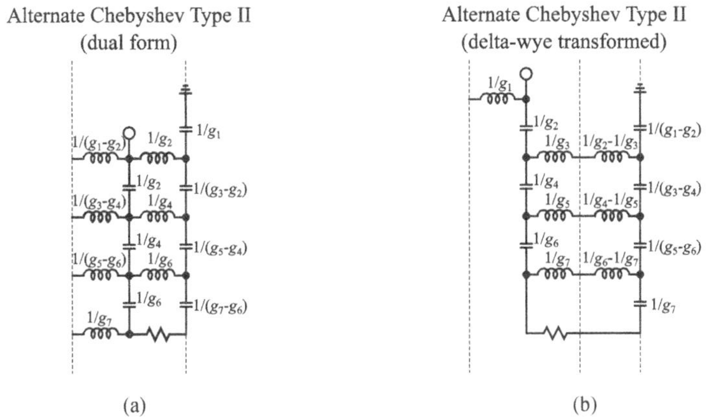

Figure 10.2 Alternate forms of the Chebyhsev Type II reflectionless filter, drawn in the split symmetry-plane representation. (a) Dual form from Figure 6.12, where $N = 7$. (b) Alternate form from Figure 6.19 where a series of delta-wye transformations was applied.

elements on the left side are open-ended and may be neglected. Although further simplification is possible by combining the capacitors in series, this is made easier to do in one's head now that the distraction of the extraneous elements has been removed.

Likewise, the odd-mode equivalent circuit is revealed in Figure 10.1(e) by covering up the components that are shorted out by the symmetry line on the right side of the schematic. The inductors that now combine in parallel are readily apparent in the drawing that remains.

10.1.2 Chebyshev Type II Alternate Forms

As the split symmetry plane is primarily a visualization tool, it is instructive to examine a number of examples using previously understood topologies. The dual form of the Chebyshev Type II reflectionless filter (which was shown first in Figure 6.12) is redrawn with a split symmetry plane in Figure 10.2(a). It is very similar to the first example, except for a slight re-ordering of the elements.

The alternate form that we derived in Figure 6.19 by applying delta-wye transformations to the inductor groups is redrawn in Figure 10.2(b). In this case, we have split the symmetry plane into three schematic lines. The boundary conditions for the even- and odd-mode equivalent circuits (an open-circuit boundary for the even-mode circuit, and a virtual short for the odd-mode) remain the same, and apply to all nodes on *any* of the three symmetry lines drawn. The cover-up method is thus applicable to this topology as well, and is illustrated in Figure 10.3.

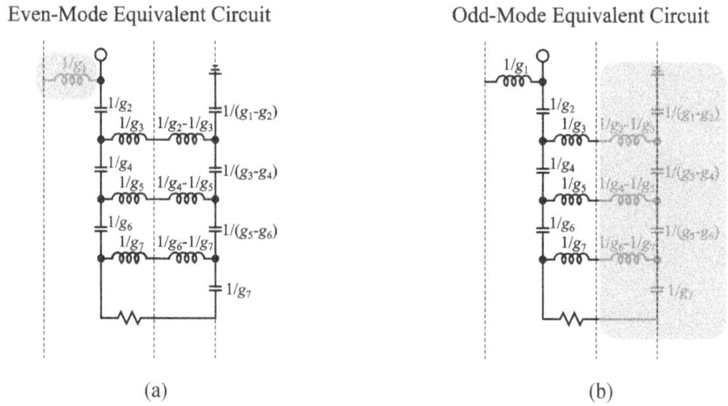

Figure 10.3 (a) Even-mode and (b) odd-mode equivalent circuits of the alternate form from Figure 10.2(b) revealed using the cover-up method.

10.1.3 A Pseudo-Elliptic Reflectionless Filter

Having demonstrated the split symmetry plane representation using previously known topologies, let us now work through an example in which we use it as an aid in the development of new reflectionless topologies. Suppose that a filter is required to have a sharper transition band than either the Chebyshev Type I or Type II transfer functions can provide. As described in Chapter 1, this extra slope can be bought by allowing ripples in both the pass-band and the stop-band. Although we do not have a reflectionless topology for a true elliptic filter function, we can approximate one by starting with our Chebyshev Type II topology and modifying it to create ripples in the pass-band.

Recall that the transmission response of our filter will be equal to the even-mode reflection coefficient. Thus, for a filter with minimum loss to have ripples in the pass-band, we must ensure that the even-mode reflection coefficient is maximum ($|\Gamma_{even}| = 1$) at finite frequencies as well as at DC. The simple high-pass ladder network of series capacitors and shunt inductors that we started with in the original derivation of Figure 3.1(a) does not achieve this for any element values. Instead, we use a modified even-mode equivalent circuit prototype in Figure 10.4(a) where a capacitor has been added in series with one of the shunt inductors. At resonance, the impedance of this arm becomes zero, leading to a peak in even-mode reflection.

Our task now is to alter this topology such that its dual appears in the odd-mode circuit. In this case, the dual form consists of a high-pass pi-network of shunt inductors and a series capacitor, followed by a parallel LC resonator in series with the termination.

First, in Figure 10.4(b), we split the first series capacitor in two and relocate

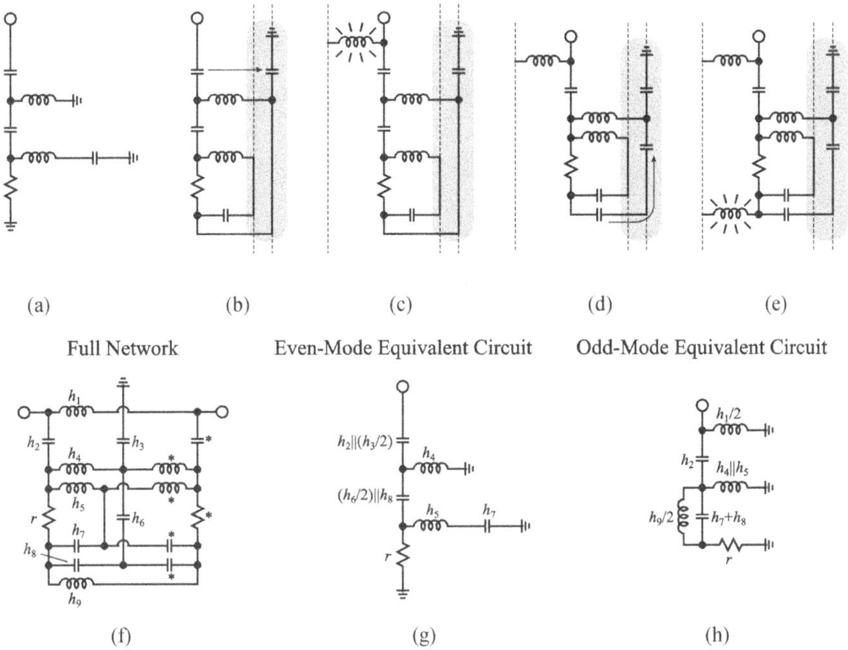

Figure 10.4 Development of a pseudo-elliptic reflectionless filter topology. (a) Initial even-mode equivalent circuit. (b) Partial relocation of the first series capacitor into the cover-up area for the odd-mode equivalent circuit. (c) Addition of shunt inductor to odd-mode circuit. (d) Partial relocation of the second series capacitor into the cover-up area. (e) Addition of another shunt inductor to odd-mode circuit. (f) Final two-port network. (g) Even-mode equivalent circuit. (h) Odd-mode equivalent circuit.

part of it into the cover-up area for the odd-mode circuit. This entails exchanging it in series with the rest of the network, and drawing a symmetry line through it. We also draw a symmetry line through the intermediate node of the series LC resonator, so that these elements become separated in the odd-mode circuit. In Figure 10.4(c) we add a shunt inductor to a symmetry line, adding it to the odd-mode circuit without modifying the even-mode circuit. In Figure 10.4(d), we split up another capacitor, moving part of into the cover-up area. Finally, the desired dual form is completed in Figure 10.4(e) by adding a second shunt inductor to the odd-mode circuit.

The final two-port network is thus shown with arbitrary element values in Figure 10.4(f) and the even- and odd-mode equivalent circuits are shown in Figure 10.4(g) and h. All that remains is to determine the constraints on the element values by equating the given elements with their dual counterparts in the even- and odd-mode equivalent circuits. This yields the following relationships:

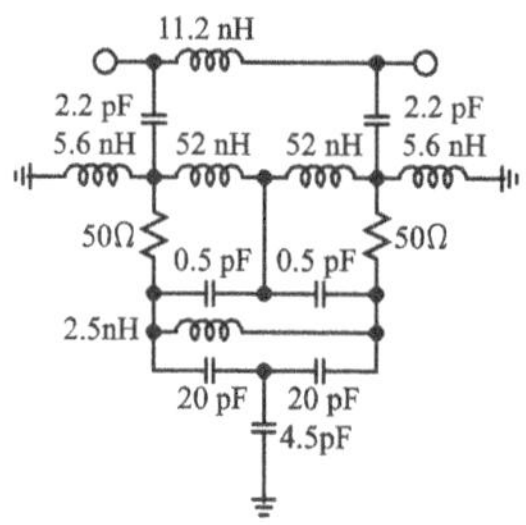

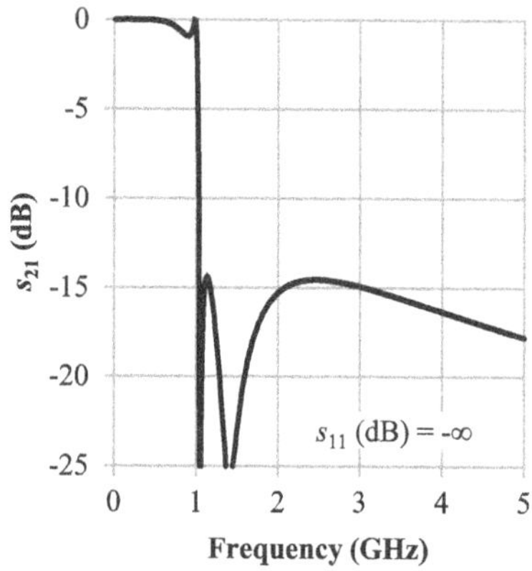

Figure 10.5 (a) Final topology and element values for pseudo-elliptic reflectionless filter. (b) Simulated frequency response.

$$\frac{\frac{1}{2}h_2 h_3}{h_2 + \frac{1}{2}h_3} = \frac{1}{2}h_1 \tag{10.1a}$$

$$h_4 = h_2 \tag{10.1b}$$

$$\frac{\frac{1}{2}h_6 h_8}{\frac{1}{2}h_6 + h_8} = \frac{h_4 h_5}{h_4 + h_5} \tag{10.1c}$$

$$h_5 = h_7 + h_8 \tag{10.1d}$$

$$h_7 = \frac{1}{2}h_9 \tag{10.1e}$$

$$r = r^{-1} = 1 \tag{10.1f}$$

Having started with ten unknowns, we wrote down six constraint equations, leaving four degrees of freedom in the choice of element values — one for frequency scaling and three for shape. Trial and error has shown that the best stop-band rejection is achieved if h_3 is allowed to become arbitrarily large, in which case the associated capacitor is omitted and by (10.1a) we find that $h_1 = 2h_2$. The final topology and a manually selected choice of element values is shown in Figure 10.5 along with the simulated performance.

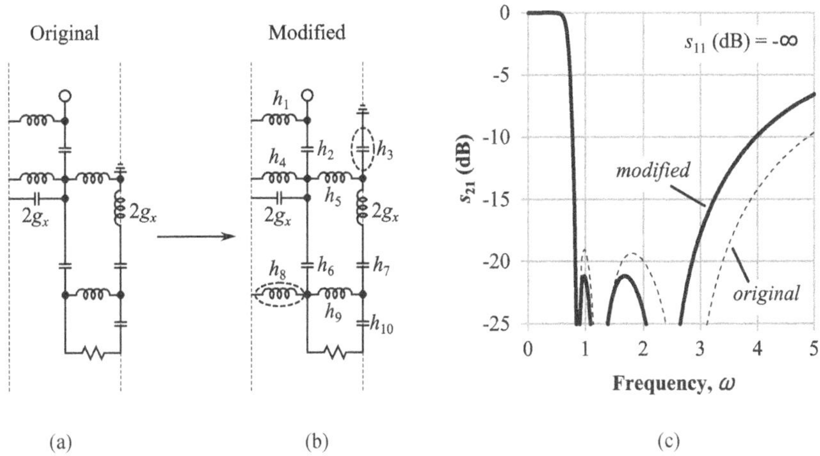

Figure 10.6 (a) Split symmetry plane representation of the filter in Figure 4.11(a). Recall that unlabeled elements have unit normalized value. (b) Modified with additional auxiliary elements. (c) Comparison of the original performance (dotted line) with one possible response of the modified circuit (solid line).

10.1.4 Modification of Prior Topologies

The split symmetry plane representation is also useful as an aid in identifying auxiliary elements that can be added to existing topologies to alter their performance without losing the reflectionless property. Take, as an example, the second-rank filter shown in Figure 4.11(a). Based on the earlier subnetwork theory, this topology had already been modified in Chapter 4 by including additional auxiliary elements, labeled g_x, which compensate one another in the dual even- and odd-mode circuits. When redrawn in the split symmetry plane representation, Figure 10.6(a), it is evident that further additions are possible to permit fine-tuned performance without loss of reflectionless-ness.

Specifically, the first series capacitor is common to both the even- and odd-mode equivalent circuits, which leads to an unnecessarily narrow constraint in its value relative to the other elements in the topology. This constraint can be lifted by inserting an additional capacitor on the symmetry line (thus appearing only in the even-mode equivalent circuit). This new capacitor is labeled h_3 in Figure 10.6(b). Similarly, an extra inductor appearing only in the odd-mode equivalent circuit, labeled h_8, adds freedom to the element values in the bottom part of the topology.

The addition of these elements allows the frequency response of the filter to be fine-tuned for specific requirements. In Figure 10.6(c) the original performance is compared with that of a possible variation using the modified topology where deeper rejection is achieved at the expense of stop-band bandwidth. Not surprisingly, the opposite goal is likewise achievable with this topology; a more extended stop-band could be obtained with lesser peak rejection levels.

10.2 UNUSUAL FORMS OF SYMMETRY

The greatest difficulty in any creative endeavor is discarding our old assumptions. Since the beginning of Chapter 3 we have seen that a fundamental aspect of reflectionless filters has been their symmetry. However, the full scope and definition of that symmetry was not fully explored. The natural assumption to make is that of topological symmetry — the geometric equivalence of a schematic drawing with itself upon a mirror transformation about a single axis. However, what is required to apply our methodology is nothing more than electrical symmetry, or the equivalence of circuit behavior from either of its ports [3]. The difference between these two definitions is best understood by the following examples.

10.2.1 Antiparallel Subnetworks

Consider an otherwise symmetric network containing an asymmetric subnetwork, R, and its mirror image connected in parallel, as shown in Figure 10.7(a). An example of the network R comprising a resistor, an inductor, and a capacitor is shown in Figure 10.7(b). Although the circuit clearly has identical electronic behavior as seen from both sides of the line of symmetry, there is no node that one can identify lying on the line of symmetry at which we can apply the usual open- and short-circuit boundary conditions associated with even- and odd-mode analysis. Instead, we must derive the even- and odd-mode equivalents of the antiparallel subnetwork pair mathematically. As R is a simple tee-network, its impedance parameters are found almost by inspection, for example using (5.2),

$$\mathbf{Z}_R = \begin{pmatrix} h_1 s + r & r \\ r & (h_2 s)^{-1} + r \end{pmatrix} \tag{10.2}$$

To combine the pair in parallel, it is useful to first convert this to admittance parameters,

$$\mathbf{Y}_R = \mathbf{Z}_R^{-1} = \frac{1}{\det(\mathbf{Z}_R)} \begin{pmatrix} (h_2 s)^{-1} + r & -r \\ -r & h_1 s + r \end{pmatrix} \tag{10.3a}$$

$$\det(\mathbf{Z}_R) = (h_1 s + r)\left((h_2 s)^{-1} + r\right) - r^2 = h_1 r s + h_1 h_2^{-1} + h_2^{-1} r s^{-1} \tag{10.3b}$$

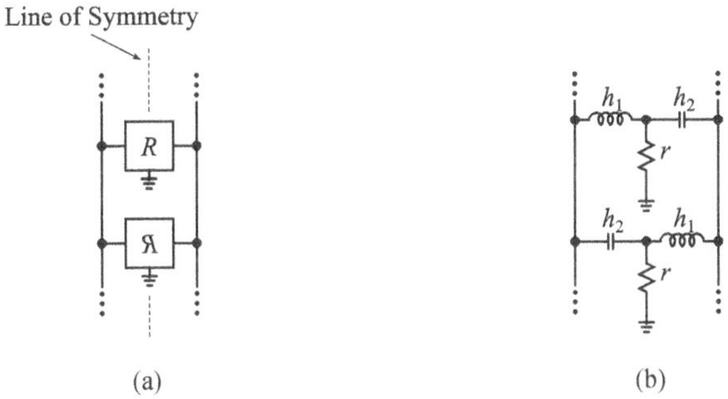

Figure 10.7 (a) Part of an electronically symmetric network incorporating antiparallel, asymmetric subnetworks. (b) Example of antiparallel, asymmetric subnetworks.

The admittance matrix of the mirrored subnetwork, $\mathbf{Y}_{R'}$, is the same but with the rows and columns exchanged.

The total admittance of the antiparallel pair is thus given by

$$\mathbf{Y}_R + \mathbf{Y}_{R'} = \frac{1}{\det(\mathbf{Z}_R)}\begin{pmatrix} (h_2 s)^{-1} + r & -r \\ -r & h_1 s + r \end{pmatrix} + \frac{1}{\det(\mathbf{Z}_R)}\begin{pmatrix} h_1 s + r & -r \\ -r & (h_2 s)^{-1} + r \end{pmatrix} \tag{10.4a}$$

$$= \frac{1}{\det(\mathbf{Z}_R)}\begin{pmatrix} h_1 s + 2r + h_2^{-1}s^{-1} & -2r \\ -2r & h_1 s + 2r + h_2^{-1}s^{-1} \end{pmatrix} \tag{10.4b}$$

which is symmetric, as expected. The even-mode equivalent admittance is given by the simultaneous excitation of both ports in-phase, therefore

$$y_{even} = \tfrac{1}{2}\begin{pmatrix}1 & 1\end{pmatrix}(\mathbf{Y}_R + \mathbf{Y}_{R'})\begin{pmatrix}1 \\ 1\end{pmatrix} = \frac{h_1 s + h_2^{-1}s^{-1}}{h_1 r s + h_1 h_2^{-1} + h_2^{-1} r s^{-1}} \tag{10.5a}$$

$$= \left(r + \frac{h_1 h_2^{-1} s}{h_1 s^2 + h_2^{-1}}\right)^{-1} = \left(r + \left(h_2 s + \frac{1}{h_1 s}\right)^{-1}\right)^{-1} \tag{10.5b}$$

which is recognizable as a resistor in series with a parallel resonator, shown in Figure 10.8(a).

The odd-mode equivalent circuit is given similarly by the antiphase excitation of the two-port network,

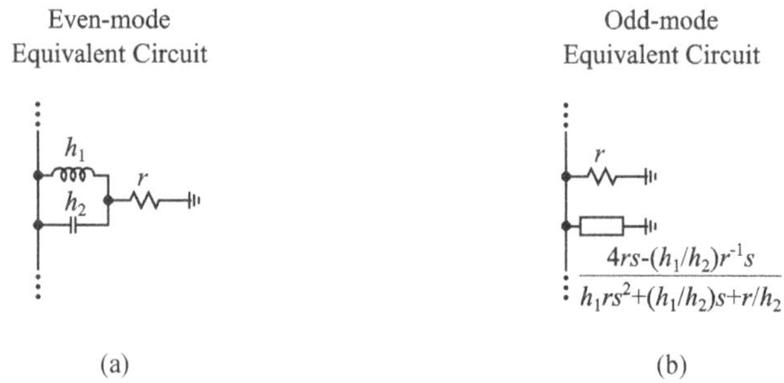

Figure 10.8 (a) Even-mode and (b) odd-mode equivalent circuits of the antiparallel subnetwork pair shown in Figure 10.7(b).

$$y_{odd} = \tfrac{1}{2}\begin{pmatrix}1 & -1\end{pmatrix}\left(\mathbf{Y}_R + \mathbf{Y}_{R'}\right)\begin{pmatrix}1\\-1\end{pmatrix} = \frac{h_1 s + 4r + h_2^{-1}s^{-1}}{h_1 rs + h_1 h_2^{-1} + h_2^{-1}rs^{-1}} \tag{10.6a}$$

$$= \frac{h_1 s^2 + 4rs + h_2^{-1}}{h_1 rs^2 + h_1 h_2^{-1}s + h_2^{-1}r} = r^{-1} + \frac{\left(4r - h_1 h_2^{-1} r^{-1}\right)s}{h_1 rs^2 + h_1 h_2^{-1}s + h_2^{-1}r} \tag{10.6b}$$

Recall that immittance functions must have Hurwitz polynomials in both the numerator and the denominator, and that the coefficients of a Hurwitz polynomial must all be positive. For the second term above to be realizable we must have

$$4r^2 \geq h_1 h_2^{-1} \tag{10.7}$$

In the case of equality above, the second admittance term becomes zero, and the odd-mode equivalent circuit is simply that of a lone resistor.

An alternative method of analyzing the subnetwork pair in Figure 10.7(b) is to apply a delta-wye transformation to both subnetworks. For simplicity, let us use prototype values that yield a lone resistor in the odd-mode circuit, namely $r = 1$, $h_1 = 2$, and $h_2 = ½$, as shown in Figure 10.9(a). The result of the delta-wye calculation is shown in Figure 10.9(b). The asymmetric shunt arms contain nonphysical elements, but become both physical and symmetric when combined in parallel,

$$z_{shunt} = \frac{(s+1)^2\left(s^{-1}+1\right)^2}{(s+1)^2 + \left(s^{-1}+1\right)^2} = \frac{(s+1)^4}{s^4 + 2s^3 + 2s^2 + 2s + 1} = \frac{(s+1)^2}{s^2+1} \tag{10.8a}$$

$$= 1 + \left(\tfrac{1}{2}s + \tfrac{1}{2}s^{-1}\right)^{-1} \tag{10.8b}$$

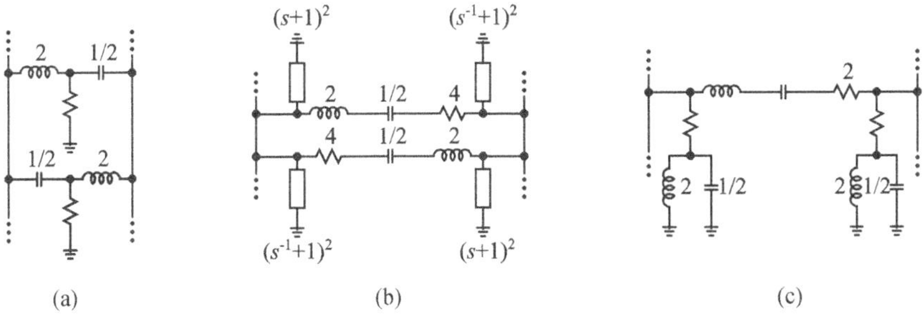

Figure 10.9 (a) Antiparallel subnetwork pair. (b) After delta-wye transformation. (c) Parallel combination of shunt arms.

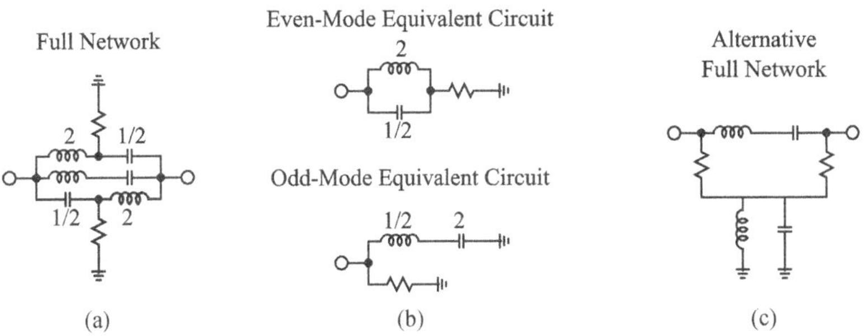

Figure 10.10 (a) Full reflectionless filter using antiparallel subnetworks. (b) Even- and odd-mode equivalent circuits. (c) Alternative reflectionless filter achieving the same even-/odd-mode equivalents.

The expression in (10.8b) corresponds to a resistor in series with a parallel LC resonator. The final topology is thus shown in Figure 10.9(c). As this network is symmetric (or can be trivially made so by splitting the series elements in two and locating each half on opposite sides of the symmetry line), the classical even-/odd-mode boundary conditions may now be applied as normal.

The even-mode circuit, where the horizontal branch is left open-circuited, is clearly the same as that in Figure 10.8(a). However it is not obvious that the odd-mode circuit reduces to a single resistor. The math, however, bears out this result,

$$y_{odd} = \left(\tfrac{1}{2}s + \tfrac{1}{2}s^{-1} + 1\right)^{-1} + \left(1 + \left(\tfrac{1}{2}s + \tfrac{1}{2}s^{-1}\right)^{-1}\right)^{-1} = \frac{2s}{(s+1)^2} + \frac{s^2+1}{(s+1)^2} = 1 \qquad (10.9)$$

Let us consider an example of this structure in a complete reflectionless topology. Perhaps the simplest such network is one that balances the parallel resonator in the even-mode circuit (Figure 10.8(a)) with a series resonator in the odd-mode circuit. The complete reflectionless structure is then shown in Figure 10.10(a). The even- and odd-mode equivalent circuits are then shown in Figure

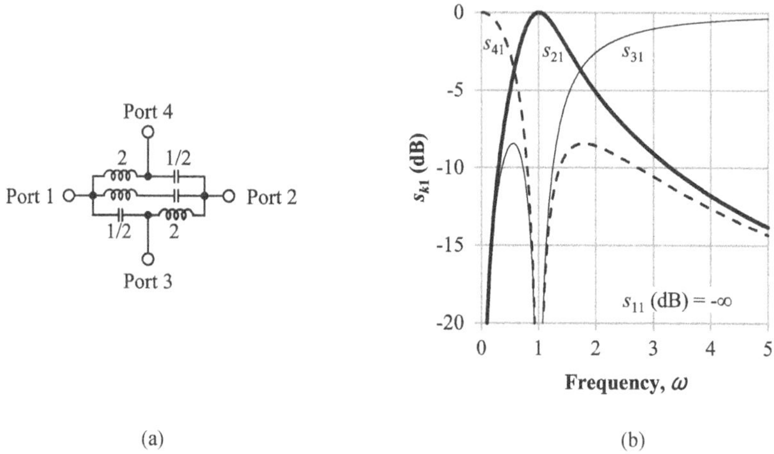

Figure 10.11 (a) Four-port reflectionless network using antiparallel subnetworks. (b) Transfer functions.

10.10(b).

It is noteworthy that the same even- and odd-mode equivalents (and therefore the same transfer function) could be achieved using the previously described techniques without employing the antiparallel subnetworks, as illustrated in Figure 10.10(c). Though equivalent in the two-port case, the first structure does have one subtle advantage, namely that the two termination resistors are grounded. This makes it easier to couple them out as external ports without needing a transformer. The resulting network behaves like a simple, low-order, reflectionless triplexer, as shown in Figure 10.11(a). The frequency responses from port 1 to the other three ports are shown in Figure 10.11(b).

This is, of course, but a single, low-order example of a wide variety of networks which may be constructed using these techniques.

10.2.2 Nonplanar Forms

Another form of electrical symmetry that was not accounted for in the prior descriptions of even-/odd-mode analysis is that of nonplanar networks. These are networks that cannot be drawn on a flat sheet of paper without crossovers [4].

Importantly, the mere presence of crossovers, like those in many of the schematics in this book, does not prove that a network is nonplanar, for it is possible to redraw them in such a way that the crossovers are avoided. See the topology of a Chebyshev Type II reflectionless filter in Figure 10.12 for an example.

For networks that are truly nonplanar, however, the diagram and boundary conditions shown in Figure 1.12 do not always apply. Instead, the situation is

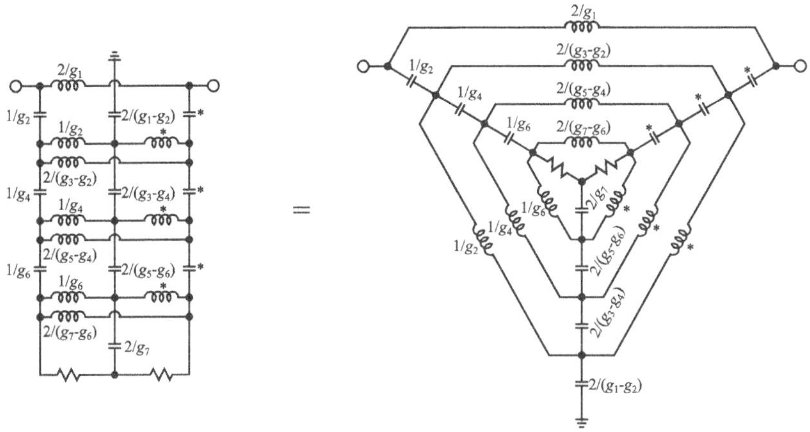

Figure 10.12 An example of a schematic having crossovers that is, in fact, planar, as it can be redrawn with the crossovers avoided.

more like that shown in Figure 10.13(a)–(b). Here, as in Chapter 1, we have bisected an electrically symmetric circuit and pulled it apart to expose the wires that cross the line of symmetry. In doing so this time, we have found that at least one pair of wires cross over one another at this midpoint in the network. The boundary conditions that applied to the noncrossing wires do not apply in this case.

Instead, we must again consider the voltages and currents that develop on these wires given the even- and odd-mode excitations. This is shown in Figure 10.13(c) for the even-mode excitation and Figure 10.13(d) for the odd-mode. In the even-mode, the currents on the noncrossing wires from both halves of the circuit are in opposing directions, and may be considered to cancel one another. This led to an open-circuit boundary condition in the conventional case. On the other hand, the currents in the crossing wires do not cancel one another. Instead, the same (or an equivalent) current enters one pin while leaving the other from the same half of the circuit. These pins are also at the same potential with respect to ground. The even-mode equivalent circuit, then, is shown to simply have these two pins linked, as illustrated in Figure 10.13(e).

The odd-mode case is somewhat more complicated. In this case, the same/equivalent current leaves both pins from one half of the circuit, while having the opposite voltage potential. The odd-mode equivalent circuit therefore requires a 1:1 transformer in order to enforce this boundary condition. Note that the dot-convention of the transformer shown in Figure 10.13(f) indicates the two coils are oriented in opposite polarity.

The presence of the transformer in the odd-mode circuit is an inconvenience, as it is difficult to match with a dual counter-part in the even-mode case; what is the dual of a 1:1 inverting transformer? Rather than trying to answer that question

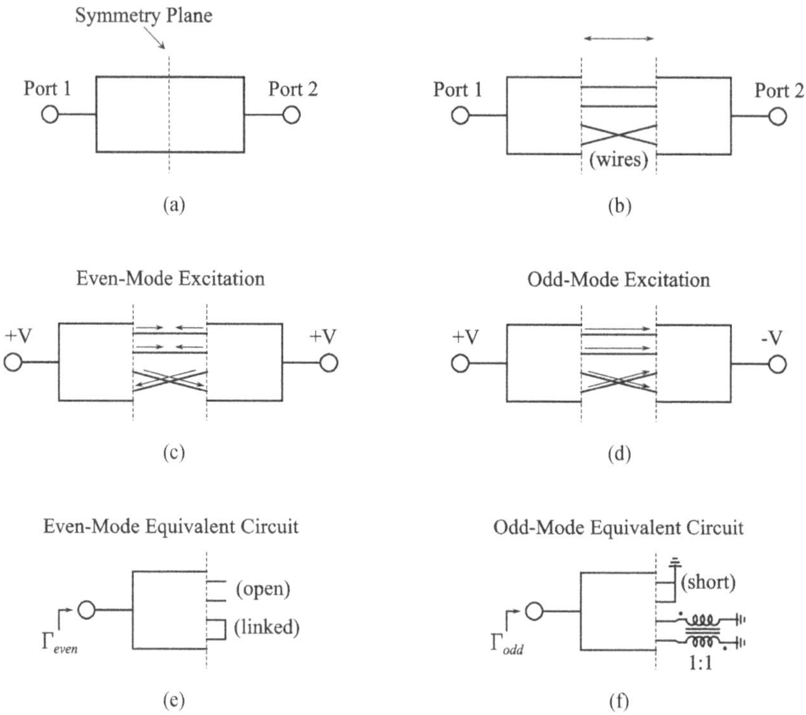

Figure 10.13 (a) A symmetric two-port network. (b) Exposed wires after separating along the symmetry plane, revealing a crossover. (c) Even-mode excitation and current flow in exposed wires. (d) Odd-mode excitation and current flow in exposed wires. (e) Even-mode equivalent circuit. (f) Odd-mode equivalent circuit.

directly (some clues may be found in [5]), it is perhaps more useful to consider how networks that contain such a transformer can be simplified.

First, consider that a crossover linking two sides of a symmetric circuit with only self-contained subnetworks between the links on either side needs not be drawn as crossover at all, as illustrated in Figure 10.14. By simply relocating the subnetworks above and below the cross-links, and then untwisting the network, we may recast the network as one containing an antiparallel subnetwork pair instead, a situation that was covered in Section 10.2.1. This assumes, of course, that there are no other hidden connections between nodes above and below the crossover, as untwisting one crossover would then twist and create another.

If elements exist inside the crossover, as shown in Figure 10.15(a), one needs to be a bit more careful. Referring back to Figure 10.13(e) and (f), we simply link the terminals of the crossover in the even-mode circuit and add the inverting transformer in the odd-mode equivalent circuit. This is shown in Figure 10.15(b) and (c), respectively. The latter may be simplified by considering the *ABCD*

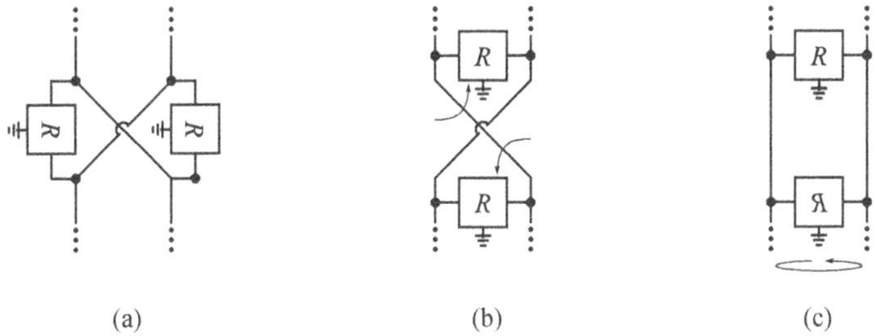

Figure 10.14 (a) Subnetworks in a symmetric circuit linked by a crossover. (b) Translation of subnetworks outside the crossover. (c) Crossover removed by untwisting the circuit.

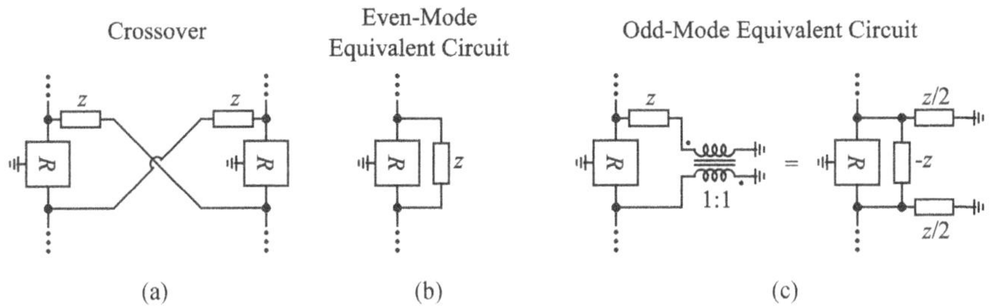

Figure 10.15 (a) Crossover with elements in the crossing arms. (b) Even-mode equivalent circuit. (c) Odd-mode equivalent circuit, first with a transformer, and then with the equivalent impedance pi-network.

parameters [6] of the series impedance element and the transformer in cascade,

$$\begin{pmatrix} A & B \\ C & D \end{pmatrix} = \begin{pmatrix} 1 & z \\ 0 & 1 \end{pmatrix} \begin{pmatrix} -1 & 0 \\ 0 & -1 \end{pmatrix} = \begin{pmatrix} -1 & -z \\ 0 & -1 \end{pmatrix} \tag{10.10}$$

and then converting to admittance parameters,

$$\mathbf{Y} = \begin{pmatrix} \frac{D}{B} & \frac{BC-AD}{B} \\ \frac{-1}{B} & \frac{A}{B} \end{pmatrix} = \begin{pmatrix} z^{-1} & z^{-1} \\ z^{-1} & z^{-1} \end{pmatrix} \tag{10.11}$$

This has the equivalent circuit of a pi-network comprising a negative z element in the series arm, and $z/2$ in the shunt arms. This has been drawn on the right side of Figure 10.15(c).

As with the transformer, the negative element does not need to be realized in practice because it arises only in the equivalent circuit as a consequence of the crossover. However, it may be difficult to match with a dual element in the even-mode circuit. Instead, it might be easier to absorb into a like-kind and larger-valued element in the subnetwork R. This virtual subtraction of physical elements

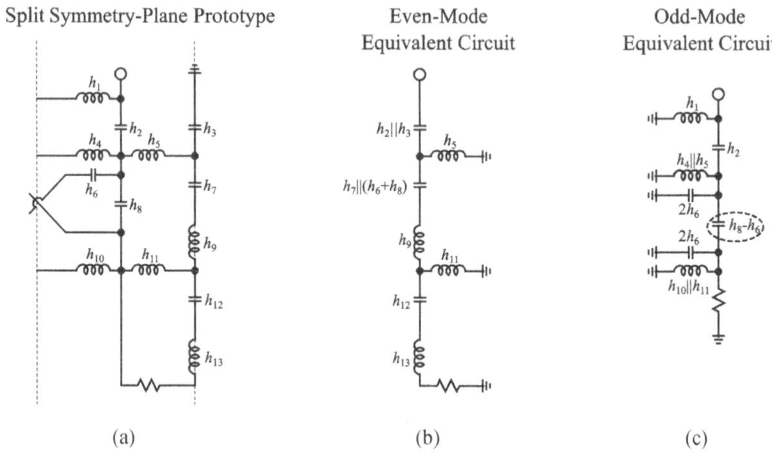

Figure 10.16 (a) Split symmetry-plane prototype of a nonplanar reflectionless filter. (b) Even-mode equivalent circuit. (c) Odd-mode equivalent circuit showing subtraction of elements.

is quite significant as it is difficult to achieve by other means.

For example, consider the nonplanar network drawn in the split symmetry-plane view in Figure 10.16(a). The cross-linkage of h_6 and h_8 causes these two elements to combine in parallel in the even-mode circuit (and in series with h_7), while leading to a capacitive pi-network in the odd-mode circuit where h_6 is subtracted from h_8. The duality constraints may be found as usual by comparing dual elements in both equivalent half-circuits, as follows,

$$\frac{h_2 h_3}{h_2 + h_3} = h_1 \tag{10.12a}$$

$$h_5 = h_2 \tag{10.12b}$$

$$\frac{h_7(h_6 + h_8)}{h_6 + h_7 + h_8} = \frac{h_4 h_5}{h_4 + h_5} \tag{10.12c}$$

$$h_9 = 2h_6 \tag{10.12d}$$

$$h_{11} = h_8 - h_6 \tag{10.12e}$$

$$h_{12} = \frac{h_{10} h_{11}}{h_{10} + h_{11}} \tag{10.12f}$$

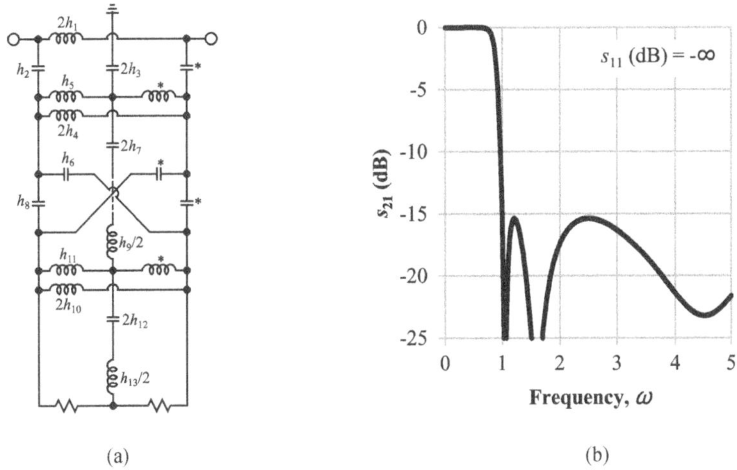

Figure 10.17 (a) Final topology of nonplanar reflectionless filter. (b) Simulated performance.

$$h_{13} = 2h_6 \tag{10.12g}$$

We have thirteen unknowns and eight equations, leaving us five degrees of freedom to tailor the frequency response. The final topology, along with the performance of one such combination that satisfies these conditions for reflectionless operation, is shown in Figure 10.17.

10.3 CONCLUSION

Reflectionless filters have the potential to change the way we think about filtering in electronic systems. This book has provided the tools to develop topologies realizing those filters, and it has demonstrated their design in a number of different regimes, from monolithic integrated circuits, to discrete lumped-elements, to transmission lines.

As stated in the introduction to this chapter, there is little doubt that many useful reflectionless filter topologies remain undiscovered at this time. To that point, I would like to posit the conjecture that any transfer function that is realizable is realizable in a reflectionless form (and without the excess flat loss that plagues the preexisting constant-resistance networks). Stated formally, I am supposing that for any transfer function that is realizable using a finite number of lumped elements (inductors, capacitors, resistors, and transformers), there exists a topology that exhibits that transfer function with an identically zero reflection coefficient at all frequencies and at all ports.

Importantly, nonphysical elements such as frequency-independent quadrature hybrids are not included in the above list, unless a way is eventually found to make one — a perfect, frequency-independent one — with only resistors, inductors, capacitors, and transformers. Had such ideal quadrature hybrids been included, the statement would be trivially true, as one could always implement a hybrid-coupled directional filter as shown in Figure 2.4.

This bold statement is not made lightly. It is the instinctive conclusion of a researcher that has found himself repeatedly surprised at what could be achieved. We have already discovered networks that realize Chebyshev responses, Type I and Type II, with variable (but limited) ripple factor and arbitrary odd orders, Achieser-Zolotarev responses, also Type I and Type II, myriad other responses without common names, some of which approximate elliptic filtering functions, and more, many of which for one reason or another did not make the cut to be listed in this book. Exposing them required pulling at a thread started decades ago with groundbreaking work on the properties of constant-resistance networks — a thread that perhaps was too quickly abandoned in favor of other approaches that at the time seemed more fruitful.

I say it is time to ask more of the filter, one of the fundamental building blocks of our electronic systems. Do not be content with the maximally poor return loss conventional filters exhibit in their stop-bands (the part of the spectrum that gives a filter its purpose) and the corruption of performance that goes along with it. It is not inevitable. Expect the same from your filters as you do every other component (a proper impedance match) and design your systems accordingly.

References

[1] Morgan, M., and Boyd, T., "Theoretical and experimental study of a new class of reflectionless filter," *IEEE Trans. Microw. Theory Tech.*, vol. 59, no. 5, pp. 1214–1221, May 2011.

[2] Morgan, M., and Boyd, T., "Reflectionless filter structures," *IEEE Trans. Microw. Theory Techn.*, vol. 63, no. 4, pp. 1263–1271, April 2015.

[3] Morgan, M., *Optimal Response Reflectionless Filters*, U.S. Patent Application No. 15/298,459, October 20, 2016, PCT Application No. PCT/US16/57829, October 20, 2016.

[4] Johnson, D., Hilburn, J., Johnson, J., and Scott, P., *Basic Electric Circuit Analysis*, 5th ed., New York: Wiley, pp. 725–730, January 1995.

[5] Bloch, A., "On methods for the construction of networks dual to nonplanar networks," *Proc. Phys. Society*, vol. 58, no. 6, pp. 677–694, November 1946.

[6] Pozar, D., *Microwave Engineering*, 4th ed., New York: Wiley, pp. 188–192, November 2011.

Appendix A: Delta-Wye Transformations

Most circuit designers are familiar with the standard delta-wye transformation formulas. Though illustrated traditionally with resistors, it is often stated that the formulae apply as well to arbitrary complex impedances; this is true, so long as one is willing to accept the nonphysical elements (typically negatively valued or quadratic reactors[1]) that often result. Consider, for example, a wye-network comprising two inductors and a resistor, having normalized values of l_1, l_2, and r, respectively. The elements of the equivalent delta-network are given as follows,

$$z_a = \frac{(sl_1)(sl_2)+(sl_1)r+(sl_2)r}{(sl_1)} = sl_2 + r\left(1+\frac{l_2}{l_1}\right) \tag{A.1a}$$

$$z_b = \frac{(sl_1)(sl_2)+(sl_1)r+(sl_2)r}{(sl_2)} = sl_1 + r\left(1+\frac{l_1}{l_2}\right) \tag{A.1b}$$

$$z_c = \frac{(sl_1)(sl_2)+(sl_1)r+(sl_2)r}{r} = s^2\frac{l_1 l_2}{r} + s(l_1 + l_2) \tag{A.1c}$$

The two equivalent networks are therefore shown below.

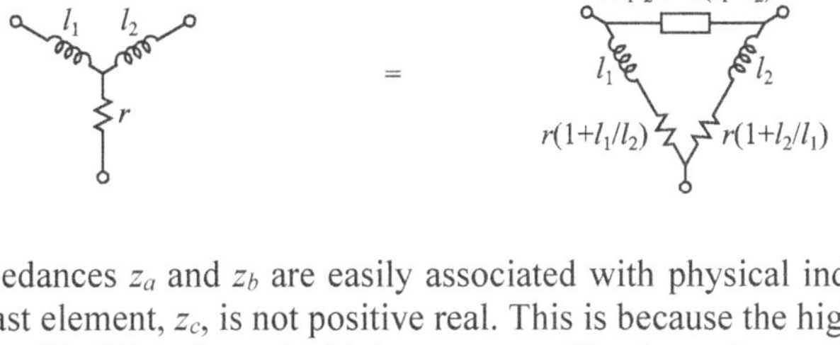

While the impedances z_a and z_b are easily associated with physical inductors and resistors, the last element, z_c, is not positive real. This is because the highest power of its numerator (2) differs from the highest power of its denominator (0) by more than unity. This element cannot be realized using physical components. The equivalent delta network, then, is a perfectly acceptable mathematical artifice for analysis (the direct calculation of a given circuit's electrical behavior) but is

[1] That is, a susceptance or reactance with a quadratic dependence on the complex frequency, s.

unacceptable for circuit synthesis.

The significance of this limitation to reflectionless filters is that the delta and wye groupings are dual topological forms, and therefore arise frequently when attempting to construct a new reflectionless topology from first principles. As explained in Section 3.7, having written down a prototype even-mode circuit, and having derived from it a dual odd-mode circuit, one is challenged to find a way of realizing them both from a common, symmetric starting point. Delta-wye transformation might at first seem to be an answer, since — topologically, at least — the even- and odd-mode equivalent circuits will look "more alike" after one of them is transformed. Or, one might assume that applying a transformation to an already completed reflectionless topology might ease its implementation or analysis.

Sometimes this is true, despite the lack of physical realizability in the intermediate steps (see Figure 10.9), but more often than not this thinking leads us down a profitless path. Forearmed with the knowledge of which networks are physically transformable, we can focus our efforts on those cases where the transformation might hold some benefit. Such was the case in Section 6.9, where the transformed network was not only realizable but yielded more moderately valued components.

For these reasons, it is worth cataloging which real combinations of resistors, inductors, and capacitors yield physically realizable networks under a delta-wye transformation. Among the class of networks comprising only three fundamental elements, of which there are 54 possibilities, only 10 meet this criterion[2] — namely, those with all elements of the same type, or those where at least two of the elements are resistors. For ease of use, these 10 cases are shown here.

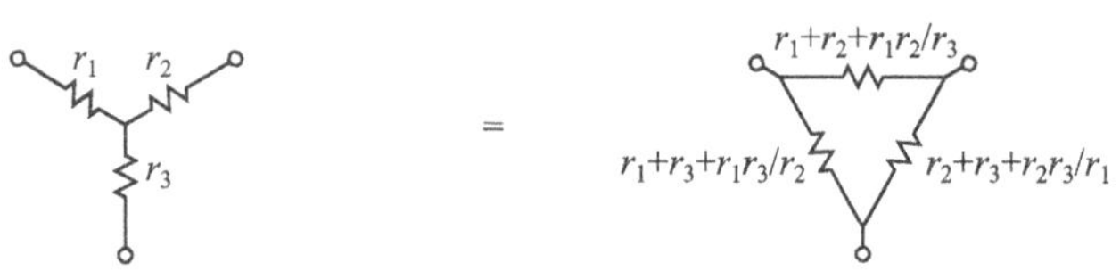

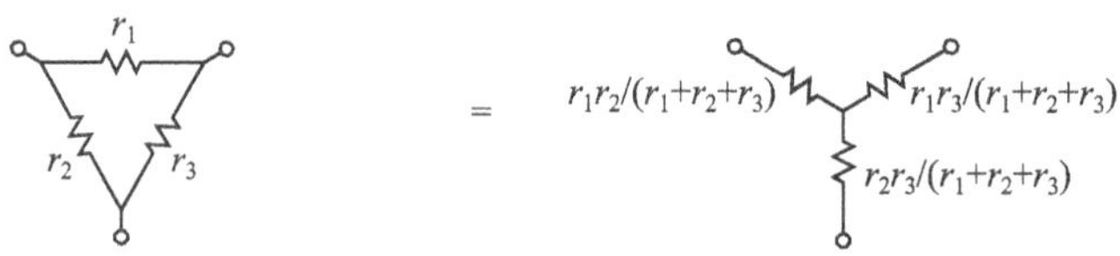

2 To my knowledge, there is no prior publication that references this fact, but having checked each of the 54 possible three-element networks individually, this is what has been found.

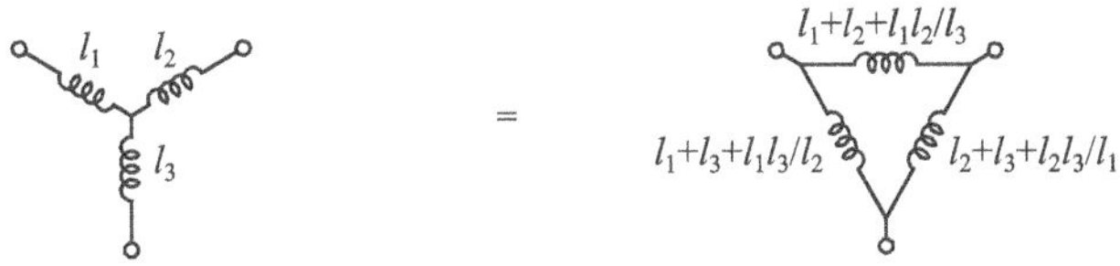
l_1
l_2
l_3
=
$l_1+l_2+l_1l_2/l_3$
$l_1+l_3+l_1l_3/l_2$
$l_2+l_3+l_2l_3/l_1$

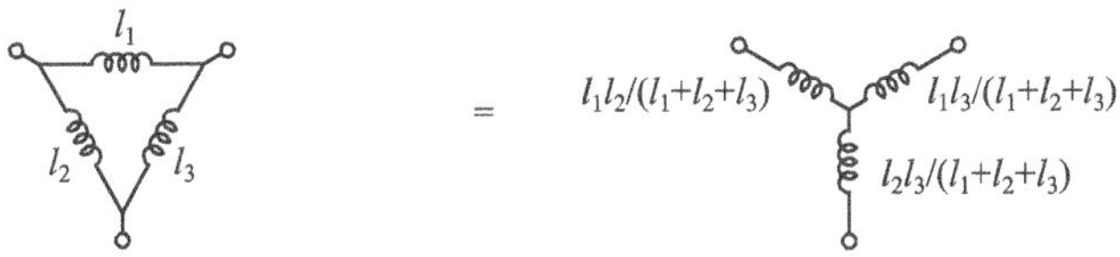
l_1
l_2
l_3
=
$l_1l_2/(l_1+l_2+l_3)$
$l_1l_3/(l_1+l_2+l_3)$
$l_2l_3/(l_1+l_2+l_3)$

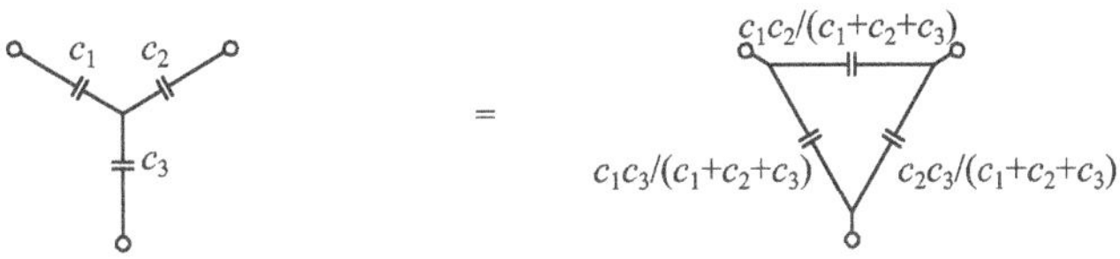
c_1
c_2
c_3
=
$c_1c_2/(c_1+c_2+c_3)$
$c_1c_3/(c_1+c_2+c_3)$
$c_2c_3/(c_1+c_2+c_3)$

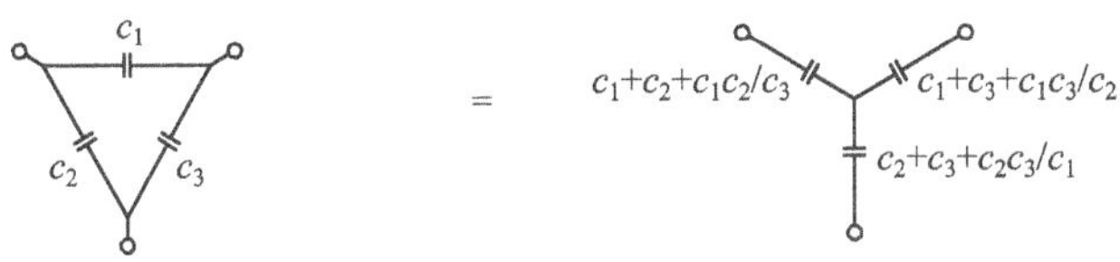
c_1
c_2
c_3
=
$c_1+c_2+c_1c_2/c_3$
$c_1+c_3+c_1c_3/c_2$
$c_2+c_3+c_2c_3/c_1$

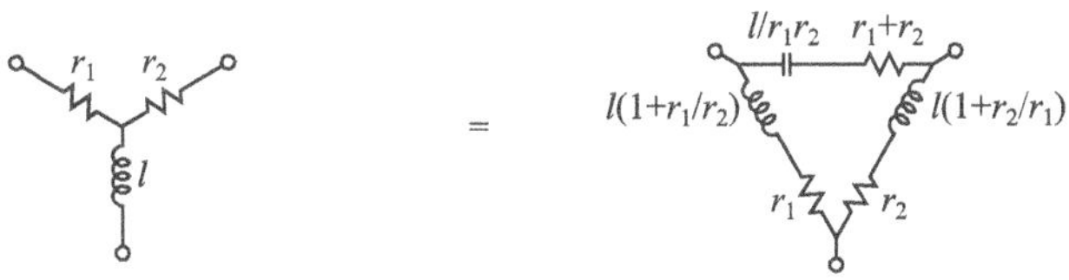
r_1
r_2
l
=
l/r_1r_2
r_1+r_2
$l(1+r_1/r_2)$
$l(1+r_2/r_1)$
r_1
r_2

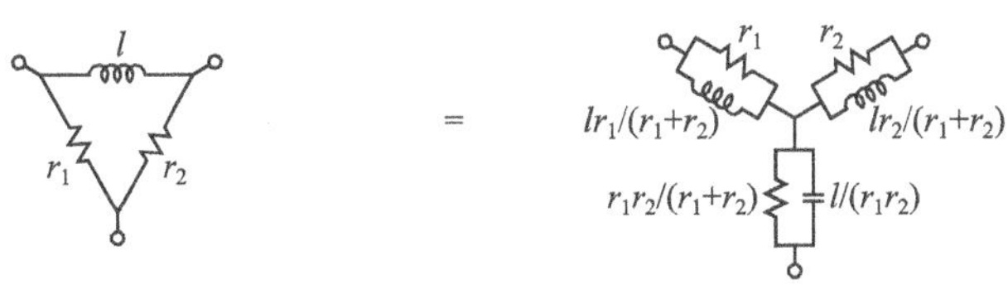
l
r_1
r_2
=
r_1
r_2
$lr_1/(r_1+r_2)$
$lr_2/(r_1+r_2)$
$r_1r_2/(r_1+r_2)$
$l/(r_1r_2)$

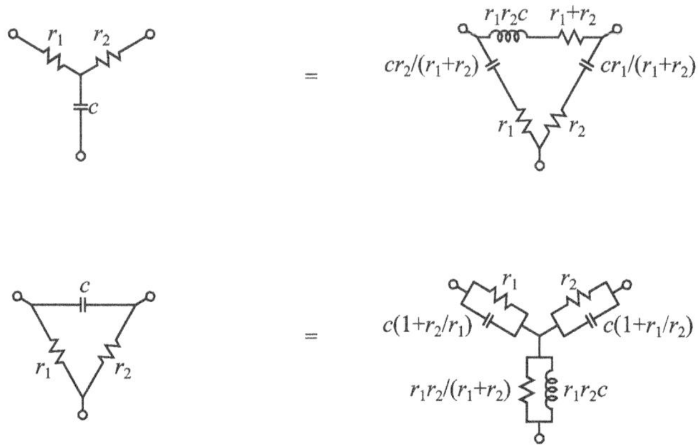

r_1
r_2
c
=
r_1r_2c
r_1+r_2
$cr_2/(r_1+r_2)$
$cr_1/(r_1+r_2)$
r_1
r_2
c
r_1
r_2
=
r_1
r_2
$c(1+r_2/r_1)$
$c(1+r_1/r_2)$
$r_1r_2/(r_1+r_2)$
r_1r_2c

Appendix B: Zolotarev Filter Prototypes

Achieser-Zolotarev filters are similar to Chebyshev Type I filters in that they approximate the ideal box filter characteristic with uniform ripples in most of the pass-band. However, a single larger ripple is permitted nearest to the origin (for a low-pass prototype). In addition to the ripple factor, ε, they have another design parameter, λ, that corresponds to the fractional bandwidth occupied by the larger-than-nominal first ripple. Larger λ for a given ripple factor, ε, yields somewhat steeper rejection with greater deflection of midfilter impedance levels (which has practical benefits in some implementations), while allowing greater amplitude of the first ripple [1, 2].

Here, we generalize the Zolotarev filter function to include an inverse or Type II Zolotarev filter, analogous to the inverse or Type II Chebyshev filters, having their ripples in the stop-band instead of the pass-band. Examples of the Type I and Type II Zolotarev filter functions are plotted in Figure B.1. Like Chebyshev filters, the nominal pass-band ripple amplitude for Type I is given by

$$a_I = \sqrt{1+\varepsilon^2} \tag{B.1}$$

and the stop-band ripple for Type II is

$$a_{II} = \frac{\varepsilon}{\sqrt{1+\varepsilon^2}} \tag{B.2}$$

The first ripple for Zolotarev filters has an enhanced ripple factor, $r_z\varepsilon$, leading to a single ripple with modified amplitude

$$a_I' = \sqrt{1+(r_z\varepsilon)^2} \tag{B.3a}$$

$$a_{II}' = \frac{r_z\varepsilon}{\sqrt{1+(r_z\varepsilon)^2}} \tag{B.3b}$$

Prototype element values for a range of ripple factors and the parameter λ are given in Tables B.1–B.5 below. The prototype elements are symmetric about the

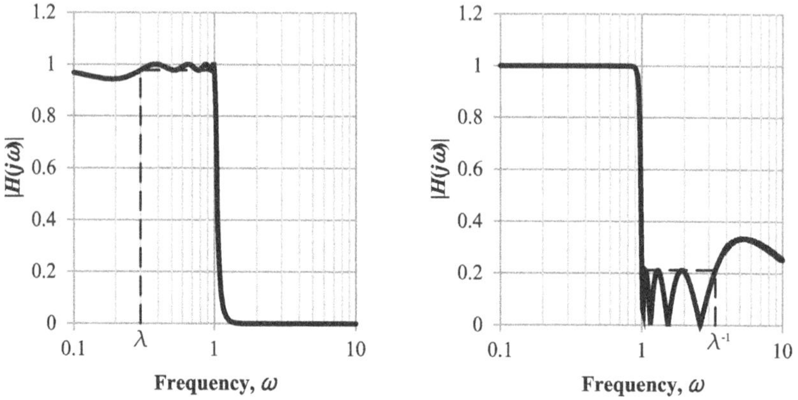

(a) (b)

Figure B.1 (a) Standard/Type I and (b) Inverse/Type II Zolotarev filters of order $N = 9$, with $\varepsilon = 0.2171$ and $\lambda = 0.3$. The element values for both filters were taken from Table B.3.

center, $g_k = g_{N-k+1}$, so the tables list only the first half of them. To use with the generalized reflectionless filter topologies of Chapter 6, we must have the odd-numbered parameters greater than the even-numbered parameters in order to avoid negative elements. For Zolotarev Type I and Type II, this condition is met if the parameters ε and λ aren't both too small. Only combinations which meet this criterion are included below. Note that when $\lambda = \sin(\pi/2N)$, the first ripple is equal in size to the rest and the prototype parameters revert to those of a standard Chebyshev filter.

It is assumed in these tables that the filter in question is equally terminated at both ends (e.g., with 50Ω). Different element values would be needed for a singly-terminated prototype, such as one might use to make a conventional-form diplexer [3].

Table B.1

Prototype Element Values for Zolotarev Filters

$\varepsilon = 0.04801$ (0.01 dB pass-band ripple or 26.4 dB stop-band ripple)

N	λ	r_z	g_1	g_2	g_3	g_4	g_5	g_6
5	0.8	11.69274	1.55433	0.93263	3.13354			
7	0.6	7.91984	1.19053	1.16532	2.58992	1.13306		
7	0.7	22.08984	1.68513	0.84489	4.76686	0.59234		
7	0.8	87.48319	2.84034	0.45395	17.39578	0.14562		
9	0.6	27.67775	1.54146	0.92204	4.62029	0.54983	7.03858	
9	0.7	109.21625	2.23378	0.59499	12.80361	0.15299	31.53560	

11	0.5	24.29549	1.33610	1.06899	3.58349	0.73261	5.91785	0.56133
11	0.6	99.47340	1.79558	0.77092	7.70305	0.26116	24.41988	0.12398

From: R. Levy, 1971 [1].

Table B.2

Prototype Element Values for Zolotarev Filters

$\varepsilon = 0.1526$ (0.1 dB pass-band ripple or 16.4 dB stop-band ripple)

N	λ	r_z	g_1	g_2	g_3	g_4	g_5	g_6
5	0.6	2.45664	1.48624	1.13729	2.61688			
5	0.7	4.72464	1.95074	0.87190	3.82242			
5	0.8	11.69274	3.14267	0.49586	8.61954			
7	0.4	1.80881	1.32520	1.30282	2.41550	1.36080		
7	0.5	3.49821	1.57660	1.10452	3.16061	1.02238		
7	0.6	7.91984	2.07606	0.79877	5.51287	0.55639		
7	0.7	22.08984	3.03669	0.48378	14.42569	0.19310		
7	0.8	87.48319	4.90272	0.25930	58.02886	0.04279		
9	0.4	3.51282	1.49803	1.17792	3.01257	1.07117	3.46259	
9	0.5	9.04492	1.91154	0.88308	5.14768	0.54578	7.34014	
9	0.6	27.67775	2.57311	0.59268	11.94679	0.18857	24.71231	
9	0.7	109.21625	3.58838	0.37951	34.88133	0.05117	115.00997	
11	0.3	2.57383	1.36840	1.29999	2.60025	1.28960	2.95225	1.24261
11	0.4	7.26933	1.67399	1.04210	3.91923	0.74857	5.75556	0.61496
11	0.5	24.29549	2.15191	0.75149	7.76332	0.29865	19.05143	0.17220
11	0.6	99.47340	2.79862	0.52602	17.84673	0.10242	84.34112	0.03538

From: R. Levy, 1971 [1].

Table B.3

Prototype Element Values for Zolotarev Filters

$\varepsilon = 0.2171$ (0.2 dB pass-band ripple or 13.5 dB stop-band ripple)

N	λ	r_z	g_1	g_2	g_3	g_4	g_5	g_6
5	*	1	1.33947	1.33700	2.16609			
5	0.4	1.10837	1.37419	1.30839	2.23281			
5	0.5	1.51975	1.51281	1.20949	2.49318			
5	0.6	2.45664	1.80067	1.01986	3.15344			
5	0.7	4.72464	2.43653	0.72233	5.07784			
5	0.8	11.69274	3.99846	0.38500	12.56353			
7	0.4	1.80881	1.56491	1.21434	2.76084	1.22301		
7	0.5	3.49821	1.90108	0.97293	3.92902	0.83414		
7	0.6	7.91984	2.53938	0.66428	7.58946	0.40356		
7	0.7	22.08984	3.69106	0.39681	20.78575	0.13310		
7	0.8	87.48319	5.88794	0.21515	83.45821	0.02963		
9	0.3	1.62586	1.50478	1.28548	2.62652	1.31770	2.79643	
9	0.4	3.51282	1.78566	1.05650	3.65326	0.87894	4.35199	

9	0.5	9.04492	2.29807	0.75520	6.83784	0.39862	10.49414	
9	0.6	27.67775	3.06627	0.50217	16.36108	0.13355	36.65254	
9	0.7	109.21625	4.24014	0.32262	47.73665	0.03667	168.57338	
11	0.3	2.57383	1.61138	1.20104	3.00251	1.11820	3.59061	1.05411
11	0.4	7.26933	1.99325	0.91641	4.94617	0.57574	7.93387	0.44563
11	0.5	24.29549	2.54407	0.65002	10.17358	0.22054	27.79834	0.11747
11	0.6	99.47340	3.28476	0.45384	23.51726	0.07618	122.15649	0.02435

*Chebyshev prototype, $\lambda = \sin(\pi/2N)$.
From: R. Levy, 1971 [1].

Table B.4

Prototype Element Values for Zolotarev Filters

$\varepsilon = 0.3493$ (0.5 dB pass-band ripple or 9.64 dB stop-band ripple)

N	λ	r_z	g_1	g_2	g_3	g_4	g_5	g_6
5	*	1	1.70582	1.22960	2.54089			
5	0.4	1.10837	1.75976	1.18745	2.65984			
5	0.5	1.51975	1.97675	1.04790	3.13007			
5	0.6	2.45664	2.42926	0.81558	4.35940			
5	0.7	4.72464	3.39896	0.52528	7.93963			
5	0.8	11.69274	5.62102	0.26783	21.13753			
7	*	1	1.73734	1.25822	2.63836	1.34430		
7	0.4	1.80881	2.03454	1.03636	3.51684	0.98510		
7	0.5	3.49821	2.53736	0.76699	5.66553	0.58307		
7	0.6	7.91984	3.41436	0.49763	12.11321	0.25125		
7	0.7	22.08984	4.90491	0.29702	34.07134	0.08056		
7	0.8	87.48319	7.72634	0.16335	135.83886	0.01813		
9	*	1	1.75049	1.26903	2.66785	1.36729	2.72397	
9	0.3	1.62586	1.93243	1.12080	3.23463	1.08449	3.49370	
9	0.4	3.51282	2.34484	0.85718	5.06951	0.62158	6.40598	
9	0.5	9.04492	3.02374	0.58726	10.37234	0.25257	17.46039	
9	0.6	27.67775	3.98681	0.38971	25.26130	0.08367	61.89998	
9	0.7	109.21625	5.46607	0.25142	73.47467	0.02334	280.02253	
11	*	1	1.75719	1.27430	2.68090	1.37593	2.74879	1.38792
11	0.3	2.57383	2.08451	1.01601	3.88152	0.85585	4.78609	0.77791
11	0.4	7.26933	2.59662	0.73499	7.07566	0.38604	12.75924	0.27604
11	0.5	24.29549	3.27935	0.51590	14.95237	0.14476	46.16127	0.07037
11	0.6	99.47340	4.20367	0.35938	34.71514	0.05053	200.30842	0.01481

*Chebyshev prototype, $\lambda = \sin(\pi/2N)$.
From: R. Levy, 1971 [1].

Table B.5

Prototype Element Values for Zolotarev Filters

$\varepsilon = 0.5088$ (1.0 dB pass-band ripple or 6.87 dB stop-band ripple)

N	λ	r_z	g_1	g_2	g_3	g_4	g_5	g_6
5	*	1	2.13495	1.09108	3.00102			

5	0.4	1.10837	2.21429	1.03962	3.19267			
5	0.5	1.51975	2.53318	0.87797	3.95533			
5	0.6	2.45664	3.18746	0.64161	5.96005			
5	0.7	4.72464	4.53201	0.39238	11.65878			
5	0.8	11.69274	7.47819	0.19801	31.82714			
7	*	1	2.16663	1.11148	3.09374	1.17348		
7	0.4	1.80881	2.59666	0.86012	4.51163	0.77714		
7	0.5	3.49821	3.28805	0.60375	7.93125	0.41653		
7	0.6	7.91984	4.41842	0.38435	17.76247	0.17039		
7	0.7	22.08984	6.29374	0.23060	50.24996	0.05437		
7	0.8	87.48319	9.84383	0.12793	199.17269	0.01234		
9	*	1	2.17980	1.11916	3.12154	1.18963	3.17474	
9	0.3	1.62586	2.44231	0.94876	4.02803	0.87331	4.42172	
9	0.4	3.51282	3.00241	0.69024	6.88557	0.44880	9.12674	
9	0.5	9.04492	3.85985	0.46506	14.67170	0.17397	26.23333	
9	0.6	27.67775	5.04944	0.30918	35.88085	0.05777	92.87708	
9	0.7	109.21625	6.88984	0.20000	104.08896	0.01628	415.57580	
11	*	1	2.18649	1.12290	3.13381	1.19567	3.19800	1.20408
11	0.3	2.57383	2.64518	0.84182	5.01550	0.65236	6.49022	0.57464
11	0.4	7.26933	3.29473	0.59283	9.65650	0.27490	18.85185	0.18620
11	0.5	24.29549	4.13183	0.41482	20.60819	0.10281	68.58980	0.04721
11	0.6	99.47340	5.27573	0.28859	47.95134	0.03613	294.98067	0.01004

*Chebyshev prototype, $\lambda = \sin(\pi/2N)$.
From: R. Levy, 1971 [1].

References

[1] Levy, R., "Characteristics and element values of equally terminated Achieser-Zolotarev quasi-low-pass filters," *IEEE Trans. Circuit Theory*, vol. 18, no. 5, September 1971.

[2] Cameron, R., Kudsia, C., and Mansour, R., *Microwave Filters for Communications Systems*, New Jersey: John Wiley & Sons, pp. 436–438, 2007.

[3] Matthaei, G., Young, L., and Jones, E., *Microwave Filters, Impedance-Matching Networks, and Coupling Structures*, Norwood MA: Artech House, pp. 104–109, 1980.

Appendix C: Transmission-Line Identities

The construction of practical transmission-line networks of any kind, including reflectionless filters, is greatly facilitated by the ability to substitute equivalent circuit structures in place of others that may be less convenient (e.g., shunt stubs in place of series stubs). In this appendix, we catalog a wide variety of known transmission-line identities to aid in that process.

An *identity* in this context refers to a pair of networks for which the port parameters (impedance, admittance, scattering, etc.) are mathematically equivalent at all frequencies. Approximate matches are not included, nor are circuits that match only at discrete frequencies. Circuit equivalents which are valid given a change of frequency variables are known as transformations (such as Richard's transformation or the well-known low-pass/high-pass/band-pass/band-stop frequency transformations given in Table 1.2) and are not included here.

This appendix is organized into identity sets — groups of two or more networks that are all equivalent — followed by the mathematical relationships between the element values under which said equivalence applies. Identities for two-port networks are given in the first section, three-port networks in the second section, and four-port networks in the third. Finally, Section C.4 identifies a special case of Kuroda's identity that is of particular value to the reflectionless filters described in Chapter 7.

C.1 TWO-PORT IDENTITIES

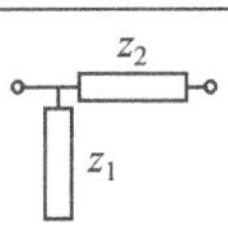	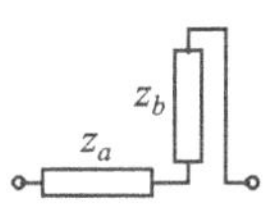	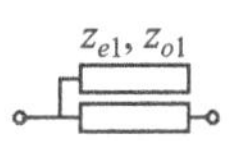	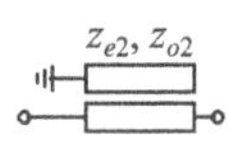
z_1 z_2	$z_1 = z_a\left(1+\frac{z_a}{z_b}\right)$ $z_2 = z_a + z_b$	$z_1 = \frac{1}{2} z_{e1}\left(1+\frac{z_{e1}}{z_{o1}}\right)$ $z_2 = \frac{1}{2}\left(z_{e1} + z_{o1}\right)$	$z_1 = 2 z_{e2} z_{o2} \frac{z_{e2} + z_{o2}}{\left(z_{e2} - z_{o2}\right)^2}$ $z_2 = \frac{1}{2}\left(z_{e2} + z_{o2}\right)$
$z_a = \frac{z_1 z_2}{z_1 + z_2}$ $z_b = \frac{z_2^2}{z_1 + z_2}$	z_a z_b	$z_a = \frac{1}{2} z_{e1}$ $z_b = \frac{1}{2} z_{o1}$	$z_a = \frac{2 z_{e2} z_{o2}}{z_{e2} + z_{o2}}$ $z_b = \frac{1}{2}\frac{\left(z_{e2} - z_{o2}\right)^2}{z_{e2} + z_{o2}}$

$z_{e1} = 2\dfrac{z_1 z_2}{z_1 + z_2}$ $z_{o1} = 2\dfrac{z_2^2}{z_1 + z_2}$	$z_{e1} = 2z_a$ $z_{o1} = 2z_b$	z_{e1} z_{o1}	$z_{e1} = \dfrac{4z_{e2}z_{o2}}{z_{e2} + z_{o2}}$ $z_{o1} = \dfrac{(z_{e2} - z_{o2})^2}{z_{e2} + z_{o2}}$
$z_{e2} = z_2\left(1 + \sqrt{\frac{z_2}{z_1 + z_2}}\right)$ $z_{o2} = z_2\left(1 - \sqrt{\frac{z_2}{z_1 + z_2}}\right)$	$z_{e2} = z_a + z_b + \sqrt{z_b(z_a + z_b)}$ $z_{o2} = z_a + z_b - \sqrt{z_b(z_a + z_b)}$	$z_{e2} = \frac{1}{2}(z_{e1} + z_{o1}) + \frac{1}{2}\sqrt{z_{o1}(z_{e1} + z_{o1})}$ $z_{o2} = \frac{1}{2}(z_{e1} + z_{o1}) - \frac{1}{2}\sqrt{z_{o1}(z_{e1} + z_{o1})}$	z_{e2} z_{o2}

The equivalence of the first two networks above is known as Kuroda's identity.

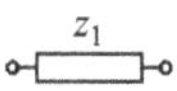	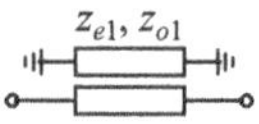	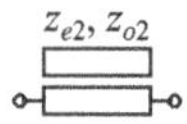	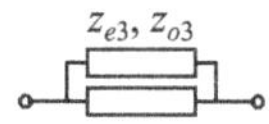
z_1	$z_1 = 2\dfrac{z_{e1}z_{o1}}{z_{e1} + z_{o1}}$	$z_1 = \frac{1}{2}(z_{e2} + z_{o2})$	$z_1 = \frac{1}{2}z_{e3}$
$z_{e1} = \frac{1}{2}z_1(1 + \rho_1)$ $z_{o1} = \frac{1}{2}z_1(1 + \rho_1^{-1})$ for any ρ_1	z_{e1} z_{o1}	$z_{e1} = \frac{1}{4}(z_{e2} + z_{o2})(1 + \rho_1)$ $z_{o1} = \frac{1}{4}(z_{e2} + z_{o2})(1 + \rho_1^{-1})$ for any ρ_1	$z_{e1} = \frac{1}{4}z_{e3}(1 + \rho_1)$ $z_{o1} = \frac{1}{4}z_{e3}(1 + \rho_1^{-1})$ for any ρ_1
$z_{e2} = 2z_1\dfrac{\rho_2}{1 + \rho_2}$ $z_{o2} = 2z_1\dfrac{1}{1 + \rho_2}$ for any ρ_2	$z_{e2} = 4\dfrac{z_{e1}z_{o1}}{z_{e1} + z_{o1}}\dfrac{\rho_2}{1 + \rho_2}$ $z_{o2} = 4\dfrac{z_{e1}z_{o1}}{z_{e1} + z_{o1}}\dfrac{1}{1 + \rho_2}$ for any ρ_2	z_{e2} z_{o2}	$z_{e2} = z_{e3}\dfrac{\rho_2}{1 + \rho_2}$ $z_{o2} = z_{e3}\dfrac{1}{1 + \rho_2}$ for any ρ_2
$z_{e3} = 2z_1$ for any z_{o3}	$z_{e3} = 4\dfrac{z_{e1}z_{o1}}{z_{e1} + z_{o1}}$ for any z_{o3}	$z_{e3} = z_{e2} + z_{o2}$ for any z_{o3}	z_{e3} z_{o3}

In the table above, ρ is the ratio of even- to odd-mode impedance of the coupled lines. In many cases above, this is a free parameter.

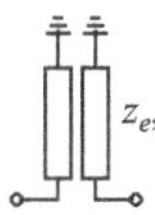	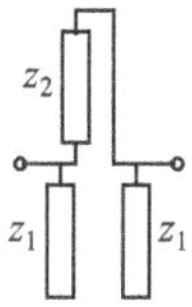	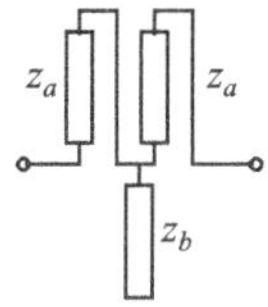	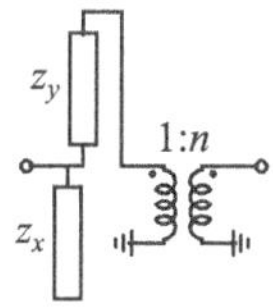 $n^2 = \frac{z_x}{z_x + z_y}$
z_e z_o	$z_e = z_1$ $z_o = \frac{z_1 z_2}{2z_1 + z_2}$	$z_e = z_a + 2z_b$ $z_o = z_a$	$z_e = z_x(n+1)$ $z_o = z_y \frac{n^2}{n+1}$
$z_1 = z_e$ $z_2 = \frac{2z_e z_o}{z_e - z_o}$	z_1 z_2	$z_1 = z_a + 2z_b$ $z_2 = (z_a + 2z_b)\frac{z_a}{z_b}$	$z_1 = z_x(n+1)$ $z_2 = z_y n$
$z_a = z_o$ $z_b = \frac{1}{2}(z_e - z_o)$	$z_a = \frac{z_1 z_2}{2z_1 + z_2}$ $z_b = \frac{z_1^2}{2z_1 + z_2}$	z_a z_b	$z_a = z_y \frac{n^2}{n+1}$ $z_b = z_x n$
$z_x = \frac{1}{2}(z_e + z_o)$ $z_y = 2z_e z_o \frac{z_e + z_o}{(z_o - z_e)^2}$ $n = \frac{z_e - z_o}{z_e + z_o}$	$z_x = \frac{z_1}{n+1}$ $z_y = \frac{z_2}{n}$ $n = \frac{z_1}{z_1 + z_2}$	$z_x = \frac{z_b}{n}$ $z_y = z_a \frac{n+1}{n^2}$ $n = \frac{z_b}{z_a + z_b}$	z_x z_y

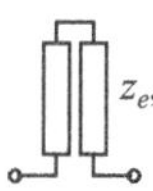	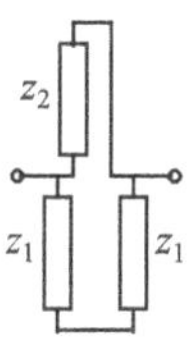	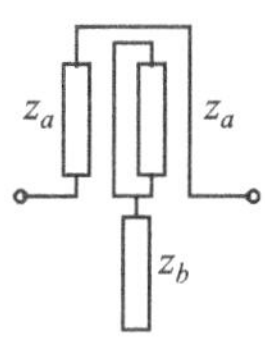	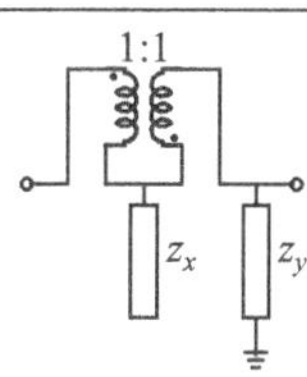
z_e z_o	$z_e = z_1$ $z_o = \frac{z_1 z_2}{2z_1 + z_2}$	$z_e = z_a + 2z_b$ $z_o = z_a$	$z_e = 2z_x$ $z_o = 2z_y$
$z_1 = z_e$ $z_2 = 2\frac{z_e z_o}{z_e - z_o}$	z_1 z_2	$z_1 = z_a + 2z_b$ $z_2 = \frac{z_a}{z_b}(z_a + 2z_b)$	$z_1 = 2z_x$ $z_2 = 4\frac{z_x z_y}{z_x - z_y}$

$z_a = z_o$ $z_b = \frac{1}{2}(z_e - z_o)$	$z_a = \dfrac{z_1 z_2}{2z_1 + z_2}$ $z_b = \dfrac{z_1^2}{2z_1 + z_2}$	z_a z_b	$z_a = 2z_x$ $z_b = z_x - z_y$
$z_x = \frac{1}{2} z_e$ $z_y = \frac{1}{2} z_o$	$z_x = \frac{1}{2} z_1$ $z_y = \frac{1}{2} \dfrac{z_1 z_2}{2z_1 + z_2}$	$z_x = \frac{1}{2}(z_a + 2z_b)$ $z_y = \frac{1}{2} z_a$	z_x z_y

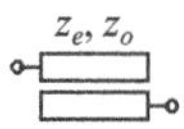	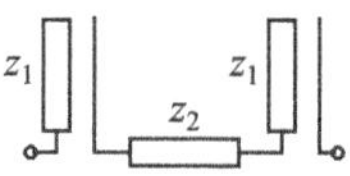	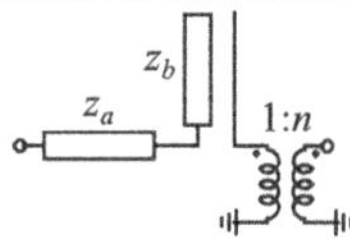 $n^2 = \dfrac{z_a}{z_a + z_b}$
z_e z_o	$z_e = z_1 + 2z_2$ $z_o = z_1$	$z_e = z_a(1+n)$ $z_o = z_a(1-n)$
$z_1 = z_o$ $z_2 = \frac{1}{2}(z_e - z_o)$	z_1 z_2	$z_1 = z_a(1-n)$ $z_2 = z_a n$
$z_a = \frac{1}{2}(z_e + z_o)$ $z_b = 2z_e z_o \dfrac{z_e + z_o}{(z_e - z_o)^2}$ $n = \dfrac{z_e - z_o}{z_e + z_o}$	$z_a = \dfrac{z_1}{1-n}$ $z_b = z_1 \dfrac{1+n}{n^2}$ $n = \dfrac{z_e - z_o}{z_e + z_o}$	z_a z_b

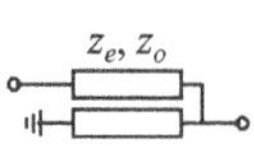	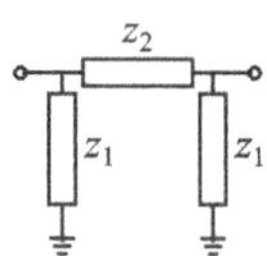	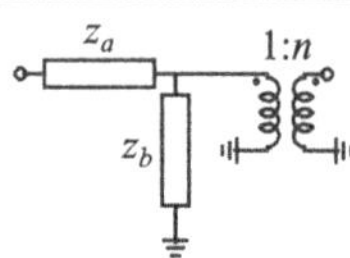 $n = \frac{1}{2}\left(1 + \dfrac{z_a}{z_b}\right)$
z_e z_o	$z_e = z_2$ $z_o = \frac{1}{2} \dfrac{z_1 z_2}{z_1 + 2z_2}$	$z_e = z_a n$ $z_o = z_b n$

$z_1 = 2\dfrac{z_e z_o}{z_e - z_o}$ $z_2 = z_e$	z_1 z_2	$z_1 = 2n\dfrac{z_a z_b}{z_a - z_b}$ $z_2 = z_a n$
$z_a = \frac{1}{n} z_e$ $z_b = \frac{1}{n} z_o$ $n = \frac{1}{2}\left(1 + \dfrac{z_e}{z_o}\right)$	$z_a = \frac{1}{n} z_2$ $z_b = \frac{1}{n}\dfrac{z_1 z_2}{z_1 + 2z_2}$ $n = 1 + \dfrac{z_2}{z_1}$	z_a z_b

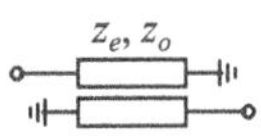	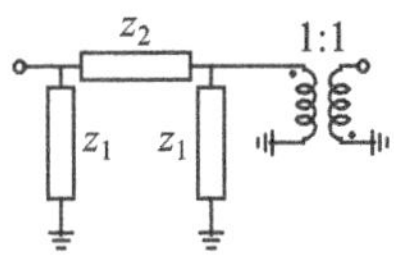	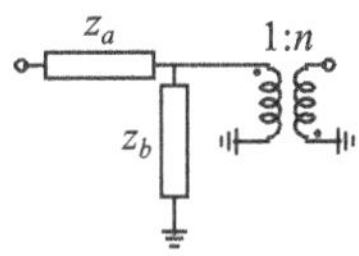 $n^2 = 1 + \dfrac{z_a}{z_b}$
z_e z_o	$z_e = z_1$ $z_o = \dfrac{z_1 z_2}{2z_1 + z_2}$	$z_e = z_a \dfrac{n}{n-1}$ $z_o = z_a \dfrac{n}{n+1}$
$z_1 = z_e$ $z_2 = 2\dfrac{z_e z_o}{z_e - z_o}$	z_1 z_2	$z_1 = z_a \dfrac{n}{n-1}$ $z_2 = z_a n$
$z_a = 2\dfrac{z_e z_o}{z_e + z_o}$ $z_b = \frac{1}{2}\dfrac{(z_e - z_o)^2}{z_e + z_o}$ $n = \dfrac{z_e + z_o}{z_e - z_o}$	$z_a = \dfrac{z_1 z_2}{z_1 + z_2}$ $z_b = z_1 \dfrac{1}{n(1+n)}$ $n = 1 + \dfrac{z_2}{z_1}$	z_a z_b

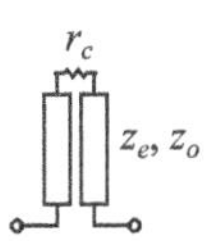	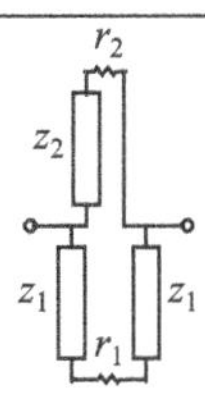 $\dfrac{r_1}{r_2} = 2\dfrac{z_1}{z_2}$	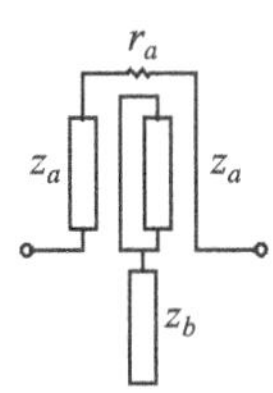

z_e z_o r_c	$z_e = z_1$ $z_o = \dfrac{z_1 z_2}{2z_1 + z_2}$ $r_c = \dfrac{r_1 r_2}{r_1 + r_2}$	$z_e = z_a + 2z_b$ $z_o = z_a$ $r_c = r_a$
$z_1 = z_e$ $z_2 = 2\dfrac{z_e z_o}{z_e - z_o}$ $r_1 = \dfrac{z_e}{z_o} r_c$ $r_2 = \dfrac{z_e}{z_e - z_o} r_c$	z_1 z_2 r_1 r_2	$z_1 = z_a + 2z_b$ $z_2 = \dfrac{z_a}{z_b}\left(z_a + 2z_b\right)$ $r_1 = \left(1 + 2\dfrac{z_b}{z_a}\right) r_a$ $r_2 = \left(1 + \dfrac{z_a}{2z_b}\right) r_a$
$z_a = z_o$ $z_b = \frac{1}{2}\left(z_e - z_o\right)$ $r_a = r_c$	$z_a = \dfrac{z_1 z_2}{2z_1 + z_2}$ $z_b = \dfrac{z_1}{z_2} \dfrac{z_1 z_2}{2z_1 + z_2}$ $r_a = \dfrac{r_1 r_2}{r_1 + r_2}$	z_a z_b r_a

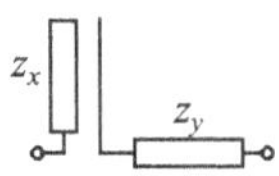	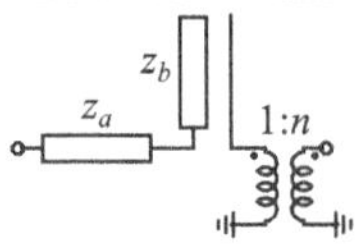 $n = \dfrac{z_a}{z_a + z_b}$
$z_x = z_b n$ $z_y = z_a n$	$z_a = z_y \frac{1}{n}$ $z_b = z_x \frac{1}{n}$ $n = \dfrac{z_x}{z_x + z_y}$

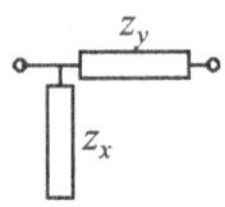	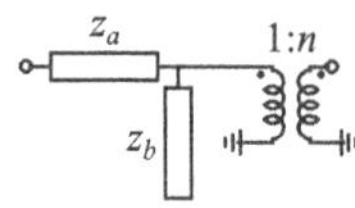 $n = 1 + \dfrac{z_a}{z_b}$
$z_x = z_b n$ $z_y = z_a n$	$z_a = z_y \frac{1}{n}$ $z_b = z_x \frac{1}{n}$ $n = 1 + \dfrac{z_y}{z_x}$

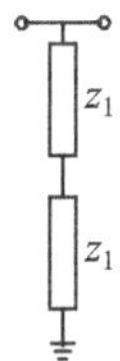	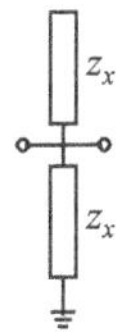
$z_1 = \frac{1}{2} z_x$	$z_x = 2z_1$

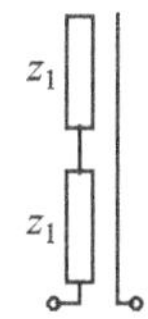	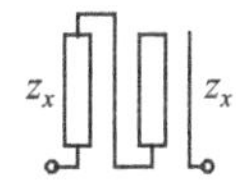
$z_1 = 2z_x$	$z_x = \frac{1}{2} z_1$

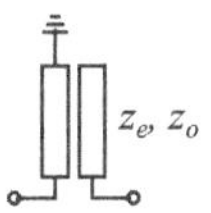	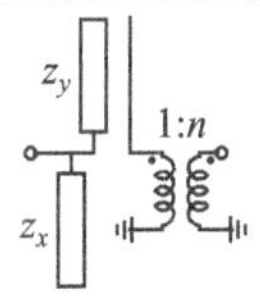 $n^2 = \dfrac{z_x}{z_x + z_y}$
$z_e = z_x(n+1)$ $z_o = z_y \dfrac{n^2}{n+1}$	$z_x = \frac{1}{2}(z_e + z_o)$ $z_y = 2z_e z_o \dfrac{z_e + z_o}{(z_e - z_o)^2}$ $n = \dfrac{z_e - z_o}{z_e + z_o}$

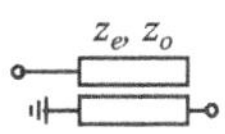	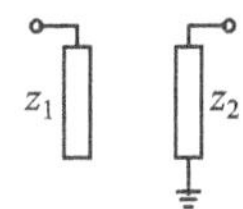 $z_2 > z_1$
$z_e = z_2 + \sqrt{z_2(z_2 - z_1)}$ $z_o = z_2 - \sqrt{z_2(z_2 - z_1)}$	$z_1 = 2\dfrac{z_e z_o}{z_e - z_o}$ $z_2 = \frac{1}{2}(z_e + z_o)$

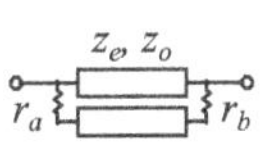	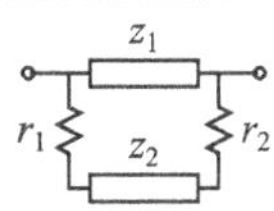
$z_e = 2\dfrac{z_1 z_2}{z_1 + z_2}$ $z_o = 2\dfrac{z_1^2}{z_1 + z_2}$ $r_a = 4r_1\left(\dfrac{z_1}{z_1 + z_2}\right)^2$ $r_b = 4r_2\left(\dfrac{z_1}{z_1 + z_2}\right)^2$	$z_1 = \frac{1}{2}(z_e + z_o)$ $z_2 = \frac{1}{2}\dfrac{z_e}{z_o}(z_e + z_o)$ $r_1 = \frac{1}{4} r_a \left(\dfrac{z_e}{z_o} + 1\right)^2$ $r_2 = \frac{1}{4} r_b \left(\dfrac{z_e}{z_o} + 1\right)^2$

C.2 THREE-PORT IDENTITIES

<table>
<tr><td>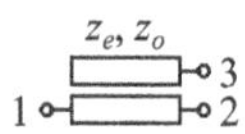
</td><td>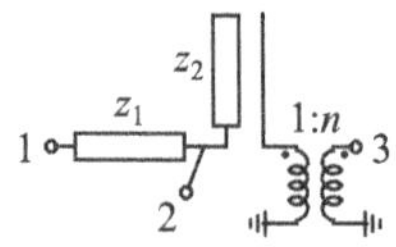

$$n^2 = \frac{z_1}{z_1 + z_2}$$</td></tr>
<tr><td>$$z_e = z_1(1+n)$$ $$z_o = z_1(1-n)$$</td><td>$$z_1 = \tfrac{1}{2}(z_e + z_o)$$ $$z_2 = 2z_e z_o \frac{z_e + z_o}{(z_e - z_o)^2}$$ $$n = \frac{z_e - z_o}{z_e + z_o}$$</td></tr>
</table>

<table>
<tr><td>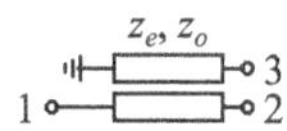
</td><td>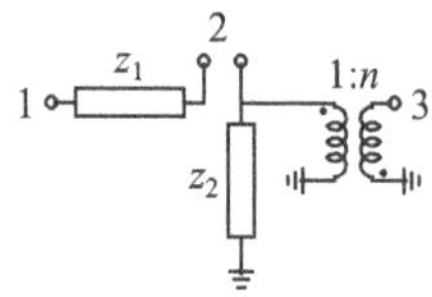

$$n^2 = 1 + \frac{z_1}{z_2}$$</td></tr>
<tr><td>$$z_e = z_2 n(n+1)$$ $$z_o = z_2 n(n-1)$$</td><td>$$z_1 = 2\frac{z_e z_o}{z_e + z_o}$$ $$z_2 = \tfrac{1}{2}\frac{(z_e - z_o)^2}{z_e + z_o}$$ $$n = \frac{z_e + z_o}{z_e - z_o}$$</td></tr>
</table>

The identity in the left table above is of critical importance to the reflectionless filters described in Chapter 7.

<table>
<tr><td>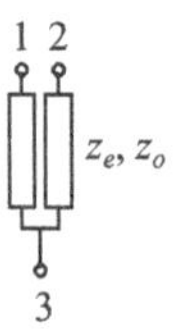
</td><td>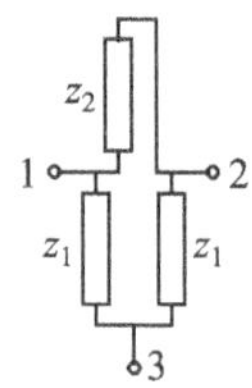
</td></tr>
<tr><td>$$z_e = z_1$$ $$z_o = \frac{z_1 z_2}{2z_1 + z_2}$$</td><td>$$z_1 = z_e$$ $$z_2 = 2\frac{z_e z_o}{z_e - z_o}$$</td></tr>
</table>

C.3 FOUR-PORT IDENTITIES

<table>
<tr><td>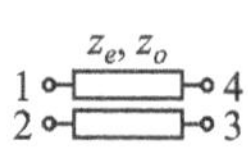
</td><td>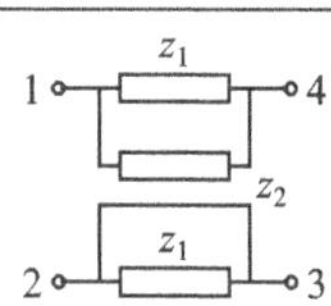
</td></tr>
<tr><td>$$z_e = z_1$$ $$z_o = \frac{z_1 z_2}{2z_1 + z_2}$$</td><td>$$z_1 = z_e$$ $$z_2 = 2\frac{z_e z_o}{z_e - z_o}$$</td></tr>
</table>

C.4 PERIODIC KURODA'S IDENTITY

A special case of the first set in Section C.1 occurs when

$$z_2 = z_a^{-1} \tag{C.1a}$$

$$z_1 = z_b^{-1} \tag{C.1b}$$

Substituting in these expressions, we have

$$z_1 = z_a\left(1+\frac{z_a}{z_b}\right) = z_2^{-1}\left(1+\frac{z_2^{-1}}{z_1^{-1}}\right) \tag{C.2a}$$

$$\therefore z_1 = \frac{1}{z_2 - z_2^{-1}} \tag{C.2b}$$

Renaming $z_2 = x$ for simplicity, we are left with the following equivalent pairs,

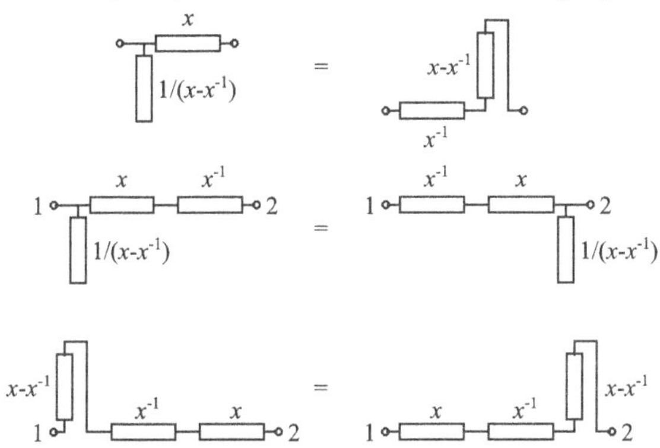

Note that the last two equivalent pairs above are simply mirror images of one another. We refer to this as the *periodic* Kuroda's identity, because repeated applications of the original Kuroda's identity to a particular stub yields stubs of the same value at every other position along a cascade of transmission lines.

Knowledge of these special cases greatly simplifies the analysis of the transmission-line reflectionless filters described in Chapter 7.

References

[1] Ozaki, H., and Ishii, J., "Synthesis of a class of strip-line filters," *IRE Trans. Circuit Theory*, pp.

104–109, June 1958.

[2] Podcameni, A., and Conrado, L., "A new transmission-line identity," *Microw. Optical Techn. Letters*, vol. 23, no. 1, October 1999.

[3] Matthaei, G., Young, L., and Jones, E., *Microwave Filters, Impedance-Matching Networks, and Coupling Structures*, Norwood MA: Artech House, pp. 219–228, 1980.

[4] Wenzel, R., "Exact design of TEM microwave networks using quarter-wave lines," *IEEE Trans. Microw. Theory Techn.*, pp. 94–111, January 1964.

[5] Morgan, M., and Boyd, T., "Reflectionless filter structures," *IEEE Trans. Microw. Theory Techn.*, vol. 63, no. 4, pp. 1263–1271, April 2015.

About the Author

Matthew A. Morgan received his B.S. in electrical engineering from the University of Virginia in 1999, and his M.S. and Ph.D. degrees in electrical engineering from the California Institute of Technology in 2001 and 2003, respectively. He has authored over 60 papers in the areas of microwave monolithic integrated circuit (MMIC) design, millimeter-wave system integration, and high-frequency packaging techniques. He holds two patents and has four pending in the areas of reflectionless filter development, high-speed serial communication, and ultrawideband millimeter-wave antennas.

During the summers of 1996 through 1998, he worked for Lockheed Martin Federal Systems in Manassas, VA, as an associate programmer, where he wrote code for acoustic signal processing, mathematical modeling, data simulation, and system performance monitoring. In 1999, he became an affiliate of NASA's Jet Propulsion Laboratory in Pasadena, CA. There, he conducted research in the development of MMICs and MMIC-based receiver components for atmospheric radiometers, laboratory instrumentation, and the deep-space communication network.

In 2003, Dr. Morgan joined the Central Development Lab (CDL) of the National Radio Astronomy Observatory (NRAO) in Charlottesville, VA, where he now holds the position of scientist/research engineer. He was project engineer for the K-Band Focal Plane Array development project on the Green Bank Telescope, and technical lead for Band 6 (211-275 GHz) cryogenic IF amplifier production, Band 6 receiver cartridge testing, and Band 3 (84-116 GHz) and Band 6 Orthomode Transducer (OMT) production for the Atacama Large Millimeter Array (ALMA). Dr. Morgan is currently the head of the CDL's Integrated Receiver Development program, and is involved in the design and development of low-noise receivers, components, and novel concepts for radio astronomy instrumentation in the cm-wave, mm-wave, and sub-mm-wave frequency ranges.

Index

T

U

V

W

Y

Z

Artech House Microwave Library

Behavioral Modeling and Linearization of RF Power Amplifiers, John Wood

Chipless RFID Reader Architecture, Nemai Chandra Karmakar, Prasanna Kalansuriya, Randika Koswatta, and Rubayet E-Azim

Control Components Using Si, GaAs, and GaN Technologies, Inder J. Bahl

Design of Linear RF Outphasing Power Amplifiers, Xuejun Zhang, Lawrence E. Larson, and Peter M. Asbeck

Design Methodology for RF CMOS Phase Locked Loops, Carlos Quemada, Guillermo Bistué, and Iñigo Adin

Design of CMOS Operational Amplifiers, Rasoul Dehghani

Design of RF and Microwave Amplifiers and Oscillators, Second Edition, Pieter L. D. Abrie

Digital Filter Design Solutions, Jolyon M. De Freitas

Discrete Oscillator Design Linear, Nonlinear, Transient, and Noise Domains, Randall W. Rhea

Distortion in RF Power Amplifiers, Joel Vuolevi and Timo Rahkonen

Distributed Power Amplifiers for RF and Microwave Communications, Narendra Kumar and Andrei Grebennikov

Electronics for Microwave Backhaul, Vittorio Camarchia, Roberto Quaglia, and Marco Pirola, editors

EMPLAN: Electromagnetic Analysis of Printed Structures in Planarly Layered Media, Software and User's Manual, Noyan Kinayman and M. I. Aksun

An Engineer's Guide to Automated Testing of High-Speed Interfaces, Second Edition, José Moreira and Hubert Werkmann

Envelope Tracking Power Amplifiers for Wireless Communications, Zhancang Wang

Essentials of RF and Microwave Grounding, Eric Holzman

FAST: Fast Amplifier Synthesis Tool—Software and User's Guide, Dale D. Henkes

Feedforward Linear Power Amplifiers, Nick Pothecary

Filter Synthesis Using Genesys S/Filter, Randall W. Rhea

Foundations of Oscillator Circuit Design, Guillermo Gonzalez

Frequency Synthesizers: Concept to Product, Alexander Chenakin

Fundamentals of Nonlinear Behavioral Modeling for RF and Microwave Design, John Wood and David E. Root, editors

Generalized Filter Design by Computer Optimization, Djuradj Budimir

Handbook of Dielectric and Thermal Properties of Materials at Microwave Frequencies, Vyacheslav V. Komarov

Handbook of RF, Microwave, and Millimeter-Wave Components, Leonid A. Belov, Sergey M. Smolskiy, and Victor N. Kochemasov

High-Linearity RF Amplifier Design, Peter B. Kenington

High-Speed Circuit Board Signal Integrity, Stephen C. Thierauf

Integrated Microwave Front-Ends with Avionics Applications, Leo G. Maloratsky

Intermodulation Distortion in Microwave and Wireless Circuits, José Carlos Pedro and Nuno Borges Carvalho

Introduction to Modeling HBTs, Matthias Rudolph

An Introduction to Packet Microwave Systems and Technologies, Paolo Volpato

Introduction to RF Design Using EM Simulators, Hiroaki Kogure, Yoshie Kogure, and James C. Rautio

Introduction to RF and Microwave Passive Components, Richard Wallace and Krister Andreasson

Klystrons, Traveling Wave Tubes, Magnetrons, Crossed-Field Amplifiers, and Gyrotrons, A. S. Gilmour, Jr.

Lumped Elements for RF and Microwave Circuits, Inder Bahl

Lumped Element Quadrature Hybrids, David Andrews

Microstrip Lines and Slotlines, Third Edition, Ramesh Garg, Inder Bahl, and Maurizio Bozzi

Microwave Circuit Modeling Using Electromagnetic Field Simulation, Daniel G. Swanson, Jr. and Wolfgang J. R. Hoefer

Microwave Component Mechanics, Harri Eskelinen and Pekka Eskelinen

Microwave Differential Circuit Design Using Mixed-Mode S-Parameters, William R. Eisenstadt, Robert Stengel, and Bruce M. Thompson

Microwave Engineers' Handbook, Two Volumes, Theodore Saad, editor

Microwave Filters, Impedance-Matching Networks, and Coupling Structures, George L. Matthaei, Leo Young, and E. M. T. Jones

Microwave Material Applications: Device Miniaturization and Integration, David B. Cruickshank

Microwave Materials and Fabrication Techniques, Second Edition, Thomas S. Laverghetta

Microwave Materials for Wireless Applications, David B. Cruickshank

Microwave Mixer Technology and Applications, Bert Henderson and Edmar Camargo

Microwave Mixers, Second Edition, Stephen A. Maas

Microwave Network Design Using the Scattering Matrix, Janusz A. Dobrowolski

Microwave Radio Transmission Design Guide, Second Edition, Trevor Manning

Microwave and RF Semiconductor Control Device Modeling, Robert H. Caverly

Microwave Transmission Line Circuits, William T. Joines, W. Devereux Palmer, and Jennifer T. Bernhard

Microwaves and Wireless Simplified, Third Edition, Thomas S. Laverghetta

Modern Microwave Circuits, Noyan Kinayman and M. I. Aksun

Modern Microwave Measurements and Techniques, Second Edition, Thomas S. Laverghetta

Modern RF and Microwave Filter Design, Protap Pramanick and Prakash Bhartia

Neural Networks for RF and Microwave Design, Q. J. Zhang and K. C. Gupta

Noise in Linear and Nonlinear Circuits, Stephen A. Maas

Nonlinear Microwave and RF Circuits, Second Edition, Stephen A. Maas

On-Wafer Microwave Measurements and De-Embedding, Errikos Lourandakis

Passive RF Component Technology: Materials, Techniques, and Applications, Guoan Wang and Bo Pan, editors

Practical Analog and Digital Filter Design, Les Thede

Practical Microstrip Design and Applications, Günter Kompa

Practical Microwave Circuits, Stephen Maas

Practical RF Circuit Design for Modern Wireless Systems, Volume I: Passive Circuits and Systems, Les Besser and Rowan Gilmore

Practical RF Circuit Design for Modern Wireless Systems, Volume II: Active Circuits and Systems, Rowan Gilmore and Les Besser

Production Testing of RF and System-on-a-Chip Devices for Wireless Communications, Keith B. Schaub and Joe Kelly

Q Factor Measurements Using MATLAB®, Darko Kajfez

QMATCH: Lumped-Element Impedance Matching, Software and User's Guide, Pieter L. D. Abrie

Radio Frequency Integrated Circuit Design, Second Edition, John W. M. Rogers and Calvin Plett

Reflectionless Filters, Matthew A. Morgan

RF Bulk Acoustic Wave Filters for Communications, Ken-ya Hashimoto

RF Design Guide: Systems, Circuits, and Equations, Peter Vizmuller

RF Linear Accelerators for Medical and Industrial Applications, Samy Hanna

RF Measurements of Die and Packages, Scott A. Wartenberg

The RF and Microwave Circuit Design Handbook, Stephen A. Maas

RF and Microwave Coupled-Line Circuits, Rajesh Mongia, Inder Bahl, and Prakash Bhartia

RF and Microwave Oscillator Design, Michal Odyniec, editor

RF Power Amplifiers for Wireless Communications, Second Edition, Steve C. Cripps

RF Systems, Components, and Circuits Handbook, Ferril A. Losee

Scattering Parameters in RF and Microwave Circuit Analysis and Design, Janusz A. Dobrowolski

The Six-Port Technique with Microwave and Wireless Applications, Fadhel M. Ghannouchi and Abbas Mohammadi

Solid-State Microwave High-Power Amplifiers, Franco Sechi and Marina Bujatti

Spin Transfer Torque Based Devices, Circuits, and Memory, Brajesh Kumar Kaushik and Shivam Verma

Stability Analysis of Nonlinear Microwave Circuits, Almudena Suárez and Raymond Quéré

Substrate Noise Coupling in Analog/RF Circuits, Stephane Bronckers, Geert Van der Plas, Gerd Vandersteen, and Yves Rolain

System-in-Package RF Design and Applications, Michael P. Gaynor

Terahertz Metrology, Mira Naftaly, editor

TRAVIS 2.0: Transmission Line Visualization Software and User's Guide, Version 2.0, Robert G. Kaires and Barton T. Hickman

Understanding Microwave Heating Cavities, Tse V. Chow Ting Chan and Howard C. Reader

Understanding Quartz Crystals and Oscillators, Ramón M. Cerda

For further information on these and other Artech House titles, including previously considered out-of-print books now available through our In-Print-Forever® (IPF®) program, contact:

Artech House Publishers
685 Canton Street
Norwood, MA 02062
Phone: 781-769-9750
Fax: 781-769-6334
e-mail: artech@artechhouse.com

Artech House Books
16 Sussex Street
London SW1V 4RW UK
Phone: +44 (0)20 7596 8750
Fax: +44 (0)20 7630 0166
e-mail: artech-uk@artechhouse.com

Find us on the World Wide Web at: www.artechhouse.com

www.ingramcontent.com/pod-product-compliance
Ingram Content Group UK Ltd.
Pitfield, Milton Keynes, MK11 3LW, UK
UKHW020951011026
475579UK00014B/20

9 781630 813277